Electronic Circuits Handbook

Design, testing and construction

MICHAEL TOOLEY

Heinemann : London

Heinemann Professional Publishing Ltd
22 Bedford Square, London WC1B 3HH

LONDON MELBOURNE JOHANNESBURG AUCKLAND

First published 1988

British Library Cataloguing in Publication Data
Tooley, Michael H.
Electronic circuits handbook: design,
testing and construction.
1. Electronic circuits
I. Title
621.3815'3 TK7867

ISBN 0 434 91968 3

Phototypesetting by Thomson Press (India) Ltd, New Delhi
Printed in England by
Robert Hartnoll Ltd, Bodmin, Cornwall

Contents

Introduction

Welcome to the world of electronic circuit design! Whether you are an electronic engineer, technician, student, or just an enthusiast working from home, this book contains something of interest to you. This book aims to explode two popular misconceptions concerning the design of electronic circuits: that only those with many years of experience should undertake circuit design and that the process relies on an understanding of advanced mathematics. Provided one is not too ambitious, neither of these popularly held beliefs is true.

Specifically, this book aims to provide the reader with a unique collection of practical working circuits together with supporting information so that circuits can be produced in the shortest possible time and without recourse to theoretical texts.

All circuits described have been thoroughly tested and, as far as possible, a range of commonly available low-cost components has been adopted. The simplest and most cost-effective solution being adopted in each case.

Furthermore, information has been included so that the circuits can readily be modified and extended by readers to meet their own individual needs. Related circuits have been grouped together and cross-referenced within the text (and also in the index) so that readers are aware of which circuits can be readily connected together to form more complex systems. As far as possible, a common range of supply voltages, signal levels and impedances has been adopted.

As a bonus, ten test gear projects have been included. These not only serve to illustrate the techniques described but also provide a range of test equipment which is useful in its own right. These items of test gear make a good starting point for those who need to get their own projects 'up and running'. The only basic item of test equipment not described, and considered essential, is a multirange meter. Such instruments can be acquired from most electronic suppliers for under £20.

As stated previously, this book differs from the vast majority of electronic textbooks in that the principles of electronics have been kept to a bare minimum. It would, however, be foolish to suggest that electronic circuit design can be accomplished without recourse to any theory or simple mathematics. This book assumes that the reader has at least an elementary understanding of electrical principles and, in particular, is familiar with common electrical units and quantities. Provided the reader has this knowledge and can perform simple arithmetic calculations, there should be no difficulty in following the mathematics presented here.

The book is organized into eleven chapters, each chapter devoted to a different topic. These chapters have been arranged in a logical sequence; each one building upon its predecessor. Despite this, the book has not been designed to be read from 'cover to cover'; rather, the aim has been that of presenting related information in as concise a manner as possible and without duplication.

Design philosophy

Electronic circuit design should be a systematic process resulting from the identification of a particular need. One should start by declaring one's overall aims and, in a larger project involving several subsystems, it will usually be necessary to define specific objectives and establish the sequence in which they should be completed.

Draft specifications should be prepared at an early stage. These will be instrumental in identifying relevant parameters as well as establishing a target performance specification. It is, however, important to remember that such specifications should remain flexible since, in practice, it may not be feasible to meet them within the available budget. Conversely, with advances in technology, specifications (particularly those relating to speed of operation, power requirements, and physical size) may well be exceeded.

Even the most 'dyed-in-the-wool' designer should be prepared to examine the widest possible range of solutions – not simply those drawn from a 'standard library' of designs. Only then should circuitry and component specifications be drawn up. With more complex projects, or those involving new and untried devices, it will usually be necessary to 'breadboard' and test specific areas of uncertainty.

Prototype circuit layouts should not be produced until one is reasonably confident that the finished design will satisfy the criteria laid down. To do otherwise would be foolhardy and is likely to be counter-productive in the long term.

Following construction, prototypes should be exhaustively tested against the draft specifications and under a variety of conditions (e.g. ambient temperature, supply voltage, etc.). Where necessary, modifications should be incorporated but note that, provided one has reasonable confidence in the design prior to prototyping, it should rarely be necessary to rebuild a prototype from scratch.

Final acceptance of a particular design involves not only verifying that the target specifications have been met but also that the originally declared aims and objectives have been fulfilled. The procedure is more meaningful if it is undertaken in conjunction with the end user.

An electronic circuit may be judged effective against a number of criteria. Most important of these is the attainment of a standard of performance which is consistent with current practice. Performance, however, can only be objectively assessed using clearly defined parameters. These must be defined or agreed at an early stage in the process.

While bearing in mind the need to attain a set of specified standards, designers should be constantly aware of the need to make their designs as cost-effective as possible. Rarely can there be any justifications for over-engineering a product (e.g. using components which are consistently overrated). Commercial constraints will inevitably dictate that the overall standard of engineering is merely sufficient to satisfy the basic design criteria.

Other factors which designers must bear in mind involve tolerance of a wide variation in component values, semiconductor characteristics (particularly transistor current gain), supply voltage, and ambient temperature. Designers should also ensure that their designs are only based on readily available components. Devices should preferably be selected from 'industry standard' ranges which have not only established a proven track record but also are available from a number of sources. New devices should not be adopted without prior evaluation nor should they find their way into production equipment without firm evidence of continuing availability.

Finally, a few words of advice and encouragement for the newcomer. Your 'success rate' will certainly improve with practice so do not be too disheartened if your first efforts do not match up fully to expectations. Keep a record of everything that you do so that you can build up a library of your own 'tried and tested' circuits. You will soon learn to recognize the circuits that you can trust and those which have pitfalls for the unwary. Last, but not least, do not be afraid to experiment – many hours of fun can be had from trying out new ideas even if they do not all work as planned.

Michael Tooley

1 Passive components

Complex electronic circuits may comprise many hundreds of individual components. These components fall into various categories but a distinction is usually made between those components which do not in themselves provide gain (such as resistors, capacitors and inductors) and those which provide amplification or switching (such as transistors and integrated circuits).

This section is devoted to common types of passive component and it aims to provide the reader with sufficient information to make an informed selection of passive components for use in the circuits which follow later.

Resistors

Resistors are essential to the functioning of almost every electronic circuit and provide us with a means of controlling the current and/or voltage present. Typical applications of resistors involve, among other things, the provision of bias potentials and currents for transistor amplifiers, changing an output current into a corresponding output voltage drop, and providing a predetermined value of attenuation.

The type and construction of a resistor are largely instrumental in determining its electrical characteristics. In any application we may be concerned with the following:

(a) *Resistance.* The required value quoted in Ω, kΩ or MΩ.

(b) *Power rating.* The maximum power dissipated by the resistor which is given by:

$$P = \frac{V^2}{R} = I^2R$$

(c) *Tolerance (or accuracy).* The maximum allowable deviation from the marked value (quoted as a percentage).

(d) *Temperature coefficient.* The change in resistance (usually expressed in parts per million, p.p.m.) per unit temperature change.

(e) *Noise performance.* The equivalent noise voltage generated by the resistor when subjected to a given set of physical and electrical conditions.

(f) *Stability.* The variation in resistance value (expressed as a percentage) which occurs under specified conditions and over a given period of time.

We shall now briefly examine each of the common types of resistor:

Type: Carbon composition.

Characteristics: Low cost. Inferior tolerance, temperature coefficient, and noise performance. Very poor long-term stability.
Construction: Moulded carbon composition (carbon, filler and resin binder) element.
Power ratings: 0.125 W, 1 W
Range of values: 2.2 Ω to 1 MΩ.
Typical tolerance: ± 10% (± 0.5 Ω for values less than 4.7 Ω).
Ambient temperature range: − 40 °C to + 105 °C.
Temperature coefficient: + 1200 p.p.m./°C.
Typical applications: General purpose resistor suitable for uncritical applications (e.g. large-signal amplifiers and power supplies).
Notes: Carbon film types should be used in preference.

Type: Carbon film.
Characteristics: Improved tolerance, noise performance and stability compared with carbon composition types.
Construction: Carbon film deposited on a ceramic former. A groove is cut into the film in order to determine the precise resistance before a protective coating is employed.
Power ratings: 0.25 W, 0.5 W, 1 W and 2 W
Range of values: 10 Ω to 10 MΩ.
Typical tolerance: ± 5%.
Ambient temperature range: − 45 °C to + 125 °C.
Temperature coefficient: − 250 p.p.m./°C.
Typical applications: General purpose applications including bias, load, and pull-up resistors.
Notes: Wirewound types should be used at power levels exceeding 2 W.

Type: Metal film.
Characteristics: Low temperature coefficient, close tolerance, high stability.
Construction: Metal alloy film deposited on a ceramic former and coated with cement.
Power ratings: 0.125 W, 0.25 W, 0.5 W.
Range of values: 10 Ω to 1 MΩ typical (resistance values as low as 0.22 Ω are also available).
Typical tolerance: ± 1% (± 10% for values below 1 Ω).
Ambient temperature range: − 55 °C to + 125 °C.
Temperature coefficient: + 50 p.p.m./°C and + 100 p.p.m./°C.

Typical applications: General purpose and low-noise circuitry ideal for bias and load resistors in low-level amplifier circuits.
Notes: Also available in leadless 'chip' encapsulation for surface mounting.

Type: Metal oxide.
Characteristics: Very low noise, high stability and reliability.
Construction: Tin oxide film bonded to an alkali-free glass former.
Power ratings: 0.5 W typical.
Range of values: 10 Ω to 1 MΩ.
Typical tolerance: ± 2%.
Ambient temperature range: − 55 °C to + 150 °C.
Temperature coefficient: + 250 p.p.m./°C.
Typical applications: General purpose resistor ideally suited for use in small-signal and low-noise amplifier stages.
Notes: Metal film resistors generally offer significantly lower temperature coefficient and closer tolerance.

Type: Aluminium clad wirewound.
Characteristics: Very high dissipation.
Construction: Resistive element coated with a silicone compound and bonded to an aluminium heatsink case.
Power ratings: 25 W, 50 W (but see note 1).
Range of values: 0.1 Ω to 1 kΩ.
Typical tolerance: ± 5%.
Ambient temperature range: − 55 °C to + 200 °C.
Temperature coefficient: 50 p.p.m./°C typical.
Maximum surface temperature: 200 °C.
Typical applications: Power supplies and high power loads.
Notes:
1 Power ratings usually assume the use of a heatsink of 4.2 °C/W (or better). 25 W and 50 W types should thus be derated to 12 W and 20 W respectively when mounted in free-air.
2 Aluminium clad wirewound types exhibit both inductive and capacitive reactance which will limit their performance at high frequencies. Above 100 kHz, or so, special high-dissipation carbon resistors will normally be preferred.

Type: Ceramic wirewound.
Characteristics: High dissipation.

Construction:	Resistive element wound on glass-fibre core and protected by a ceramic body.
Power ratings:	4 W, 7 W, 11 W and 17 W.
Range of values:	0.47 Ω to 22 kΩ.
Typical tolerance:	± 5%.
Ambient temperature range:	− 55 °C to + 200 °C.
Maximum surface temperature:	310 °C.
Typical applications:	Power supplies.
Notes:	In common with most wirewound types, these resistors exhibit self-inductance which will limit their performance at high frequencies.

Type:	Silicon and vitreous enamel wirewound.
Characteristics:	High dissipation.
Construction:	Resistive element wound on ceramic core and coated with silicon or vitreous enamel.
Power ratings:	2.5 W typical.
Range of values:	0.1 Ω to 22 kΩ.
Typical tolerance:	± 5%.
Ambient temperature range:	− 55 °C to + 200 °C.
Maximum surface temperature:	350 °C.
Typical applications:	Power supplies, power amplifiers and drivers.
Notes:	The insulation resistance of the vitreous enamel coating falls with increasing temperature. If the resistor is likely to be operated near the maximum permitted dissipation, it should be mounted on insulated stand-off pillars and should not be in contact with any conducting surface or any other component.

Variable resistors

Variable resistors are available in a variety of forms, the most popular of which are the carbon track and wirewound potentiometers. Carbon types are suitable for low power applications (generally less than 1 W) and are available at relatively low cost with values which range from around 5 kΩ to 1 MΩ. They do, however, suffer noise and track wear which may make them unacceptable for critical applications (such as instrumentation, low-noise amplifiers etc.)

Carbon potentiometers are available with linear and semilogarithmic law tracks (the latter being preferred for such applications as volume controls) and in rotary or linear (slider) formats. Wirewound potentiometers are generally only available with linear law tracks, power ratings up to 3 W, and values which range from 10 Ω to 100 kΩ.

In addition, ganged controls are available in which several carbon potentiometers are mechanically connected to a common control shaft. Very fine adjustment is made possible with the use of precision multi-turn wirewound potentiometers. These devices are, naturally, rather expensive and their use tends to be confined to precision measuring and test equipment.

Preset resistors

Various forms of preset resistor are in common use. Such components generally use carbon tracks and are available for both horizontal and vertical mounting. The lowest cost types (skeleton presets) use open carbon tracks. Multi-turn presets are also available and, like their fully variable counterparts, are considerably more expensive than their conventional counterparts.

Preferred values

It is important to realize that the resistance marked on the body of a resistor is only a guide to its actual value of resistance. The accuracy of the marked value (in terms of the maximum percentage deviation allowable) is known as its tolerance. A resistor marked 100 Ω and having a tolerance of 10% will, for example, have a value which falls within the range 90 Ω to 110 Ω. If our circuit requires a resistance of 95 Ω, then a resistor of this type should prove to be quite adequate. If, on the other hand, we require a value of 99 Ω, then it would be necessary to obtain a resistor having a marked value of 100 Ω and a tolerance of 1%.

Resistors are available in several series of fixed decade values, the number of values provided with each series being governed by the tolerance involved. In order to cover the full range of resistance values using resistors having a 20% tolerance, for example, the following six decade values (known as the 'E6 series') will be required:

1.0, 1.5, 2.2, 3.3, 4.7, 6.8

The 'E12 series' for 10% tolerance resistors is:

1.0, 1.2, 1.5, 1.8, 2.2, 2.7, 3.3, 3.9, 4.7, 5.6, 6.8, 8.2

Whilst the 'E24 series' for resistors of 5% tolerance is:

1.0, 1.1, 1.2, 1.3, 1.5, 1.6, 1.8, 2.0, 2.2, 2.4, 2.7, 3.0
3.3, 3.6, 3.9, 4.3, 4.7, 5.1, 5.6, 6.2, 6.8, 7.5, 8.2, 9.1

Less values are available for variable resistors (for obvious reasons) and these are usually supplied in the following decade ranges:

1.0, 2.2 or 2.5, 4.7 or 5.0

Colour codes

Carbon and metal oxide resistors are invariably marked with colour codes which indicate their value and tolerance. Two methods of colour coding

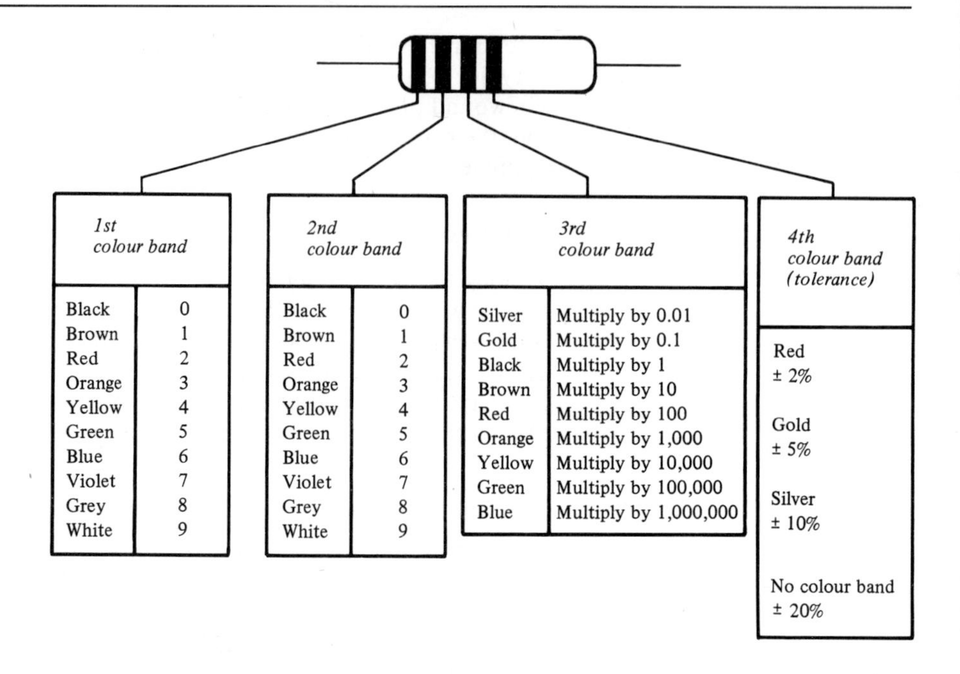

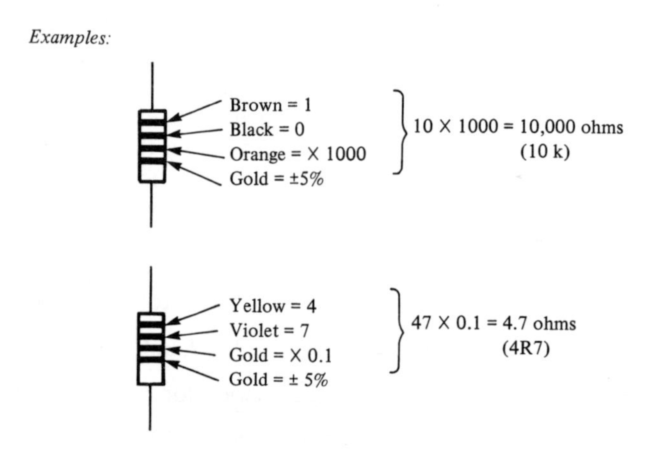

Figure 1.1 *Colour code for resistors with four bands*

are in common use: one involves four coloured bands (see Figure 1.1) while the other uses five colour bands (see Figure 1.2).

Other types of resistor have values marked using the system of coding defined in BS 1852. This system involves marking the position of the decimal point with a letter to indicate the multiplier concerned as follows:

Letter	*Multiplier*
R	1
K	1000
M	1000000

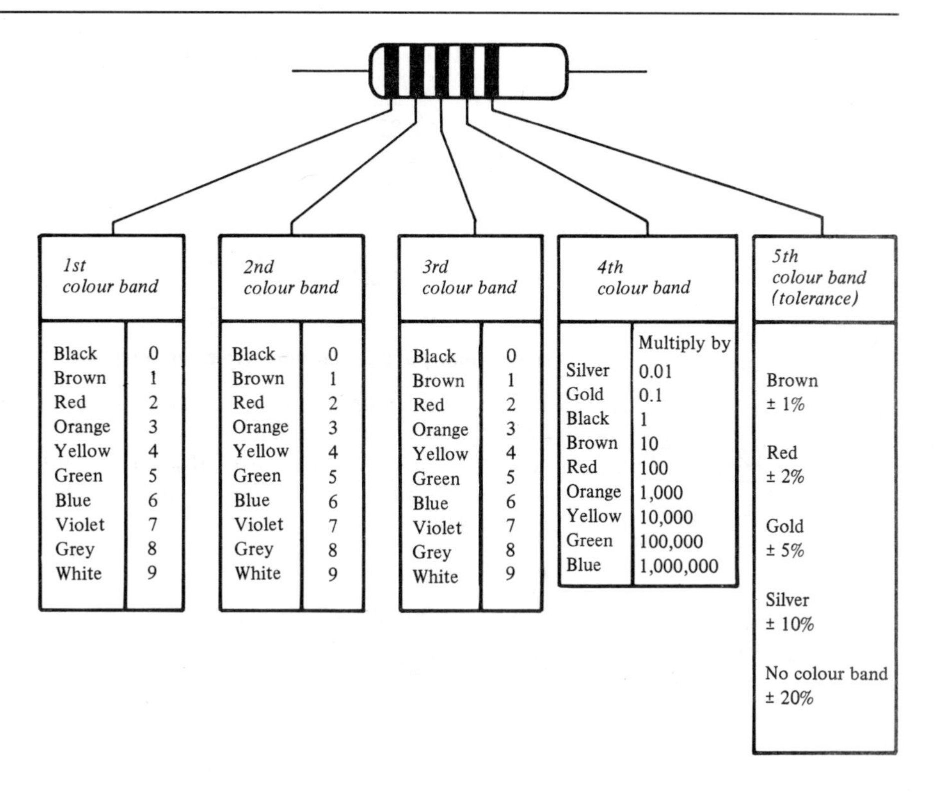

Examples:

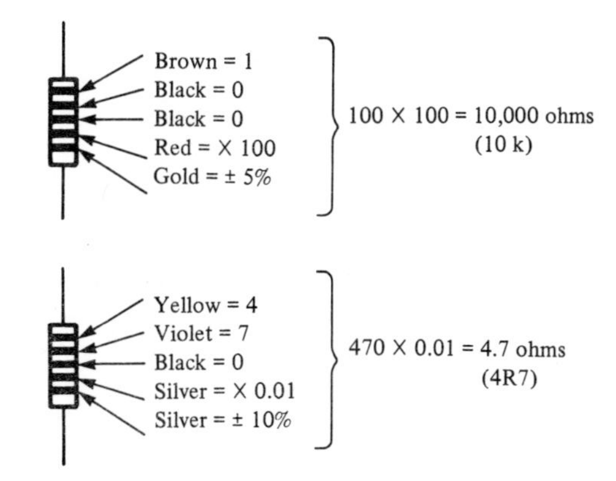

Figure 1.2 *Colour code for resistors with five bands*

A further letter is then appended to indicate the tolerance on the following basis:

Letter	*Tolerance*
F	±1%
G	±2%
J	±5%
K	±10%
M	±20%

The following examples should help make this method of coding clear:

Coding	*Value*	*Tolerance*
R22M	0.22 Ω	±20%
4R7K	4.7 Ω	±10%
68RJ	68 Ω	±5%
330RG	330 Ω	±2%
1M0F	1 MΩ	±1%
5M6M	5.6 MΩ	±20%

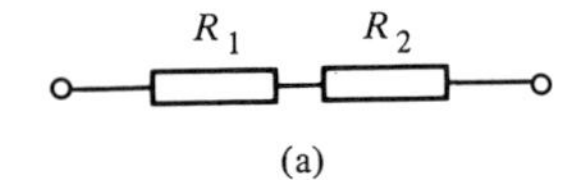

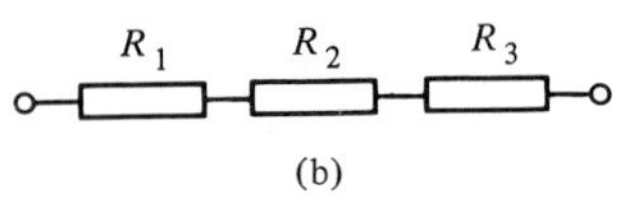

Figure 1.3 *Series connected resistors*

Series and parallel combinations of resistors

In order to obtain a particular value of resistance, fixed resistors may be arranged in either series or parallel as shown in Figures 1.3 and 1.4. The effective resistance of each of the series networks shown in Figure 1.3 is simply equal to the sum of the individual resistances. Hence, for Figure 1.3(a).

$$R = R_1 + R_2$$

Whilst for Figure 1.3(b).

$$R = R_1 + R_2 + R_3$$

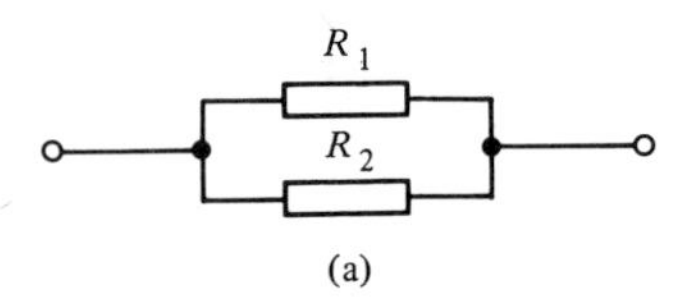

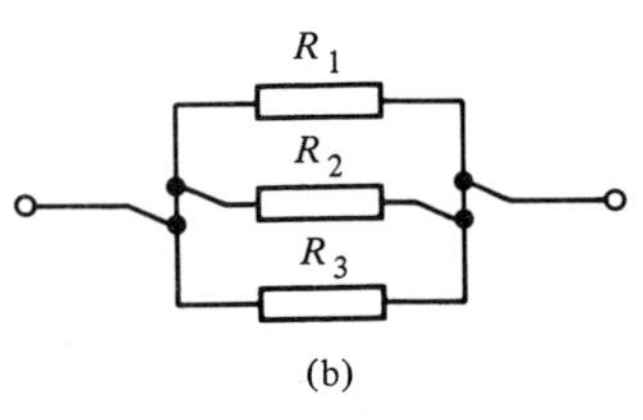

Figure 1.4 *Parallel connected resistors*

Turning to the parallel resistors shown in Figure 1.4, the reciprocal of the effective resistance of each network is equal to the sum of the reciprocals of the individual resistances. Hence, for Figure 1.4(a):

$$\frac{1}{R} = \frac{1}{R_1} + \frac{1}{R_2}$$

While for Figure 1.4(b):

$$\frac{1}{R} = \frac{1}{R_1} + \frac{1}{R_2} + \frac{1}{R_3}$$

In the former case, the formula can be more conveniently rearranged as follows:

$$R = \frac{R_1 \times R_2}{R_1 + R_2}$$

(This can be simply remembered as the product of the two resistor values divided by the sum of the two values.)

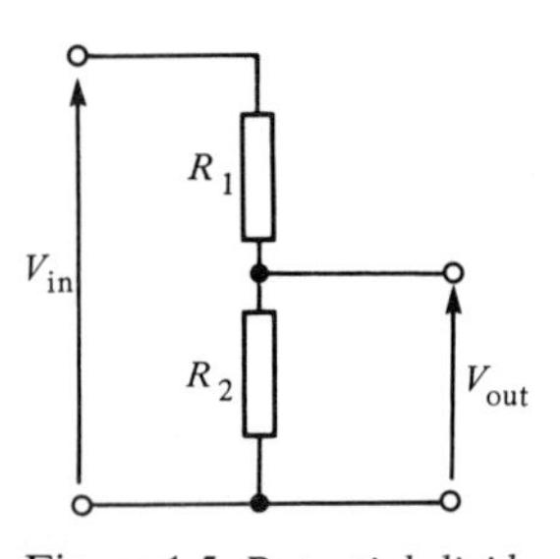

Figure 1.5 *Potential divider*

The potential divider

A common application of resistors is that of providing a simple form of attenuator in which an input voltage is reduced by a factor determined by the resistor values present. Figure 1.5 shows a typical potential divider arrangement. The output voltage produced by the circuit is given by:

$$V_{out} = V_{in} \times \frac{R_2}{R_1 + R_2}.$$

It is, however, important to note that the output voltage will fall when current is drawn from the arrangement by a load connected to its output terminals. A general rule of thumb is that the resistance of the load (ideally infinite) should be at least ten times the value of R_2.

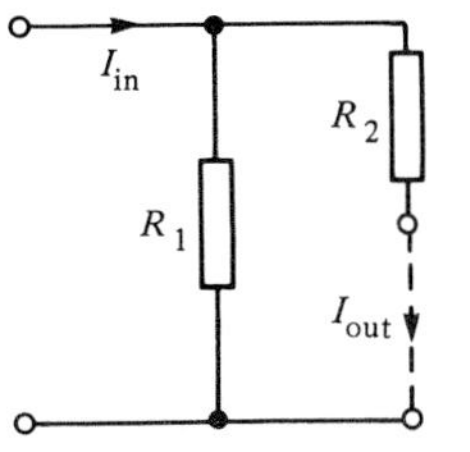

Figure 1.6 *Current divider*

The current divider

Another application of resistors is that of providing a means of diverting current from one branch of a network to another. In this case, the input current is divided by a factor which is determined by the resistor values present. Figure 1.6 shows a typical current divider arrangement. The current flowing between the output terminals is given by:

$$I_{out} = I_{in} \times \frac{R_1}{R_1 + R_2}$$

It is, however, important to note that the output current will fall when the load connected between the output terminals has any appreciable resistance. A general rule of thumb is that the resistance of the load (ideally zero) should be no more than one tenth of the value of R_2.

High frequency effects

We have already mentioned that the inductive nature of wirewound resistors will generally preclude their use at frequencies in excess of about 100 kHz. Other types of resistor also suffer from stray inductance which is primarily attributable to their axial connecting leads. These usually possess a self-inductance of around 0.004 μH/mm. A 10 Ω carbon film resistor having leads, each of length 10 mm, will thus exhibit a total self-inductance of approximately 0.08 μH. Furthermore, the total stray capacitance which exists between the two ends of the resistor will be in the region of 1 pF. Hence, we have a component which, although purporting to be a resistor, also exhibits capacitance and inductance. The consequence of all this is simply that impedance of the component will vary in a complex manner with frequency.

Happily, for many applications these relatively small values of capacitance and inductance will not prove too problematic. Difficulties do, however, arise with wirewound components for which the values of self-inductance and stray capacitance are significantly larger. For low values of wirewound resistor (470 Ω or less), the lead inductance has the effect of increasing the impedance of the resistor at frequencies in excess of several hundred kHz.

For larger values of resistor (2.2 kΩ and above) the shunt capacitance becomes dominant and the impedance generally falls with frequency above 500 kHz. Intermediate values of resistance (in the region of 1 kΩ) are, fortunately, less prone to variation of impedance with frequency as the effect of increasing inductive reactance is largely counteracted by falling capacitive reactance.

Thermistors

One of the desirable properties of an ordinary resistor is that its value should not be susceptible to too much variation over a fairly wide range of temperatures. Thermistors (thermally sensitive resistors), on the other hand, are deliberately chosen to have a resistance characteristic which varies markedly with temperature. They can thus be used as temperature sensing and compensating elements.

There are two basic types of thermistor: negative temperature coefficient (n.t.c.) and positive temperature coefficient (p.t.c.). Typical n.t.c. thermistors have resistances which vary from a few hundred (or thousand) ohms at 25 °C to a few tens (or hundreds) of ohms at 100 °C. P.t.c. thermistors, on the other hand, usually have a resistance-temperature characteristic which remains substantially flat (typically at around 100 Ω) over the range 0 °C to around 75 °C. Above this, and at a critical temperature (usually in the range 80 °C to 120 °C) their resistance rises very rapidly to values of up to, and beyond, 10 kΩ.

Capacitors

Like resistors, capacitors are also crucial to the correct working of nearly every electronic circuit and provide us with a means of storing electrical energy in the form of an electric field. Capacitors have numerous applications including storage capacitors in power supplies, coupling of a.c. signals between the stages of an amplifier, and decoupling power supply rails so that, as far as a.c. signal components are concerned, the supply rails are indistinguishable from 0 V.

As with resistors, the type and construction of a capacitor will be instrumental in determining its electrical characteristics. In any application we may be concerned with the following:

(a) *Capacitance.* The required value quoted in μF, nF or pF.

(b) *Voltage rating.* The working voltage rating for a capacitor is the maximum voltage which can be continuously applied to the capacitor. Exceeding this value can have dire consequences and the dielectric may eventually rupture and the capacitor will suffer irreversible damage. Where the voltage rating is expressed in terms of a direct voltage (e.g. 250 V d.c.) this is normally related to the maximum working temperature. It is, however, wise to operate capacitors with a considerable margin for safety. This helps promote long-term reliability. As a rule of thumb, the working d.c. voltage should be limited to no more than 50% of the rated d.c. voltage.

Where an a.c. voltage rating is specified this is normally for sinusoidal operation at either 50 Hz or 60 Hz. Performance will not be significantly affected over the audio frequency range (up to 100 kHz) but, above this, or when non-sinusoidal (e.g. pulse) waveforms are involved the capacitor must be derated in order to minimize dielectric losses which in turn may result in excessive internal heating.

(c) *Tolerance (or accuracy)*. The maximum allowable deviation from the marked value (quoted as a percentage).

(d) *Temperature coefficient*. The change in capacitance (usually expressed in parts per million, p.p.m.) per unit temperature change.

(e) *Leakage current*. The d.c. current flowing in the dielectric when the rated d.c. voltage is applied (normally quoted at a given temperature).

(f) *Insulation resistance*. The resistance of the dielectric when the rated d.c. voltage is applied (normally quoted at a given temperature).

(g) *Stability*. The variation in capacitance value (expressed as a percentage) which occurs under specified conditions and over a given period of time.

We shall now briefly examine each of the common types of capacitor:

Type:	Ceramic.
Characteristics:	Small size, low inductance.
Construction:	Metallized ceramic plates with hard lacquer or epoxy coating. Available as monolithic and multilayer types.
Voltage rating:	100 V d.c. typical.
Range of values:	2.2 pF to 220 pF (plate). 10 pF to 1 μF (multilayer). 1 nF to 100 nF (disk).
Typical tolerance:	± 10%, ± 20%.
Ambient temperature range:	− 85 °C to + 85 °C typical.
Temperature coefficient:	+ 100 p.p.m./d °C to − 4700 p.p.m./°C
Insulation/leakage:	$> 10^{10}\,\Omega$.
Typical applications:	Medium and high frequency decoupling. Timing. Temperature compensation of oscillator circuits.
Notes:	1 Monolithic types are also available in 'chip' form for surface mounting. 2 The following markings are commonly used for temperature compensating capacitors:

Marking	*Temperature coefficient*	*Tip colour*
NP0	0 p.p.m./°C	Black
N030	− 30 p.p.m./°C	Brown
N080	− 80 p.p.m./°C	Red
N150	− 150 p.p.m./°C	Orange
N220	− 220 p.p.m./°C	Yellow
N330	− 330 p.p.m./°C	Green
N470	− 470 p.p.m./°C	Blue
N750	− 750 p.p.m./°C	Violet
N1500	− 1500 p.p.m./°C	Orange/orange

N2200	-2200 p.p.m./°C	Yellow/orange
N3300	-3300 p.p.m./°C	Green/orange
N4700	-4700 p.p.m./°C	Blue/orange

3 Ceramic feed-through capacitors are available for very high frequency decoupling. Such components are soldered directly into a bulkhead or screening enclosure and are typically rated at 1 nF 350 V.
4 Modern monolithic ceramic capacitors are marked with a three-digit code. The first two digits correspond to the first two digits of the value while the third digit is a multiplier which gives the number of zeros to be added to give the value in pF. Monolithic ceramics marked '222' and '103' will have values of 2200 pF (2.2 nF) and 10000 pF (10 nF) respectively.

Type: Electrolytic.

Characteristics: Relatively large values of capacitance are possible. Electrolytic capacitors are polarized and require the application of a d.c. polarizing voltage.

Voltage rating: 6.3 V to 400 V.

Range of values: 0.1 μF to 68 nF.

Typical tolerance: ±20% (large 'can' types exhibit poorer tolerance, -10% to $+50\%$ being typical).

Ambient temperature range: -40 °C to $+85$ °C.

Temperature coefficient: $+1000$ p.p.m./°C typical.

Leakage current: Typically equal to $0.01\,C\,V$ (where C is the capacitance in μF and V is the applied d.c. polarizing voltage in V).

Typical applications: Power supply 'reservoir' capacitors. Low frequency decoupling.

Notes:
1 Non-polarized electrolytic capacitors are available for special applications (e.g. loudspeaker cross-over networks) in which a d.c. polarizing voltage cannot be readily obtained. Typical values for non-polarized electrolytics range from 1 μF to 100 μF and such devices are normally rated for operation at up to 50 V r.m.s. (These capacitors are, however, more expensive than conventional polarized electrolytics.)
2 When selecting a reservoir capacitor, care must be taken to ensure that the device has an adequate ripple current rating.

3 Care must be exercised when using electrolytic capacitors for interstage coupling as the leakage current may significantly affect the bias conditions.
4 Electrolytic capacitors exhibit a fairly wide tolerance and hence it is usually quite permissible to substitute one value for another provided, of course, one selects a working voltage of the same, or higher, value. As an example, a circuit which specifies a 50 μF capacitor rated at 25 V will invariably operate successfully with a 47 μF component rated at 35 V.

Type: Metallized film.
Characteristics: Medium values of capacitance suitable for high voltage applications. Expensive and somewhat bulky.
Construction: Metallized film dielectric encapsulated in tubular or rectangular plated steel housing.
Voltage rating: 600 V d.c. (250 V a.c.) typical.
Range of values: 2 μF, 4 μF, 8 μF, 16 μF typical.
Typical tolerance: ±20%.
Ambient temperature range: −25 °C to +85 °C.
Temperature coefficient: +100 p.p.m./°C to +200 p.p.m./°C.
Insulation resistance: $> 10^{10}\ \Omega$.
Typical applications: Reservoir capacitors in high voltage d.c. power supplies. Power factor correction in a.c. circuits.
Notes:
1 Working voltages should be derated at high temperatures.
2 Capacitors of this type used in high-voltage circuits may retain an appreciable charge for some considerable time. In such cases, a carbon film 'bleed' resistor (of typically 1 MΩ 0.5 W) should be connected in parallel with the capacitor.

Type: Mica.
Characteristics: Stable, low temperature coefficient, close tolerance.
Construction: Cement-coated silvered mica plate.
Voltage rating: 350 V d.c. typical.
Range of values: 2.2 pF to 10 nF.
Typical tolerance: ±1% (or +0.5 pF for values below 50 pF).
Ambient temperature range: −40 °C to +85 °C.
Temperature coefficient: +75 p.p.m./°C for values below 50 pF.
+35 p.p.m./°C for values above 50 pF.

Insulation resistance: Typically $> 5 \times 10^{10}\,\Omega$.
Typical applications: High frequency oscillators, timing, filters, pulse applications.
Notes: Lower cost polystyrene types should be used in low voltage circuits where tolerance and stability are relatively unimportant.

Type: Polycarbonate.
Characteristics: High stability and excellent temperature characteristics coupled with small physical size.
Construction: Metallized polycarbonate film sealed in epoxy resin.
Voltage rating: 63 V d.c. (45 V a.c.), 160 V d.c. (100 V a.c.), 630 V d.c. (300 V a.c.)
Range of values: 10 nF to 10 μF.
Typical tolerance: $\pm 20\%$.
Ambient temperature range: $-55\,^{\circ}$C to $+100\,^{\circ}$C.
Temperature coefficient: $+60$ p.p.m./°C.
Insulation resistance: Typically $> 10^{10}/C$ (where C is in μF).
Typical applications: Timing and filter circuits.
Notes:
1. Not recommended for use in conjunction with a.c. mains – use polypropylene instead.
2. PCB mounting polycarbonate 'block' types have extremely low inductance.

Type: Polyester.
Characteristics: General purpose.
Construction: Metallized polyester film dielectric. The capacitor element is either resin dipped or encapsulated in a tubular plastic sleeve.
Voltage rating: 250 V d.c. (125 V a.c.), 400 V d.c. (200 V a.c.)
Range of values: 10 nF to 2.2 μF.
Typical tolerance: $\pm 20\%$.
Ambient temperature range: $-40\,^{\circ}$C to $+100\,^{\circ}$C.
Temperature coefficient: $+200$ p.p.m./°C typical.
Insulation resistance: Typically $> 10^{10}/C$ (where C is in μF).
Typically applications: General purpose coupling and decoupling.
Notes:
1. Polyester capacitors are generally unsuitable for critical timing applications due to their relatively large temperature coefficient.
2. Resin dipped types exhibit much lower values of self-inductance than tubular types and are thus preferred for decoupling applications.
3. The colour code shown in Figure 1.7 was, until relatively recently, commonly used

to mark resin dipped polyester capacitors.

4 The modern method of marking resin dipped polyester capacitors involves the following system of coding:
   First line: capacitance (in pF or μF) and tolerance (K = 10%, M = 20%).
   Second line: rated d.c. voltage and code for dielectric material.

Type: Polypropylene.
Characteristics: Very low loss dielectric suitable for high a.c., d.c. voltages and pulse applications.
Construction: Epoxy cased foil and polypropylene film
Voltage rating: 1 kV d.c. (350 V a.c.), 1.5 kV d.c. (450 V a.c.).
Range of values: 1 nF to 470 nF.
Typical tolerance: ±20%.
Ambient temperature range: −55 °C to +100 °C.
Temperature coefficient: −200 p.p.m./°C.
Insulation resistance: $> 10^{11}\,\Omega$.
Typical applications: Coupling and decoupling in high voltage circuits, mains filters.
Notes: Polypropylene capacitors are relatively expensive and should be reserved for use in applications in which polyester capacitors would be inadequately rated.

Type: Polystyrene.
Characteristics: Low-cost, close tolerance capacitor available only in relatively low values and only suitable for low voltage applications.
Construction: Polystyrene dielectric rolled into tubular form.
Voltage rating: 160 V d.c. (40 V a.c.).
Range of values: 10 pF to 10 nF.
Typical tolerance: ±1%, ±2.5% and ±5%.
Ambient temperature range: −40 °C to +70 °C.
Temperature coefficient: −150 to +80 p.p.m./°C typical.
Insulation resistance: $> 10^{12}\,\Omega$.
Typical applications: Timing, filters, oscillators and discriminators (see note 2).
Notes:
1 The outer foil connection is usually coded with a red stripe and, where appropriate, this connection should be returned to 0 V.
2 Mica capacitors may be peferred for applications in which stability is important.

Type: Tantalum.
Characteristics: Relatively large values coupled with very small physical size.

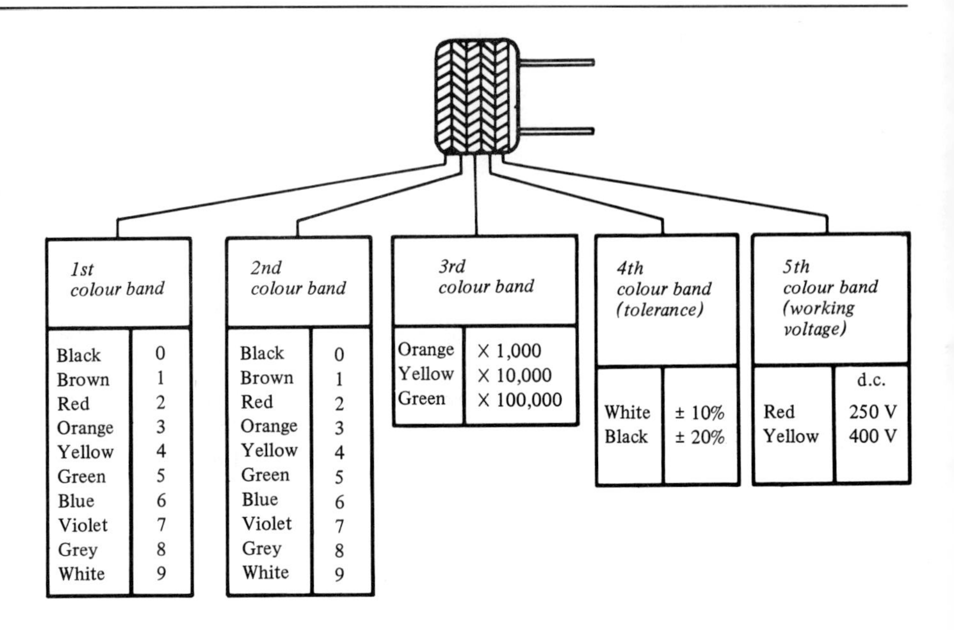

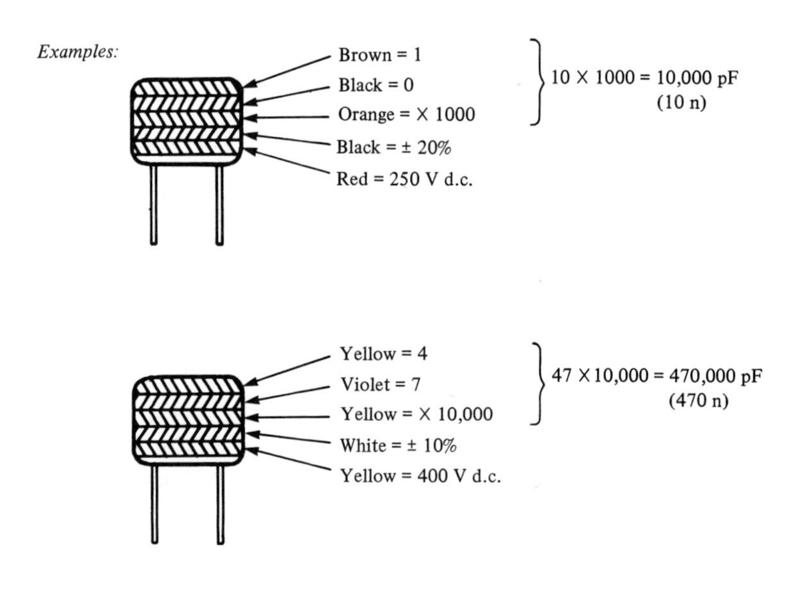

Figure 1.7 *Capacitor colour code*

Construction:	Porous tantalum pellet with electrolyte filling.
Voltage rating:	6.3 V to 35 V.
Range of values:	0.1 μF to 100 μF.
Typical tolerance:	$\pm 20\%$.
Ambient temperature range:	-55 °C to $+85$ °C.

Temperature coefficient:	+ 100 p.p.m./°C to + 250 p.p.m./°C.
Leakage current:	Typically less than 1 μA.
Typical applications:	Coupling and decoupling.
Notes:	Large values of tantalum capacitor are relatively expensive and modern radial lead electrolytics can often be substituted at much lower cost.

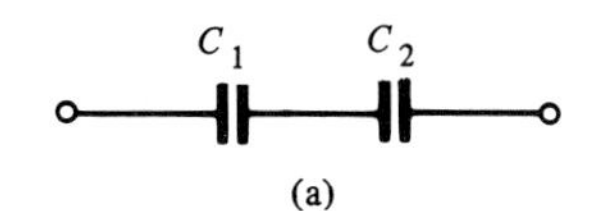

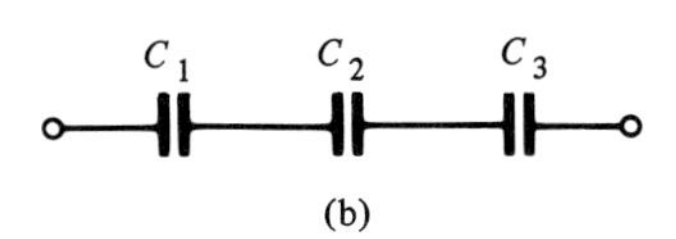

Figure 1.8 *Series connected capacitors*

Series and parallel combination of capacitors

In order to obtain a particular value of capacitance, fixed capacitors may be arranged in either series or parallel as shown in Figures 1.8 and 1.9. The reciprocal of the effective capacitance of each of the series networks shown in Figure 1.7 is equal to the sum of the reciprocals of the individual capacitances. Hence, for Figure 1.8(a):

$$\frac{1}{C}=\frac{1}{C_1}+\frac{1}{C_2}$$

While for Figure 1.8(b):

$$\frac{1}{C}=\frac{1}{C_1}+\frac{1}{C_2}+\frac{1}{C_3}$$

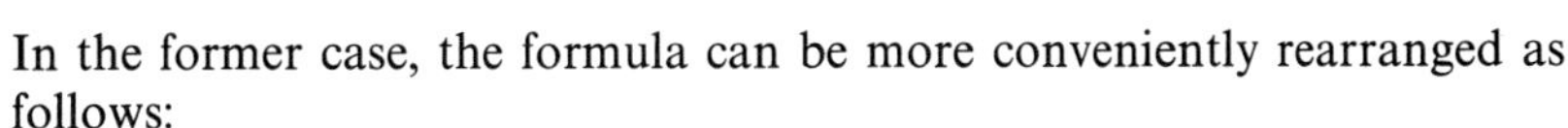

In the former case, the formula can be more conveniently rearranged as follows:

$$C=\frac{C_1 \times C_2}{C_1+C_2}$$

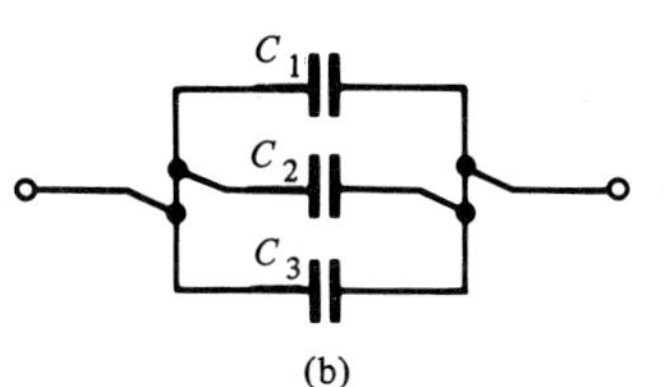

Figure 1.9 *Parallel connected capacitors*

(This can be simply remembered as the product of the two capacitor values divided by the sum of the two values.)

For parallel arrangements of capacitors, the effective capacitance of the network is simply equal to the sum of the individual capacitances. Hence, for Figure 1.9(a):

$$C=C_1+C_2$$

Whilst for Figure 1.9(b):

$$C=C_1+C_2+C_3$$

Capacitors in a.c. circuits

The reactance of a capacitor is defined as the ratio of applied voltage to current and it is measured in Ω. The reactance of a capacitor depends not only upon the value of capacitance but also upon the frequency of the applied voltage. The following formula can be used to determine the capacitive reactance:

$$X_C=\frac{V_C}{I_C}=\frac{1}{2\pi fC}$$

Where X_c is the reactance in Ω, f is the frequency in Hz, and C is the capacitance in F.

As an example, consider the case of a 100 nF capacitor which forms part of a filter connected across a 240 V 50 Hz mains supply. The reactance of the capacitor will be:

$$X_C = \frac{1}{2\pi 50 \times 100 \times 10} = 31.8\,\text{k}\Omega$$

The r.m.s. current flowing in the capacitor will thus be:

$$I_C = \frac{240}{31.8} = 7.55\,\text{mA}$$

High frequency effects

Like resistors, capacitors possess stray lead inductance. This has the effect of presenting a resonant L-C circuit in which the stray inductance of the leads becomes series resonant with its own capacitance. The self-resonant frequency depends upon several factors including the construction of the capacitor, the lengths of its leads, and the value of capacitance. Typical values of self-resonant frcquency for a variety of capacitor types are given in Table 1.1:

Table 1.1

Capacitor type	*Value*	*Lead length*	*Self-resonant frequency*
Tubular polycarbonate	1 μF	30 mm	750 kHz
Tubular polyester	100 nF	20 mm	4 MHz
Tubular polyester	10 nF	20 mm	10 MHz
Plate ceramic	1 nF	20 mm	30 MHz
Plate ceramic	1 nF	10 mm	45 MHz

Inductors

Inductors are less commonly used than resistors or capacitors but are important in such applications as high frequency filters and radio frequency amplifiers. Fixed radio frequency inductors are available from several manufacturers in a range of preferred values (E6 series) for values in the range 1 μH to 10 mH. Where the inductor is required to form part of a resonant circuit (in conjunction with a series or parallel connected capacitor) the requirement is usually that the inductance value should be adjustable to allow precise tuning of the arrangement. In such cases one can make use of a range of adjustable inductors produced by manufacturers (such as Toko) or wind ones own value of inductance using a suitable coil former fitted with an adjustable ferrite core.

Compact inductors for low and medium frequency use, suitable for PCB mounting and having values ranging from around 100 μH to around 100 mH, can be readily manufactured using one of the range of the RM series ferrite cores. The core material of these inductors is available in one of three grades: A13, Q3, and N28.

The RM series ferrite cores are usually supplied in kit form comprising a pair of matched core halves, a single-section bobbin (having integral 0.1 inch pitch pin connections), a pair of retaining clips, and an adjuster.

The following data refers to the RM series of ferrite cores:

Table 1.2

Core type	*RM6*	*RM6*	*RM7*	*RM10*	*RM10*
Inductance factor, A (nH/turns^2)	160	250	250	250	400
Turns factor (turns for 1 mH)	79	63.3	63.3	63.3	50
Adjustment range	+20%	+14%	+15%	+17%	+20%
Effective permeability, μ	110	171	146	100	160
Temperature coefficient (p.p.m./°C)	51–154	80–241	73–219	50–149	80–239
Frequency range (kHz)	5.5–800	3.5–700	3–650	2–650	1.2–500
Saturation flux density (mT)	250	250	250	250	250
Maximum turns on bobbin					
(using 0.2 mm diameter wire)	205	205	306	612	612
(using 0.5 mm diameter wire)	36	36	50	98	98
(using 1.0 mm diameter wire)	9	9	11	25	25

To determine the number of turns necessary for a particular inductance the following formula is used:

$$\text{Number of turns, } n = \sqrt{\frac{L}{A}}$$

Where L is the desired inductance (expressed in nH) and A is the inductance factor quoted in Table 1.2.

Several factors need to be taken into account when deciding which type of core is to be used. Paramount among these is the number of turns of wire which can be accommodated on the bobbin bearing in mind that the resistance of the winding will increase as the diameter of the wire decreases.

As a general rule of thumb, to avoid appreciable temperature rise due to I^2R losses within the winding, wires of 0.2 mm diameter should be used for r.m.s. currents not exceeding 100 mA. Wires of 0.5 mm diameter may be used for r.m.s. currents of less than 750 mA while wires of 1 mm diameter may be used with r.m.s. currents of up to 4 A.

As an example, let us assume that we need to construct a 1.5 mH (1,500,000 nH) inductor which will operate with an r.m.s. current of between 400 and 600 mA at a frequency of 2.5 kHz.

First, use the table to eliminate the cores for which the operating frequency lies outside the recommended working range. From this we would exclude the RM6 core having an inductance factor of 160.

Second, select an appropriate wire diameter, bearing in mind the working current. We would choose a wire diameter of 0.5 mm (or thereabouts).

Third, determine the number of turns required using an inductance factor of 250 (this would allow the use of either RM6, RM7 or RM10 cores):

$$\text{Hence } n = \frac{1,500,000}{250} = 77.5 \text{ turns}$$

Finally, use the table to check that this number of turns of 0.5 mm diameter can be accommodated on the bobbin. The only acceptable bobbin type is the RM10 (which can accommodate up to 98 turns of 0.5 mm diameter wire).

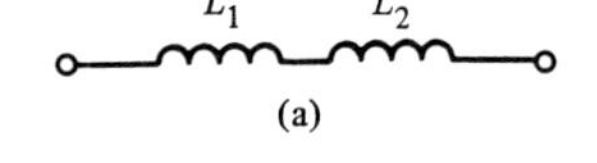

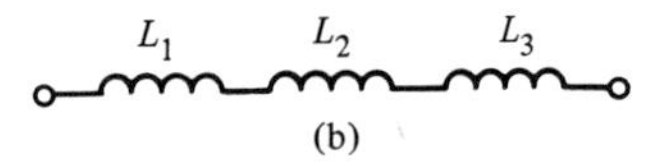

Figure 1.10 *Series connected inductors*

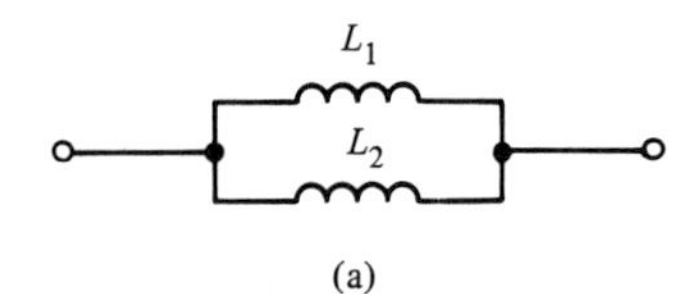

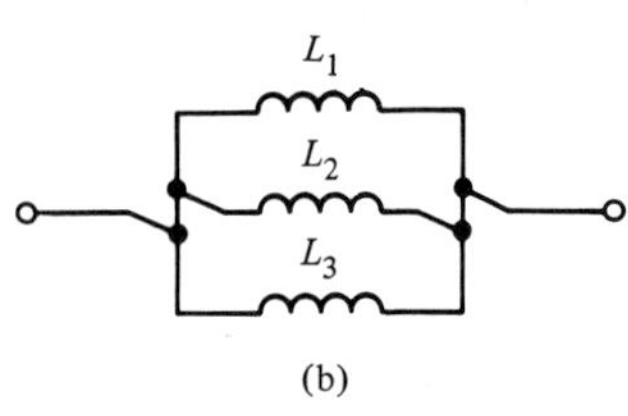

Figure 1.11 *Parallel connected inductors*

Series and parallel combinations of inductors

In order to obtain a particular value of inductance, fixed inductors may be arranged in either series or parallel as shown in Figures 1.10 and 1.11. The effective inductance of each of the series networks shown in Figure 1.10 are simply equal to the sum of the individual inductance. Hence, for Figure 1.10(a):

$$L = L_1 + L_2$$

While for Figure 1.10(b):

$$L = L_1 + L_2 + L_3$$

Turning to the parallel inductors shown in Figure 1.11, the reciprocal of the effective inductance of each of network is equal to the sum of the reciprocals of the individual inductances. Hence, for Figure 1.11(a):

$$\frac{1}{L} = \frac{1}{L_1} + \frac{1}{L_2}$$

While for Figure 1.11(b):

$$\frac{1}{L}=\frac{1}{L_1}+\frac{1}{L_2}+\frac{1}{L_3}$$

In the former case, the formula can be more conveniently rearranged as follows:

$$L=\frac{L_1 \times L_2}{L_1+L_2}$$

(This can be simply remembered as the product of the two inductance values divided by the sum of the two values.)

Inductors on a.c.

The reactance of an inductor is defined as the ratio of applied voltage to current and it is measured in Ω. The reactance of an inductor depends not only upon the value of inductance but also upon the frequency of the applied voltage. The following formulae can be used to determine the inductive reactance:

$$X_{\mathrm{L}}=\frac{V_{\mathrm{L}}}{I_{\mathrm{L}}}=2\pi fL$$

Where X_{L} is the reactance in Ω, f is the frequency in Hz, and L is the inductance in H.

As an example, consider the case of a 100 mH inductor which forms part of a filter connected in series with a 240 V 50 Hz mains supply. The reactance of the inductor will be:

$$X_{\mathrm{L}}=2\pi \times 50 \times 0.1=31.4\,\Omega$$

If the r.m.s. current flowing is 50 mA, the voltage dropped across the inductor will be:

$$V_{\mathrm{L}}=50 \times 31.4=1570\,\mathrm{mV}=1.57\,\mathrm{V}$$

Transformers

Transformers tend to fall into one of the four general categories listed below:

(a) *Mains transformers.* Such devices are designed for operation at 50 Hz or 60 Hz and occasionally at 400 Hz for aircraft applications. Cores are laminated silicon iron (usually E and I section) but torodial construction is also used.

(b) *Audio frequency transformers.* Transformers for coupling signals in the frequency range 20 Hz to 20 kHz, or thereabouts. Cores are laminated silicon iron (usually E and I sections).

(c) *High frequency transformers.* Transformers designed for operation at radio frequencies (100 kHz and above). Such devices often form part of an L-C resonant circuit and thus provide selectivity as well as impedance matching. Cores are often adjustable and take the form of threaded ferrite slugs.

(d) *Pulse transformers.* Transformers designed for pulse and wideband operation (the p.r.f. of such pulses generally falls within the range 1 kHz to 100 kHz). Ferrite cores are almost exclusively used in such applications.

When selecting a transformer, the following items should be considered:

Primary and secondary voltage ratios

The relationship between turns ratio and voltage ratio is:

$$\frac{N_p}{N_s} = \frac{V_p}{V_s}$$

Where N_p and N_s are the number of primary and secondary turns respectively and V_p and V_s are the primary and secondary voltages respectively.

Primary and secondary current ratings

The relationship between primary and secondary currents of a transformer is given by:

$$\frac{N_p}{N_s} = \frac{I_s}{I_p}$$

Where N_p and N_s are the number of primary and secondary turns respectively and I_p and I_s are the primary and secondary voltages respectively.

The current transfer ratio of a transformer is the inverse of its voltage transfer ratio and hence:

$$\frac{I_s}{I_p} = \frac{V_p}{V_s}$$

Power rating

Ratings of power transformers usually quote a maximum rating in terms of volt-amperes (VA). The working VA for a power transformer can be estimated by calculating the total power consumed by each secondary load and multiplying this by 1.1 (to allow for losses within the transformer when on-load).

Saturation flux density

In order to maintain high efficiency, it is important to avoid saturation within the transformer. In the case of sinusoidal operation and provided one keeps within the ratings specified by the manufacturer, saturation

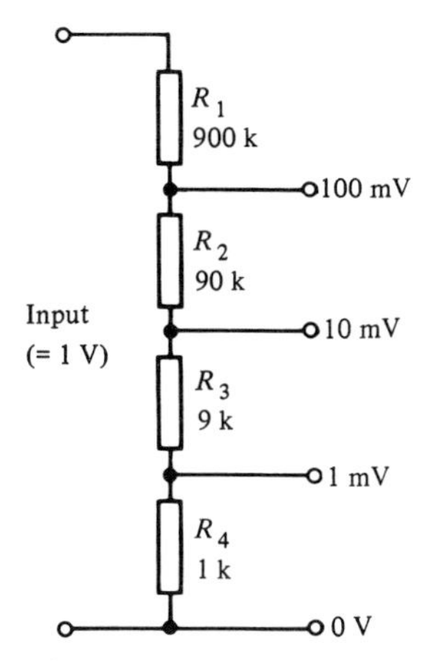

Figure 1.12 *Simple potential divider attenuator* ($R_{in} = 1\ \mathrm{M\Omega}$)

should not occur. For pulse operation, or when an appreciable direct current is applied to a transformer, it may be necessary to make an estimation of the peak flux density in order to check that saturation is not occurring.

Regulation

The regulation of a transformer is a measure of the ability of a transformer to maintain its rated output voltage under load. This subject is dealt with more fully in Chapter 3.

It is not usually practicable to manufacture one's own power transformers and one should select from the large range available from various manufacturers. Some manufacturers will supply transformers to individual customers on a 'one-off' basis though it is well worth 'shopping around' for a stock item before taking this particular course. An alternative solution (particularly where one may require an unusual combination of secondary voltages) is that of making use of a proprietary transformer 'kit'. These are available in several sizes (including 20 VA, 50 VA, 100 VA and 200 VA) and usually comprise a double-section bobbin complete with pre-wound mains primary, insulating shrouds, laminations and end mounting frames.

Attenuators

Simple attenuators can be constructed using nothing more than a potential divider arrangement. Figures 1.12 and 1.13 show typical potential divider arrangements (the latter is based on commonly available preferred values). Each of these circuits has a fixed input resistance (1 MΩ for the circuit of Figure 1.12 and 910 Ω in the case of Figure 1.13). In both cases, the output load circuit must have a comparatively high impedance in order that the accuracy is maintained. High stability 1% tolerance resistors should be used in both circuits.

The simple potential divider type attenuators described in Figures 1.12 and 1.13 do not, unfortunately, possess equal input and output resistances.

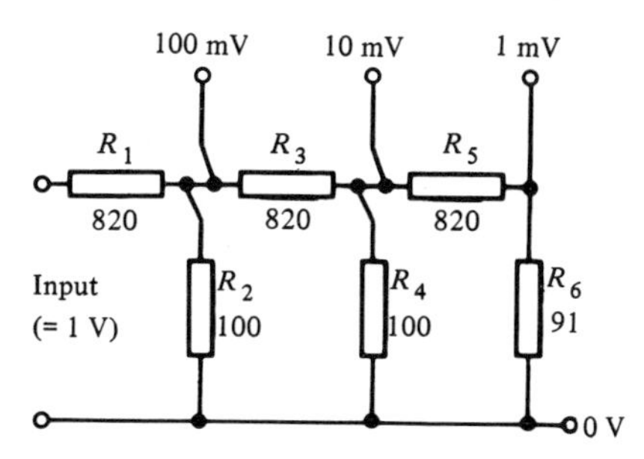

Figure 1.13 *Improved potential divider attenuator* ($R_{in} = 910\ \Omega$)

This is a desirable feature whenever an attenuator is to form part of a matched system. Figures 1.14, 1.15, and 1.16 show typical 'constant impedance' attenuators based on T, π and H arrangements of resistors.

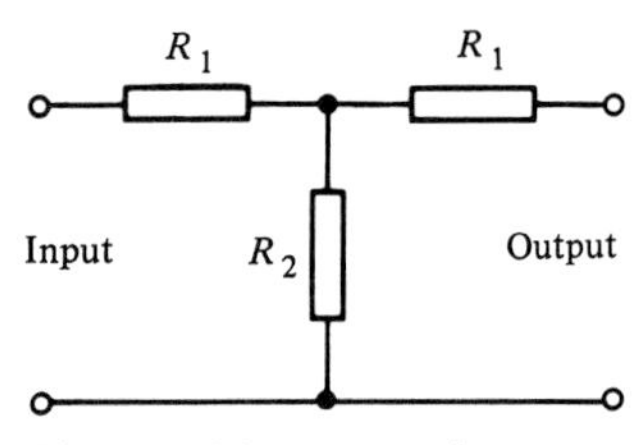

Figure 1.14 *T-network attenuator (unbalanced)*

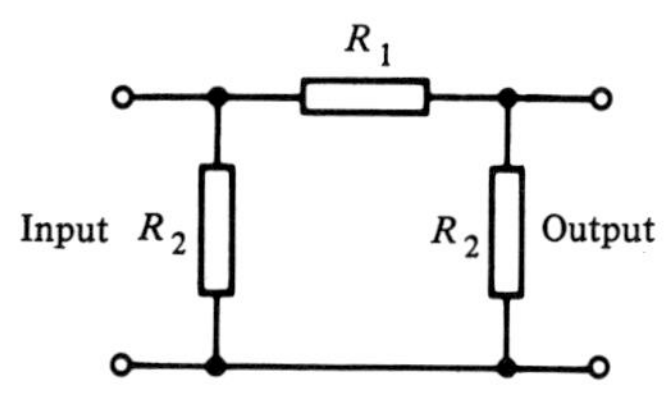

Figure 1.15 *π-network attenuator (unbalanced)*

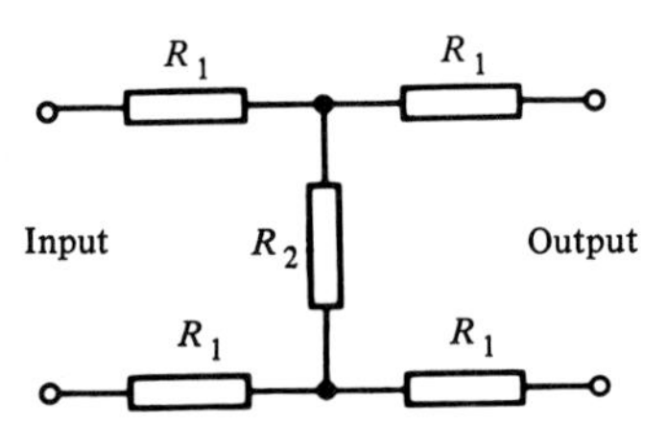

Figure 1.16 *H-network attenuator (balanced)*

The design formulae for these circuits are as follows:

Table 1.3

	Filter type		
Resistor	*T*	*π*	*H*
R_1	$Z\frac{(x-1)}{(x+1)}$	$Z\frac{(x^2-1)}{2x}$	$\frac{Z(x-1)}{2(x+1)}$
R_2	$Z\frac{2x}{(x^2-1)}$	$Z\frac{(x+1)}{(x-1)}$	$Z\frac{2x}{(x^2-1)}$

Where Z is the design impedance of the network (often 600 Ω for line matching purposes) and x is the desired ratio of input to output voltage (i.e. $x = V_{in}/V_{out}$).

Where attenuation is to be expressed in decibels (dB) the table given in Chapter 4 may be used.

For radio frequency circuits, the design impedance will usually be 50 Ω or 75 Ω rather than 600 Ω. In practice it will be necessary to use the nearest preferred values but, with care, it should be possible to realize values of attenuation which are very close to the desired values. As an example, the following preferred values are all within ±2% of the desired values for a π-network attenuator of the type shown in Figure 1.15 having a design impedance of 75 Ω:

	Attenuation (in dB)			
Resistor	*3*	*6*	*10*	*20*
R_1	27	56	100	390
R_2	470	220	390	100

Finally, readers should note that attenuators may be connected in cascade in order to produce an attenuation which is equal to the product of the voltage ratios or sum of the decibel values. As an example, an attenuation of 19 dB can be produced by cascading 3 dB, 6 dB, and 10 dB attenuators.

CR timing and waveshaping networks

Networks of capacitors and resistors (known as C-R networks) form the basis of many timing and pulse shaping circuits and are thus found in many practical electronic circuits. Figure 1.17 shows a simple timing network

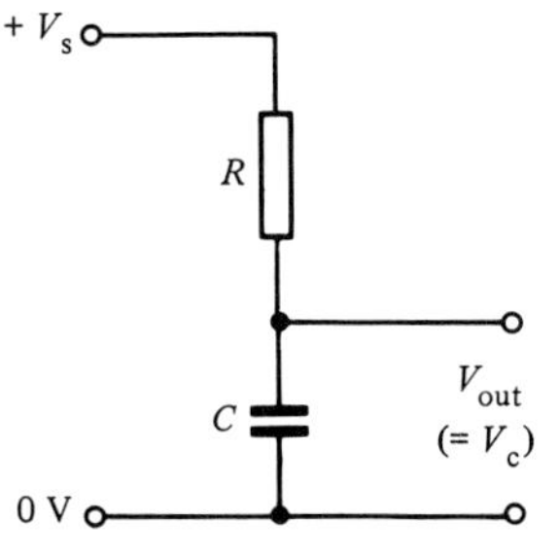

Figure 1.17 *Basic C-R timing network*

in which a capacitor is charged from a constant voltage supply. When the supply is connected, the output voltage will rise exponentially with time (as shown in Figure 1.18). At the same time, the current will fall (as shown in Figure 1.19). The rate of charge will depend upon the product of

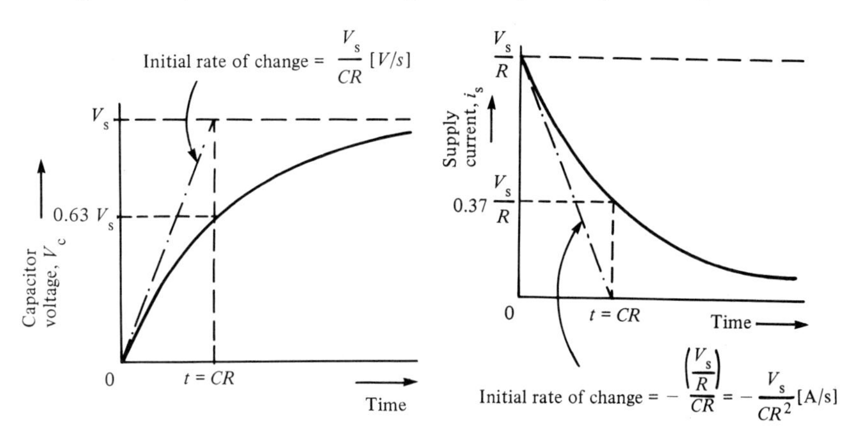

Figure 1.18 *Output voltage versus time for the circuit of Figure 1.17*

Figure 1.19 *Current versus time for the circuit of Figure 1.17*

capacitance and resistance and is known as the 'time constant'. Hence:

Time constant, $\tau = C \times R$ seconds

Where C is in farads and R is in ohms.

To avoid some fairly complex mathematics (which, for the curious, can be found in most electronic textbooks) it is only necessary to remember that the capacitor voltage will rise to 63% of the supply voltage in a time interval equal to the time constant. Furthermore, at the end of the next interval of time equal to the time constant (i.e. after a total time of 2τ has elapsed) the voltage will have risen by 63% of the remainder, and so on.

If the components are reversed, as shown in Figure 1.20, the output voltage will fall exponentially with time (as shown in Figure 1.21). In a

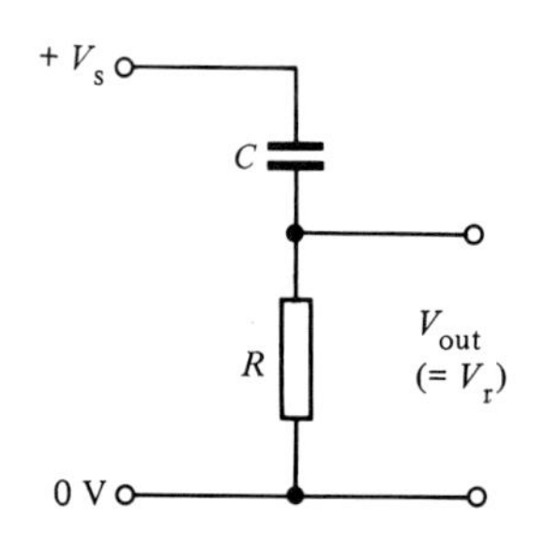

Figure 1.20 *C-R timing network with components reversed*

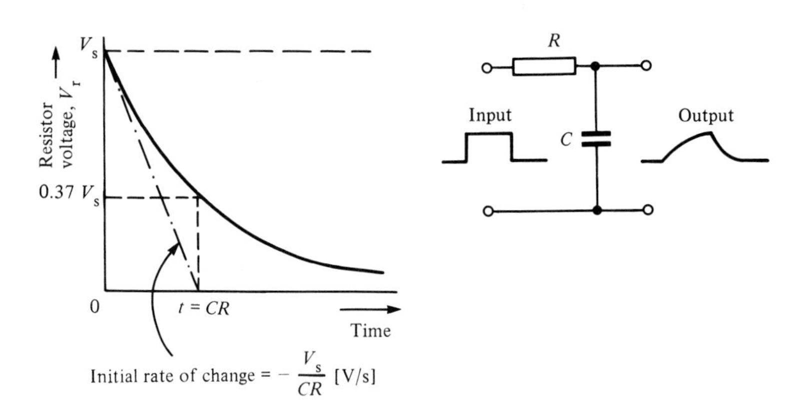

Figure 1.21 *Output voltage versus time for the circuit of Figure 1.20*

Figure 1.22 *C-R 'integrator'*

time equal to the time constant, the resistor voltage will have fallen to 37% of the supply voltage, and so on.

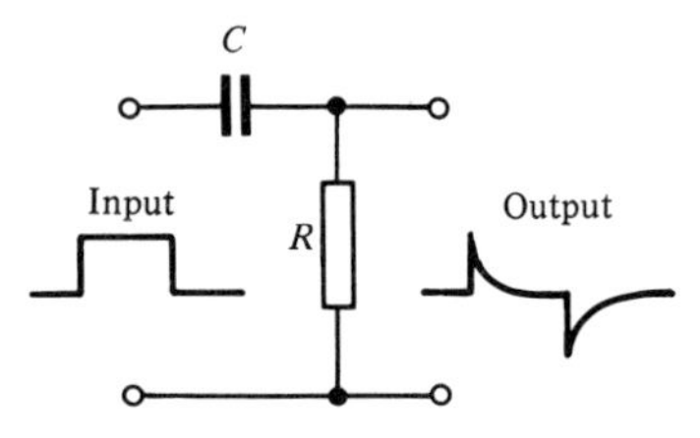

Figure 1.23 *C-R 'differentiator'*

Either of these circuits can be used in conjunction with some form of voltage sensitive circuit to provide a reasonably accurate time delay. Where continuous (or 'astable') operation is required, it will be necessary to discharge the capacitor at the end of each timing period (see Chapter 7 for further details).

Another application of simple CR networks which is directly attributable to their charge/discharge characteristics are the simple waveshaping circuits shown in Figures 1.22 and 1.23. These circuits function as 'square-to-triangle' and 'square-to-pulse' converters by respectively 'integrating' and 'differentiating' their inputs.

In order to produce effective 'integration' of the input waveform in the simple integrator circuit of Figure 1.22, the time constant should be very much larger than the periodic time of the input. Conversely, in order to produce effective 'differentiation' of the input waveform in the simple differentiator shown in Figure 1.23, the time constant should be very much smaller than the periodic time of the input.

Passive filters

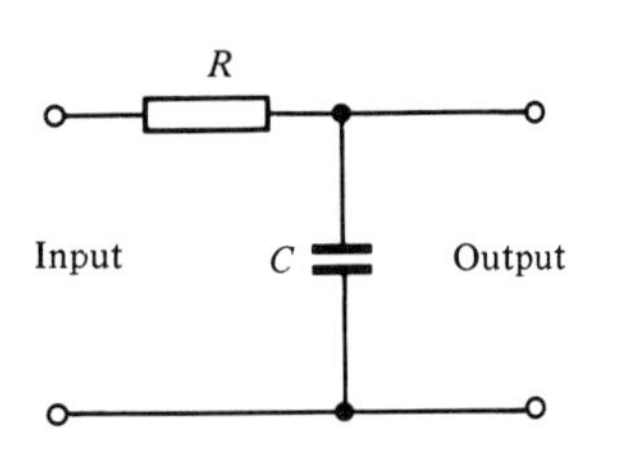

Figure 1.24 *C-R low-pass filter*

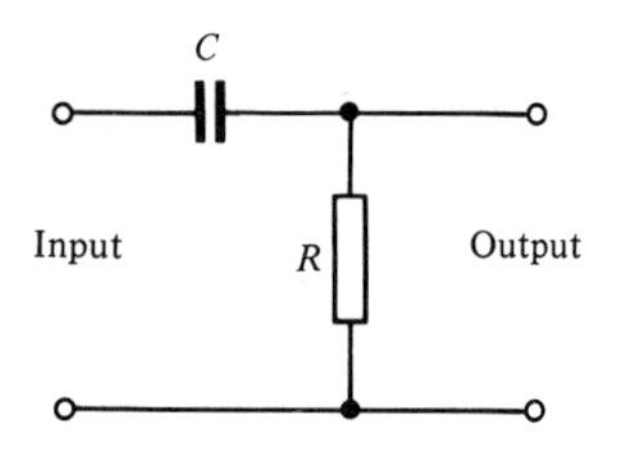

Figure 1.25 *C-R high-pass filter*

C-R networks can also be used as simple low-pass and high-pass filters when fed with sinusoidal signals. Simple single-section CR low-pass and high-pass filters are shown in Figures 1.24 and 1.25 together with representative frequency response characteristics shown in Figures 1.26 and 1.27. The low-pass arrangement has a response which, while initially flat, falls progressively as the frequency increases beyond cut-off. The high pass arrangement, on the other hand, provides a response which rises progressively with frequency and then remains flat as the frequency increases above cut-off.

In both of these circuits, the cut-off frequency is the frequency at which the output voltage falls to 0.707 of the input voltage (this is equivalent to a reduction of 3 dB – see Chapter 4). The cut-off frequency is given by:

$$f_c = \frac{1}{2\pi CR}$$

Where f_c is in Hz, C is in F, and R is in Ω.

The rate of fall-off beyond cut-off is constant at 6 dB per octave (or 20 dB/decade). It should be noted that a roll-off of 6 dB per octave is equivalent to a halving of output voltage each time the frequency doubles. Hence, taking the case of a low-pass filter beyond cut-off, if the output of the filter is 2 V at 1 kHz, it will produce 1 V at 2 kHz, 500 mV at 4 kHz, and so on.

To produce steeper rates of roll-off, two or more single-section filters can be cascaded, as shown in Figure 1.28. A two-section passive C-R filter will provide an attenuation of 12 dB/octave (40 dB/decade) while a three-section filter will provide 18 dB/octave (60 dB/decade). In practice, active filter arrangements should be used in preference to multi-section passive filters. Representative circuits are shown in Chapter 5.

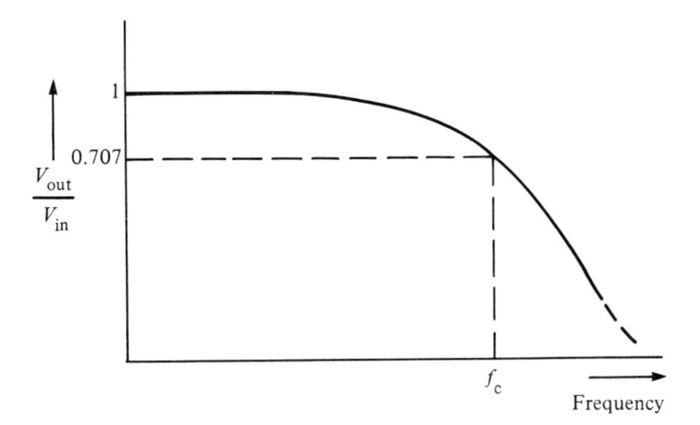

Figure 1.26 *Frequency response for the circuit of Figure 1.24*

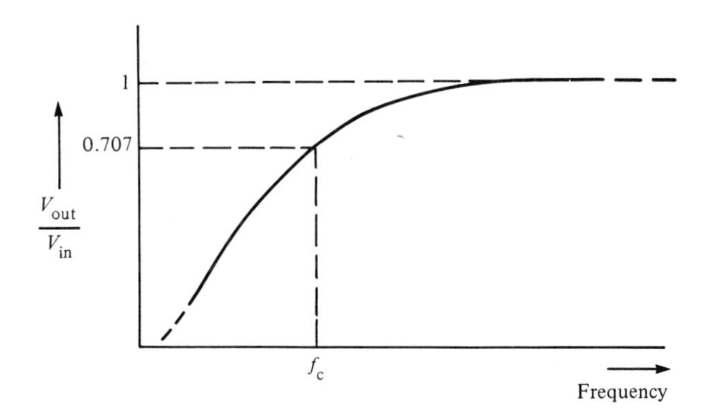

Figure 1.27 *Frequency response for the circuit of Figure 1.25*

A simple form of bandpass filter can be produced by cascading a single-section low-pass filter with a single-section high-pass filter, as shown in Figure 1.29. The frequency response of this circuit is shown in Figure 1.30. To avoid appreciable attenuation at mid-band, the ratio of upper to lower cut-off frequency should be at least 10:1. Where a high degree of selectivity is required coupled with low mid-band attenuation, an active filter should be employed along the lines suggested in Chapter 5.

Two other simple filters of note are the Wien bridge (Figure 1.31) and twin-T (Figure 1.32). Both of these filters are band-stop types having a frequency response characteristic like that shown in Figure 1.33.

The Wien bridge filter comprises a combination of series and parallel C-R networks. The two resistive arms of the bridge have values such that one-third of the input voltage appears across the lower resistor and two-thirds appears across the upper resistor. The voltage appearing at the junction of the series and parallel arms will vary according to the frequency of the input voltage. At a certain critical frequency, one-third of the input voltage will be developed across the parallel C-R arm. This voltage will be exactly equal and in phase with the voltage appearing across the lower resistor and hence the output voltage will fall to zero. The frequency at which balance is obtained (infinite attenuation) is given by:

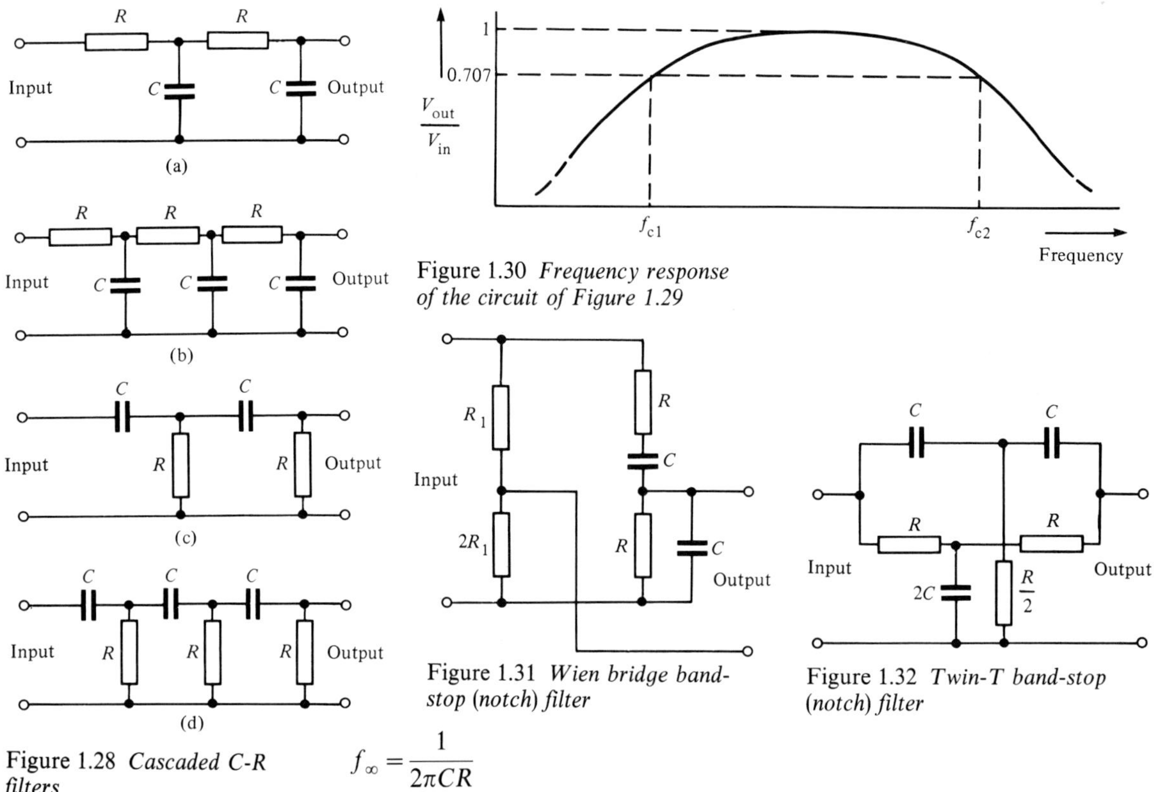

Figure 1.28 *Cascaded C-R filters*
(a) *Two-section low-pass*
(b) *Three-section low-pass*
(c) *Two-section high-pass*
(d) *Three-section high-pass*

Figure 1.30 *Frequency response of the circuit of Figure 1.29*

Figure 1.31 *Wien bridge band-stop (notch) filter*

Figure 1.32 *Twin-T band-stop (notch) filter*

$$f_{\infty} = \frac{1}{2\pi CR}$$

Where f_{∞} is in Hz, C is in F, and R is in Ω.

In order to vary the frequency of operation, it is necessary to simultaneously vary both resistors or both capacitors. To achieve a wide range of operating frequencies it is often convenient to switch decade values of capacitor and vary the resistive element over a 10:1 range, as shown in Figure 1.34.

One important disadvantage of the Wien bridge circuit is that there is no common connection between the input and output. Another limitation is that the maximum output from a Wien bridge filter is only equal to one-third of the input voltage. The twin-T overcomes both of these problems at the cost of two additional passive components (one resistor and one capacitor).

The balance equation for the twin-T network is the same as that of its Wien bridge counterpart. The principal disadvantage of the twin-T circuit is that it is more difficult to vary the frequency of operation since, ideally, all three resistors (or capacitors) should be simultaneously varied.

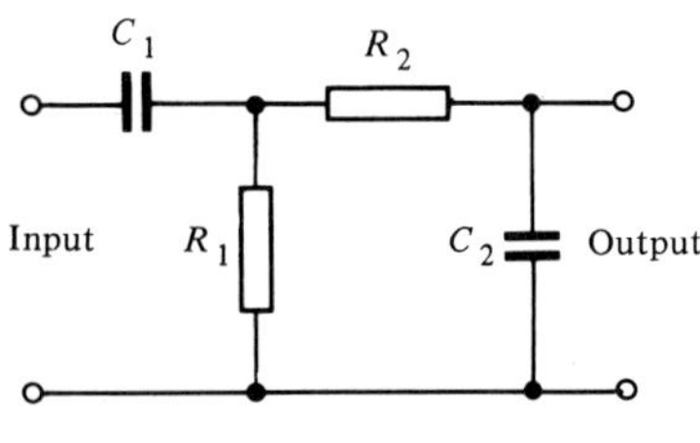

Note: $R_1C_1 \gg R_2C_2$

Figure 1.29 *C-R band-pass filter*

L-C filters

At high frequencies (e.g. above 100 kHz), C-R filters have serious shortcomings and L-C filters are usually preferred. Simple constant-impedance

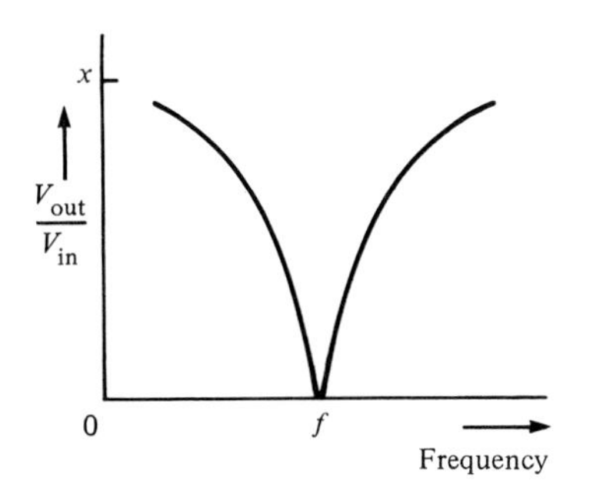

Note: $x = \frac{1}{3}$ for Wien bridge filter
$x = 1$ for twin-T filter

Figure 1.33 *Frequency response of the circuits shown in Figures 1.31 and 1.32*

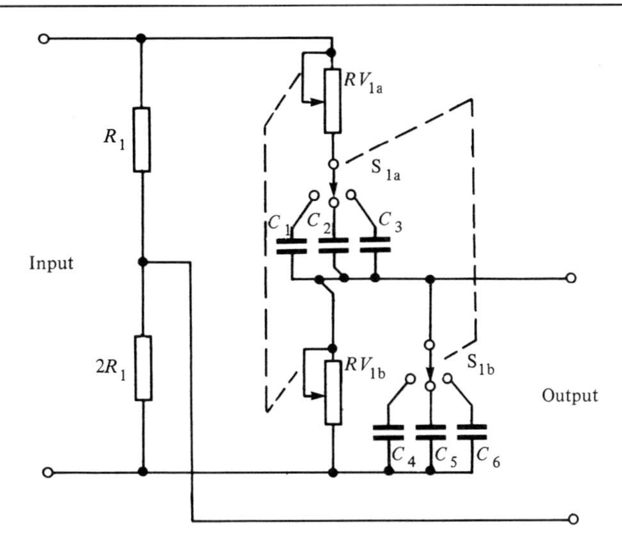

Figure 1.34 *Variable frequency Wien bridge filter*

L-C low-pass and high-pass filters are shown in Figure 1.35. The design equations for these filters are as follows:

Characteristic impedance, $Z_0 = \sqrt{\frac{L}{C}}$

Cut-off frequency, $f_c = \frac{1}{2\pi\sqrt{LC}}$

Inductance, $L = \frac{Z_0}{2\pi f_c}$

Capacitance, $C = \frac{1}{2\pi f_c Z_0}$

Where Z is in Ω, f is in Hz, L is in H, and C is in F.

	T - *section*	π - *section*
Low-pass	L, L, $2C$	$2L$, C, C
High-pass	C, C, $\frac{L}{2}$	$\frac{C}{2}$, L, L

Figure 1.35 *L-C low-pass and high-pass filters*

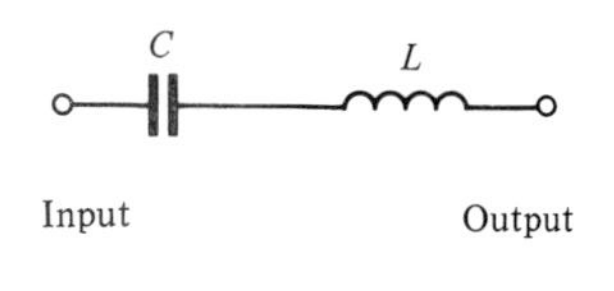

Figure 1.36 *Series L-C band-pass filter*

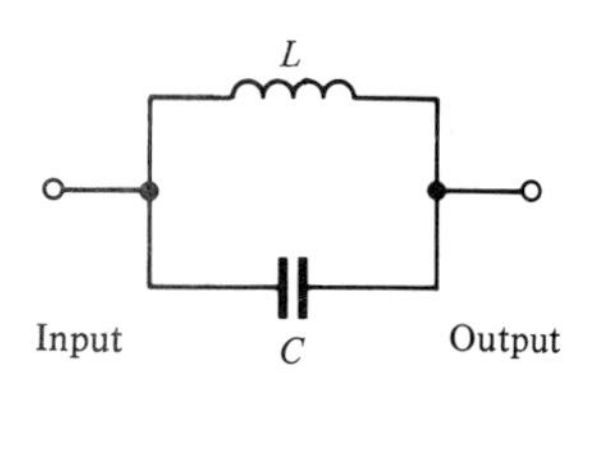

Figure 1.37 *Parallel L-C band-stop filter*

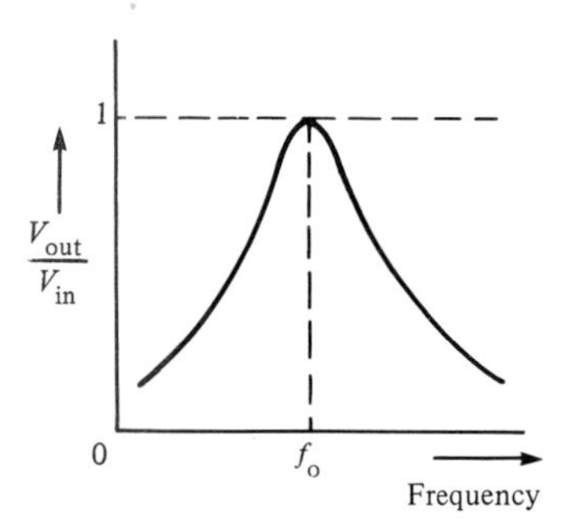

Figure 1.38 *Frequency response of the circuit shown in Figure 1.36*

Simple L-C band-pass (series tuned circuit) and band-stop (parallel tuned circuit) filters are shown in Figures 1.36 and 1.37. Neither of these arrangements has a constant impedance and both operate best when the source and load impedances are relatively low. Representative frequency response characteristics for the two circuits are shown in Figures 1.38 and 1.39.

Asssuming that losses are negligible, the centre-frequency of both of the previous circuits is given by:

$$f_0 = \frac{1}{2\pi\sqrt{LC}}$$

Figure 1.39 *Frequency response of the circuit shown in Figure 1.37*

Where f_0 is in Hz, L is in H, and C is in F.

The performance of simple filters can be greatly improved by adding shunt elements along the lines shown in Figures 1.40 and 1.41. Note that,

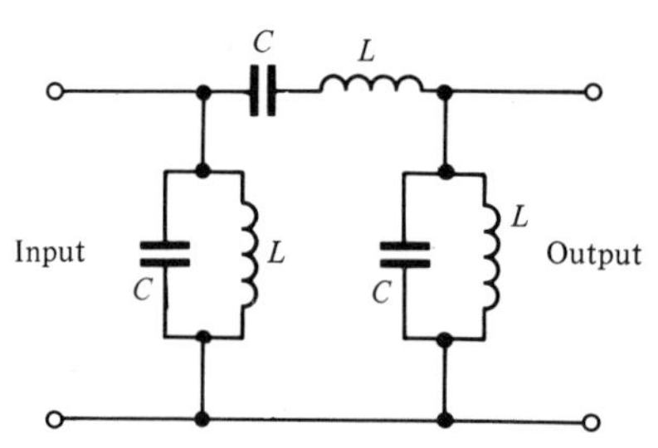

Figure 1.40 *Improved L-C band-pass filter*

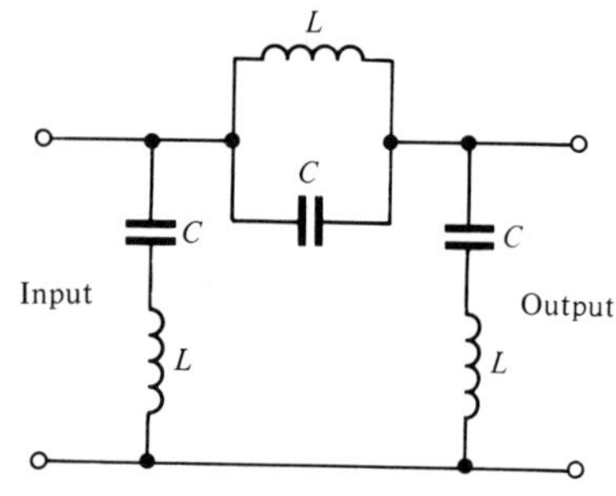

Figure 1.41 *Improved L-C band-stop filter*

in each case, the shunt element takes the opposite form to that of the series element. Performance is improved if spurious coupling is avoided; inductors, in particular, should be arranged so that electromagnetic coupling is avoided.

2 Semiconductors

The term semiconductor covers many devices, from the humblest diode to the very latest VLSI integrated circuit. Modern electronic systems rely on such devices to fulfil an immense variety of applications; from power rectification to digital frequency synthesis. Despite the obvious differences in cost and complexity, it is important to remember that all such devices rely on the same fundamental principles.

This section provides an introduction to a variety of the most commonly used semiconductor devices. However, as with other sections, we shall not concern ourselves too much with principles of operation (the necessary theory can be found in any modern electronics textbook) but simply deal with those characteristics which relate specifically to their use in practical circuits.

Diodes

Diodes are two-terminal devices which exhibit low resistance to current flow in one direction and high resistance to current flow in the other. The direction in which current flows is often referred to as the 'forward' direction while that in which negligible current flows is known as the 'reverse' direction. When a diode is conducting, a small voltage is dropped across it. This voltage is known as the 'forward voltage drop' and typical values for representative silicon and germanium diodes are shown in Table 2.1:

Table 2.1

	Forward voltage drop	
Current flowing	*Silicon (1N4148)*	*Germanium (OA91)*
10 μA	0.43 V	0.12 V
100 μA	0.58 V	0.26 V
1 mA	0.65 V	0.32 V
10 mA	0.75 V	0.43 V

Semiconductor diodes generally comprise a single p-n junction. The p-type connection is known as the 'anode' while the n-type connection is known as the 'cathode'. Conventional current flows from anode to cathode when the diode is conducting. Negligible current flows in the reverse direction (cathode to anode). The semiconductor material used is generally either germanium (Ge) or silicon (Si).

Germanium diodes conduct at lower forward voltages than their silicon counterparts (typically 100 mV as compared with 600 mV) but they tend to exhibit considerably more reverse leakage current (1 μA as compared with 10 nA for an applied reverse voltage of 50 V). Furthermore, the forward resistance of a conducting silicon diode is much lower than that of a comparable germanium type. Hence germanium diodes are used primarily for signal detection purposes whereas silicon devices are used for rectification and for general purpose applications. Typical forward and reverse characteristics for comparable germanium and silicon diodes are shown in Figure 2.1.

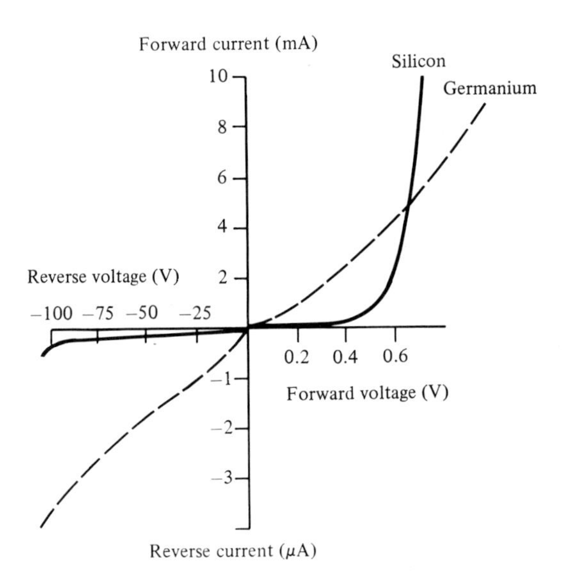

Figure 2.1 *Characteristics of representative silicon and germanium diodes*

A general measure of the efficiency of a diode is the ratio of forward to reverse current under a given set of conditions. It is important to note that the reverse leakage current for a diode increases markedly as the junction temperature increases hence diodes are less efficient at high temperatures.

The maximum reverse voltage that a diode can tolerate is usually specified in terms of its 'reverse repetitive maximum' voltage (V_{RRM}) or peak inverse voltage (PIV). In either case, operating the diode beyond this voltage can result in a high risk of breakdown. Indeed, to avoid chances of diode failure it is advisible to operate the diode well within this limit.

Typical characteristics of some representative diode types are summarized in Table 2.2:

Table 2.2

Type	*OA47*	*OA91*	*1N4148*	*1N4007*	*1N5402*
Material	Ge	Ge	Si	Si	Si
Maximum forward voltage (V_F)	0.6 V	2.1 V	1.0 V	1.6 V	1 V
Maximum forward current (I_F)	50 mA	50 mA	100 mA	1 A	3 A
Maximum reverse voltage (V_{RRM})	25 V	115 V	75 V	1 kV	200 V
Maximum reverse current (I_R)	30 μA	75 μA	25 nA	10 μA	10 μA
Application	Low level detector	General purposes		High voltage rectifier	Low voltage rectifier

Diodes are often divided into 'signal' and 'rectifier' types according to their principal field of application. The requirements of diodes in each of these categories are quite different. Signal diodes, for example, require consistent forward characteristics with low forward voltage drop being paramount. Rectifier diodes, on the other hand, need to be able to cope with high values of reverse voltage and large values of forward current. Consistency of characteristics is usually unimportant in such applications.

Wire ended diodes are commonly available with maximum forward current ratings of up to 6 A and maximum reverse voltage ratings of up to 1250 V. Stud-mounting diodes (designed for bolting to a heatsink) are available with forward current ratings ranging typically from 16 A to 75 A.

Silicon diodes are also available encapsulated in bridge form. Bridge arrangements are useful as full-wave rectifiers (see Chapter 3 for details). The following types of encapsulation are employed:

Table 2.3

Type	*Encapsulation*	*Mounting*	*Maximum forward current rating (typical)*
Vm series	4-pin DIL	PCB	0.9 A
WO series	Cylindrical	PCB	1A
SKB2 series	In-line	PCB	1.6 A
KBPC series	Square	PCB	2 A to 6 A
SKB25 series	Epoxy-potted	Heatsink	6 A to 35 A

Diode applications

Figure 2.2 shows some typical applications for diodes. Figure 2.2(a) and (b) show simple half-wave and full-wave rectifiers (this subject is covered in greater detail in Chapter 3). Figure 2.2(c) shows a simple diode detector

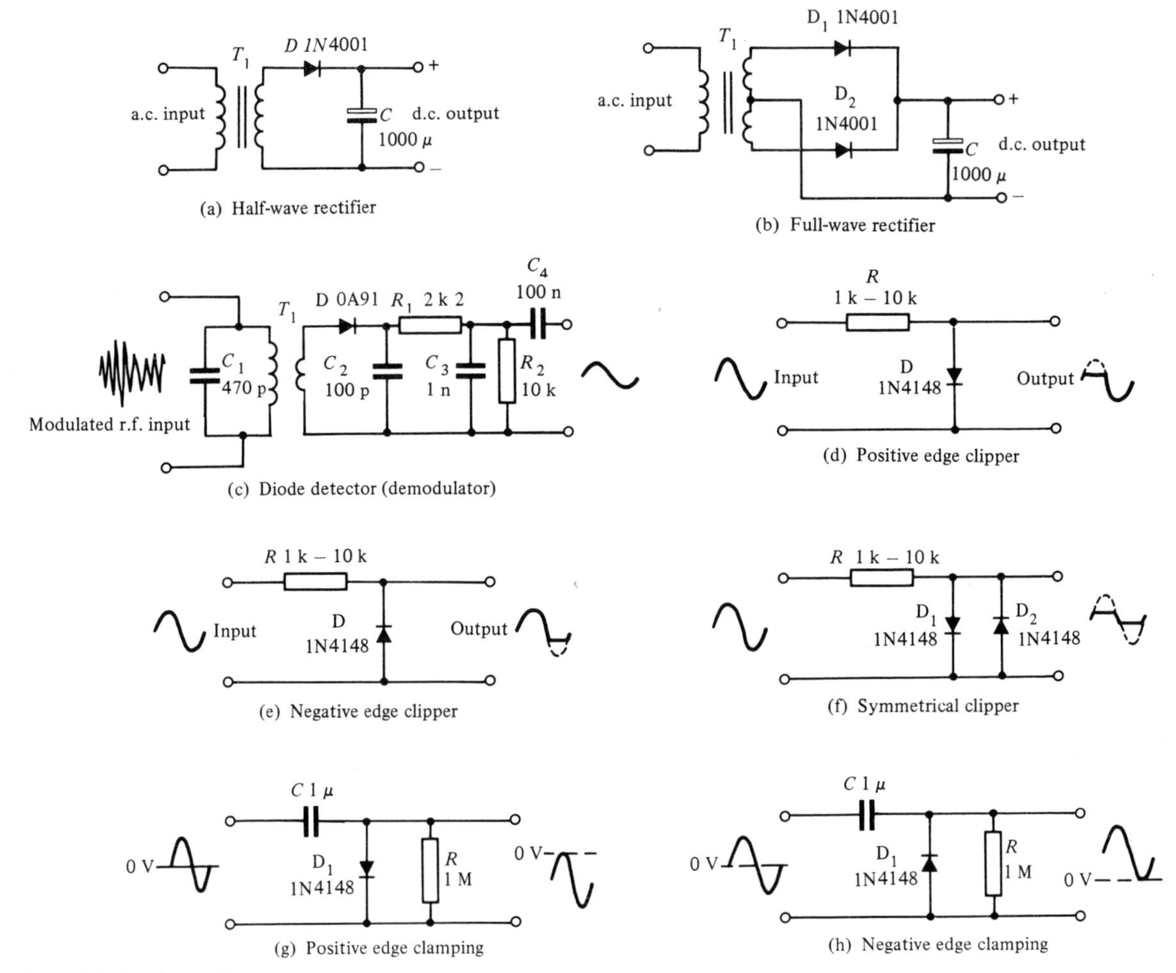

Figure 2.2 *Diode applications*

which can be used to recover the modulation envelope from an amplitude modulated wave. Figure 2.2(d) and (e) show positive and negative clippers. These circuits respectively remove the positive half-cycle and negative half-cycle of an applied a.c. input. In practice, the forward voltage drop of the diodes should be small compared with the peak value of the applied signal. Figure 2.2(f) shows a symmetrical clipping arrangement. This circuit will remove both positive and negative going peaks and the output voltage will have a peak-to-peak value which is equal to twice the forward voltage drop of the diodes used.

Figure 2.2(g) and (h) show positive and negative clamping circuits. The input to both of these circuits should be derived from a low-impedance source and is assumed to swing uniformly either side of 0 V. The output voltage will take the same shape as the input but will be shifted so that it all appears either positive or negative. It should be noted that clamping circuits will only be effective where:

(a) The source has a relatively low impedance.
(b) The input signal is significantly larger than the forward voltage drop associated with the diode.
(c) The time constant of the coupling network (C and R) is long in comparison with the lowest input signal frequency.

Zener diodes

Zener diodes are silicon diodes which are specially designed to exhibit consistent reverse breakdown characteristics. Zener diodes are available in various families (according to their general characteristics, encapsulation and power ratings) with reverse breakdown (zener) voltages in the E12 and E24 series (ranging from 2.7 V to around 68 V).

The following families are commonly available:

BZY88 series Miniature glass encapsulated diodes rated at 500 mW (at 25 °C). Zener voltages range from 2.7 V to 15 V (voltages are quoted for 5 mA reverse current at 25 °C).

BZX85 series Miniature glass encapsulated diodes rated at 1.3 W (at 25 °C). Zener voltages range from 2.7 V to 6.8 V.

BZX61 series Encapsulated alloy junction rated at 1.3 W (25 °C ambient). Zener voltages range from 7.5 V to 72 V.

BZY93 series High power diodes in stud mounting encapsulation. Rated at 20 W for ambient temperatures up to 74 °C. Zener voltages range from 9.1 V to 75 V.

1N5333 series Plastic encapsulated diodes rated at 5 W. Zener voltages range from 3.3 V to 24 V.

The temperature coefficient of the zener voltage of a zener diode tends to vary with the zener voltage rating. The following are typical values of temperature coefficient for low-voltage, low-power zener diodes:

Table 2.4

Zener voltage (V)	*Temperature coefficient* (mV/°C)
2.7	−2.0
3.3	−1.8
3.9	−1.4
4.3	−1.0
4.7	+0.3
5.1	+1.0
5.6	+1.5
6.2	+2.0
6.8	+2.7

From Table 2.4 it should be obvious that the minimum value of temperature coefficient occurs at zener voltages of approximately 4.5 V. Hence, circuits which use zener diodes as voltage references should be designed for operation with 4.7 V devices for optimum temperature performance.

Schottky diodes

A forward conducting diode requires a finite time to return to the non-conducting state. This is due to internal charge storage which can cause problems when diodes are used in pulse circuits. Charge storage tends to be a more significant problem with larger power diodes. However, special 'fast-recovery' diodes are available for use in such applications as switched mode power supplies.

Schottky diodes exhibit a forward voltage drop, which is approximately half that of conventional silicon diodes, and very fast reverse recovery times. They are thus preferred for use in switched mode power supplies and in applications where very low forward voltage drop is a prime consideration.

Thyristors

Thyristors (also known as silicon controlled rectifiers) are three-terminal devices which can be used for switching and a.c. power control. Thyristors can switch very rapidly from a conducting to a non-conducting state depending upon the conditions which exist at the gate terminal. In the 'off' state, the thyristor exhibits negligible leakage current, while in the 'on' state the device exhibits very low resistance. This results in very little power loss within the thyristor even when appreciable power levels are being controlled. Once switched into the conducting state, the thyristor will remain conducting (i.e. it 'latches') until the forward current is removed from the device.

In d.c. applications (such as 'crowbar protection' described in Chapter 3), this means that the supply must be interrupted or disconnected before the device can be reset into its non-conducting state. Where the device is used with an alternating supply, the device will automatically become reset whenever the main supply reverses. The device can then be triggered on the next half-cycle having correct polarity to permit conduction. Like their conventional silicon diode counterparts, thyristors have 'anode' and 'cathode' connections but they also have a 'gate' terminal and can be triggered into conduction by applying a current pulse to this electrode.

Figure 2.3 shows a simple a.c. power controller based on a thyristor and simple C-R phase shifting network. This circuit has a number of important limitations not the least of which is that it only provides half-wave control. Furthermore, triggering can be much improved using electronically generated pulses (using, for example, an astable timer).

The general rules for ensuring satisfactory triggering of a thyristor involve:

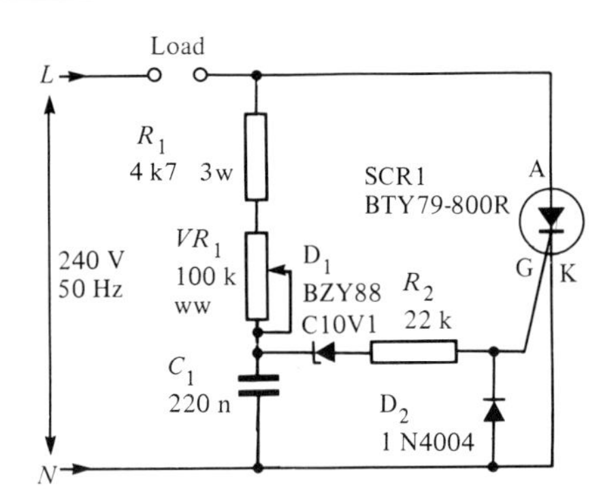

Figure 2.3 *Simple thyristor a.c. power controller*

(a) Wherever possible, use a trigger pulse having a fast rise time – avoid triggering using sinusoidal signals.
(b) Ensure that sufficient gate current is available. Thyristors will turn on faster (and less power will be dissipated within the device) for large values of gate current.
(c) Keep the pulse width of the gate signal fairly short. Failure to observe this precaution may result in excessive gate power dissipation.
(d) Avoid negative going gate voltages – these will also result in unwanted power loss.
(e) For a.c. power control applications make sure that the device can be triggered over a sufficiently wide angle of the supply. Failure to observe this rule will result in inadequate control range.

Triacs

Triacs are a development of thyristors which exhibit bidirectional characteristics. Triacs have three terminals known as 'main terminal one' (MT_1), 'main terminal two' (MT_2) and 'gate' (G) and can be triggered by both positive and negative voltages applied between G and MT_1 with either positive or negative voltage present at MT_2. Triacs are thus suitable for full-wave control of a.c. circuits and can be used in a similar manner to their half-wave thyristor equivalents. Figure 2.4 shows an a.c. mains power controller based on a triac and diac (roughly equivalent to a bi-directional zener diode).

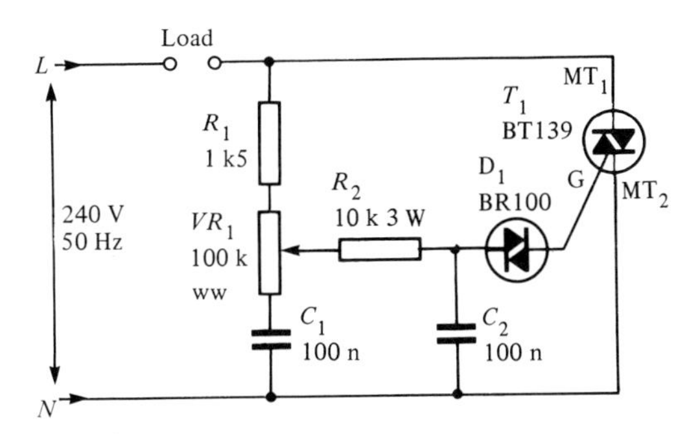

Figure 2.4 *Triac a.c. power controller*

Thyristors and triacs both switch 'on' and 'off' very rapidly. This results in very rapid transients which way be conveyed over some considerable distance along the a.c. mains supply. To prevent radiation of such noise, simple L-C filters will normally be required. A typical filter is shown in Figure 2.5.

Light emitting diodes

Light emitting diodes provide a useful means of indicating the state of a circuit and offer several advantages over conventional filament lamps, important among which are significantly lower operating current requirements and much greater reliability.

Light emitting diodes are based on gallium phosphide and gallium arsenide phosphide and generally provide adequate light output at forward currents of between 5 mA and 30 mA. (It should, however, be noted that green and yellow types are less efficient and may require higher levels of forward current to achieve the same light output.)

Light emitting diodes (LEDs) are available in various formats with the round types being most popular. Round LEDs are commonly available in the 3 mm and 5 mm (0.2 in) diameter plastic packages and also in 5 mm × 2 mm rectangular format. The viewing angle for round LEDs tends to be in the region of 20° to 40° whereas, for rectangular types this is increased to around 100°.

Typical characteristics for commonly available red LEDs are given in Table 2.5:

Table 2.5

Type	*Standard*	*Standard*	*High efficiency*	*High intensity*
Diameter (mm)	3	5	5	5
Maximum forward current (mA)	40	30	30	30
Typical forward current (mA)	12	10	7	10
Typical forward voltage drop (V)	2.1	2.0	1.8	2.2
Maximum reverse voltage (V)	5	3	5	5
Maximum power dissipation (mW)	150	100	27	135
Peak wavelength (nm)	690	635	635	635

In order to limit the forward current to an appropriate value, it is usually necessary to include a fixed resistor in series with an LED indicator, as shown in Figure 2.6. The value of the resistor is given by:

$$R_S = \frac{V_S - V_F}{I}$$

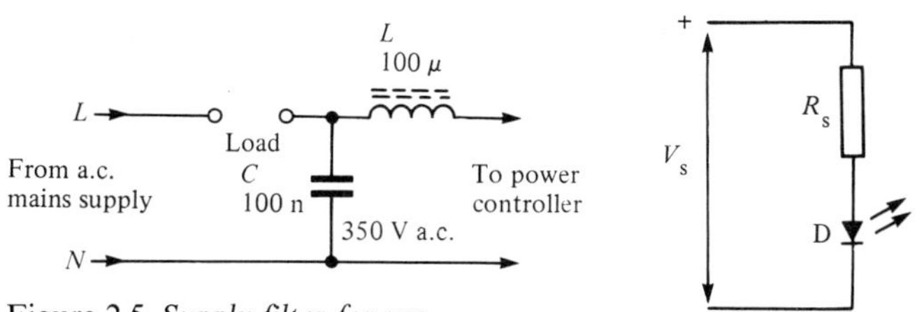

Figure 2.5 *Supply filter for use in conjunction with a.c. power controllers*

Figure 2.6 *LED indicator*

Where V_F is the forward voltage drop produced by the LED and V_S is the applied voltage. It is usually safe to assume that V_F will be 2 V and choose the nearest preferred value for R_S. Typical values for R_S in order to produce an LED current to 10 mA are given in Table 2.6:

Table 2.6

Supply voltage, V_S (*V*)	*Series resistance, R_S* (Ω)
3	180
5	270
6	390
9	680
12	1 k
15	1.2 k
18	1.5 k
24	2.2 k

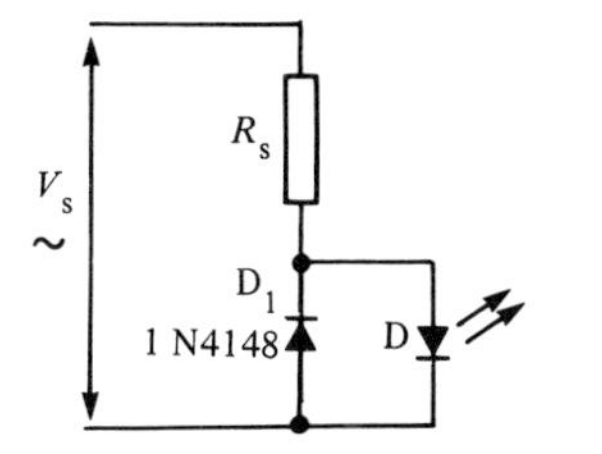

Figure 2.7 *Diode protection of an LED for low-voltage a.c. applications*

It is important to be aware of a serious limitation of LED devices as regards the maximum value of reverse voltage that may be applied. Exceeding these very modest reverse voltages will result in destruction of the LED junction; the penalty for misconnection of the device will invariably result in having to replace the device. Therefore, it is essential to connect LED devices with the correct polarity. Furthermore, in applications involving a.c. voltages, it is essential to wire a conventional low-current silicon diode in shunt with an LED, as shown in Figure 2.7.

Transistors

Modern transistors are invariably silicon types and are available in either n-p-n or p-n-p forms. Transistors tend to have diverse applications but the following categories cover most situations:

Linear: Transistors designed for linear applications (such as low-level voltage amplification).

Switching: Transistors designed for switching applications.

Power: Transistors which operate at significant power levels (such devices are often subdivided into audio frequency at radio frequency power types).

Radio-frequency: Transistors designed specifically for high-frequency applications.

High voltage: Transistors designed specifically to handle high voltages.

Typical characteristics of some representative transistors are summarized in Table 2.7.

Table 2.7

Type	*BC109*	*BC184L*	*BC212L*	*TIP31A*	*TIP3055*
Material	Silicon	Silicon	Silicon	Silicon	Silicon
Construction	n-p-n	n-p-n	p-n-p	n-p-n	n-p-n
Case style	TO18	TO92	TO92	TO220	TAB
Maximum collector power dissipation (P_C)	360 mW	300 mW	300 mW	40 W	90 W
Maximum collector current (I_C)	100 mA	200 mA	−200 mA	3 A	15 A
Maximum collector – emitter voltage (V_{CEO})	20 V	30 V	−50 V	60 V	60 V
Maximum collector – base voltage (V_{CBO})	30 V	45 V	−60 V	60 V	100 V
Current gain (h_{FE})	200–800	250 minimum	60–300	10–60	5–30
Transition frequency	250 MHz	150 MHz	200 MHz	8 MHz	8 MHz

Transistor amplifiers

Transistors are three-terminal devices. Amplifiers, on the other hand, have four terminals. It should thus be readily apparent that one terminal of a transistor must be 'common' to both the input and output of any practical amplifier circuit.

There are, therefore, three different circuit configurations in which transistors may be operated. For bipolar transistors these are:

(a) Common emitter.
(b) Common collector (also known as emitter follower).
(c) Common base.

These three basic configurations are shown in Figure 2.8 and their characteristics are summarized in Table 2.8 (typical values are shown in brackets):

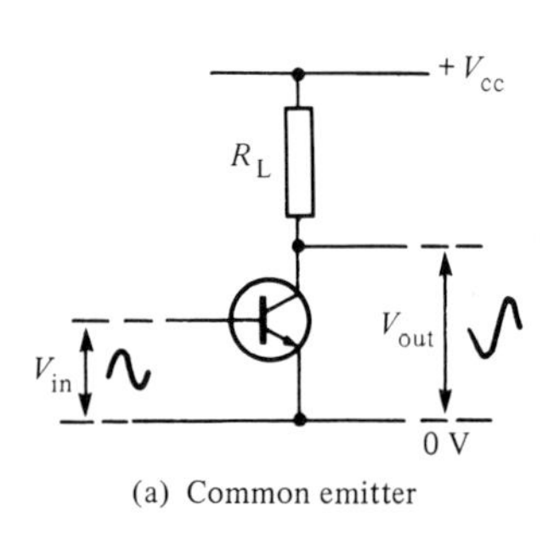

(a) Common emitter

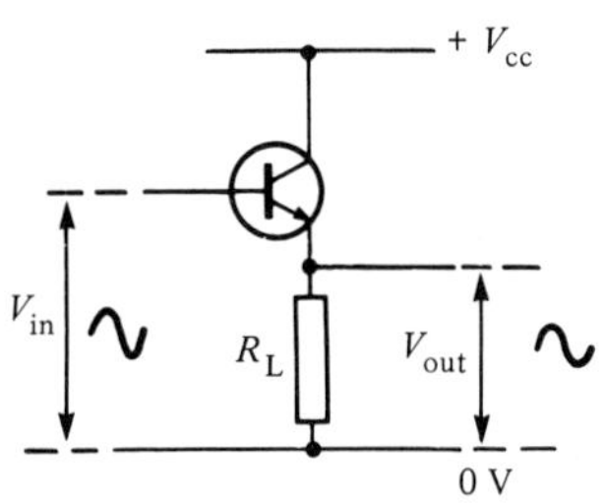

(b) Common collector (emitter follower)

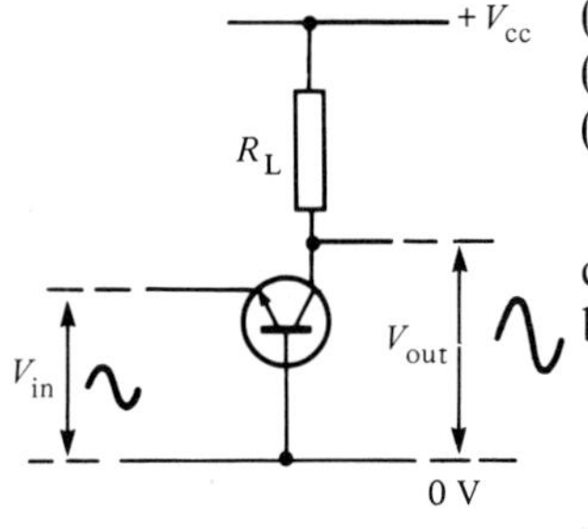

(c) Common base

Figure 2.8 *Transistor amplifier configurations*

For field effect transistors, the three amplifier configurations are:

(a) Common source.
(b) Common drain (also known as source follower).
(c) Common gate.

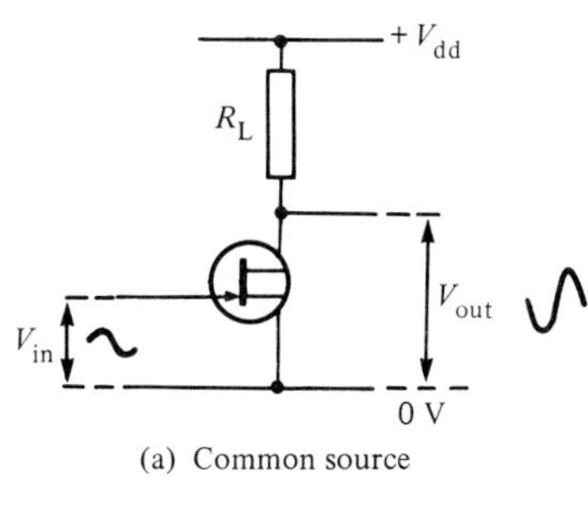

(a) Common source

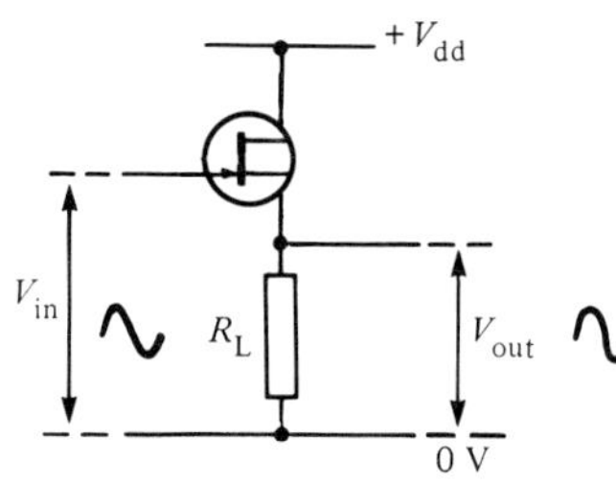

(b) Common drain (source follower)

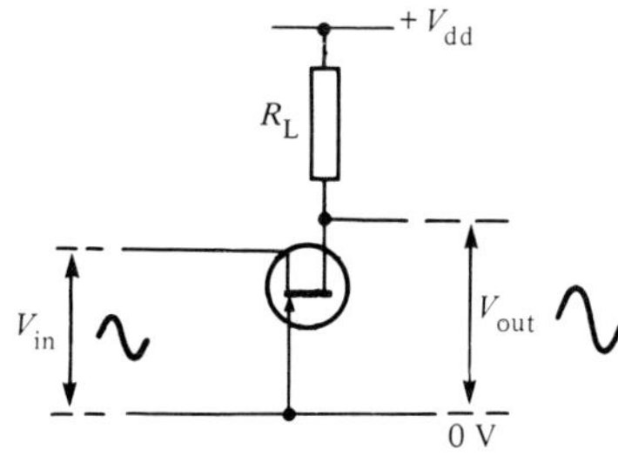

(c) Common gate

Figure 2.9 *Field effect transistor amplifier configurations*

Table 2.8

	Mode of operation		
Parameter	*Common emitter*	*Common collector*	*Common base*
Voltage gain	Medium/high (50)	Unity (1)	High (500)
Current gain	High (200)	High (200)	Unity (1)
Power gain	Very high (10,000)	High (200)	High (500)
Input resistance	Medium (2.5 kΩ)	High (100 kΩ)	Low (200 Ω)
Output resistance	Medium/high (20 kΩ)	Low (100 Ω)	High (100 kΩ)
Phase shift	180°	0°	0°

These three basic configurations are shown in Figure 2.9 and their characteristics are summarized in Table 2.9 (typical values are again shown in brackets):

Table 2.9

	Mode of operation		
Parameter	*Common source*	*Common drain*	*Common gate*
Voltage gain	Medium (40)	Unity (1)	High (250)
Current gain	Very high (200,000)	Very high (200,000)	Unity (1)
Power gain	Very high (8,000,000)	Very high (200,000)	High (250)
Input resistance	Very high (1 MΩ)	Very high (1 MΩ)	Low (500 Ω)
Output resistance	Medium/high (50 kΩ)	Low (200 Ω)	High (150 kΩ)
Phase shift	180°	0°	0°

Transistors as switches

Transistors, both bipolar and unipolar (FET), are ideal for use as switches. A conducting transistor is equivalent to a switch in the 'on' state while a non-conducting transistor is equivalent to a switch in the 'off' state, as shown in Figure 2.10. The transistor is operated under saturated switching conditions; in the conducting state sufficient base current must be applied

Transistor conducting

Switch 'on'

Transistor non-conducting

Switch 'off'

Figure 2.10 *Transistor switch*

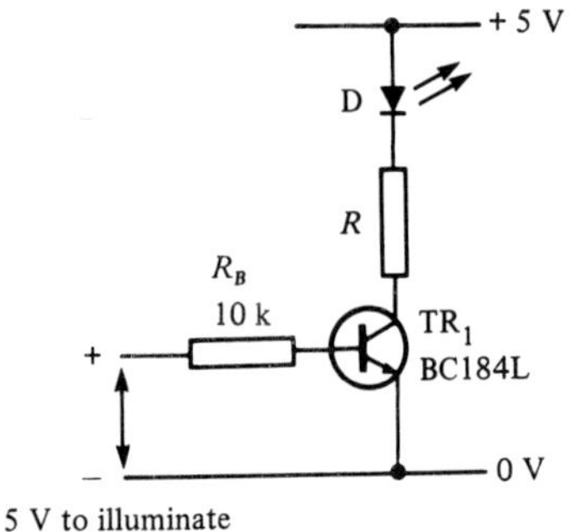

5 V to illuminate
0 V to extinguish

Figure 2.11 *Transistor LED driver*

for the collector voltage to fall to approximately 0 V (a typical value of V_{CE} for a saturated transistor is less than 200 mV.)

Provided the transistor operates under saturated conditions, dissipation within the transistor is kept to a relatively low level. In the conducting state V_{CE} is very low and in the non-conducting state I_C is very low; the collector power dissipation is the product of I_C and V_{CE} and thus, in both cases, the dissipation will be negligible.

As a rule of thumb, the value of base resistance should be chosen so that the current in the base circuit is between one tenth and one twentieth of the current in the collector circuit. Figure 2.11 shows a simple application of a transistor switch in the form of an LED driver; the LED will become illuminated whenever the input is taken to +5 V and will be extinguished when the input is taken to 0 V. Where operation involves supply voltages other than 5 V, R_1 should be selected from the table of resistor values shown in Table 2.1.

Transistor circuit configurations

Quite apart from the basic transistor amplifier configurations described earlier, a number of other transistor circuits are commonly used. Several of these are depicted in Figure 2.12.

Figure 2.12(a) shows a Darlington pair. This compound arrangement behaves essentially as one transistor having a current gain equal to the product of the individual current gains. It is thus possible to achieve an overall current gain of several thousand times with such an arrangement. The second transistor will operate with a higher value of collector current than the first and, in many practical applications, it is necessary to choose a high-gain small signal transistor for the first stage, TR_1, and a medium gain transistor capable of handling an appreciable collector current in the second stage, TR_2. Typical applications of the Darlington arrangement are in the filament lamp and relay drivers shown in Figures 2.13 and 2.14.

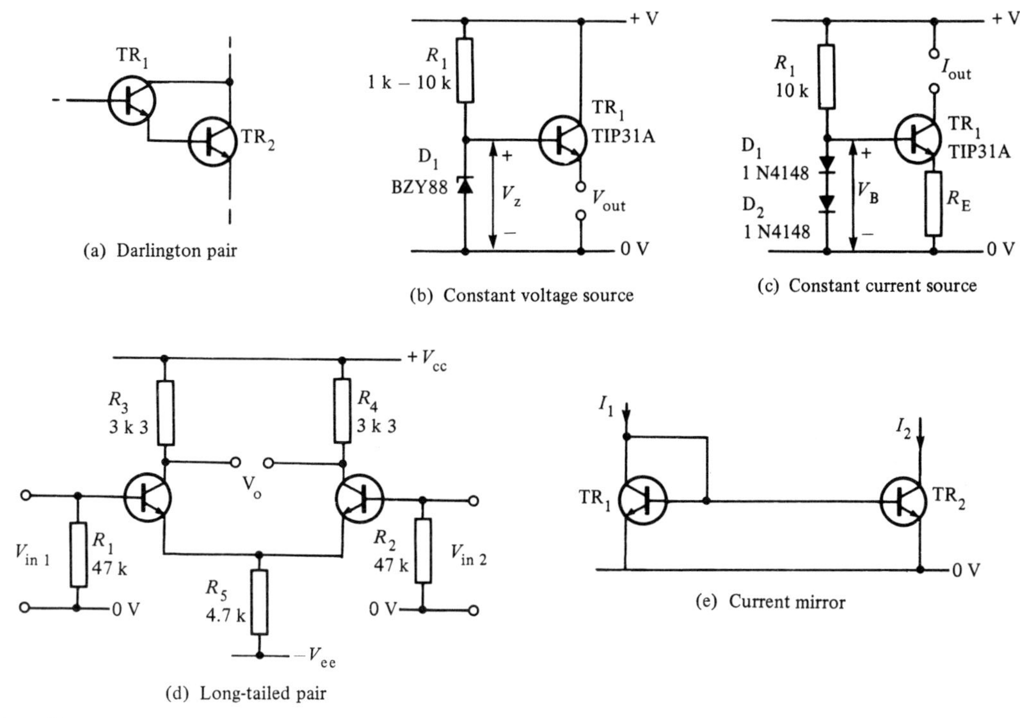

Figure 2.12 *Transistor applications*

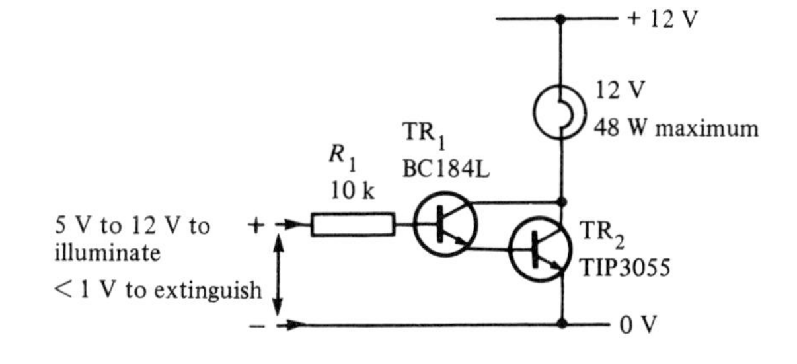

Figure 2.13 *Darlington lamp driver*

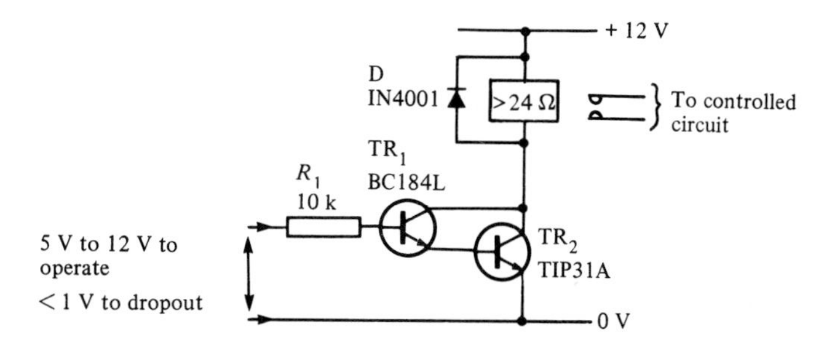

Figure 2.14 *Darlington relay driver*

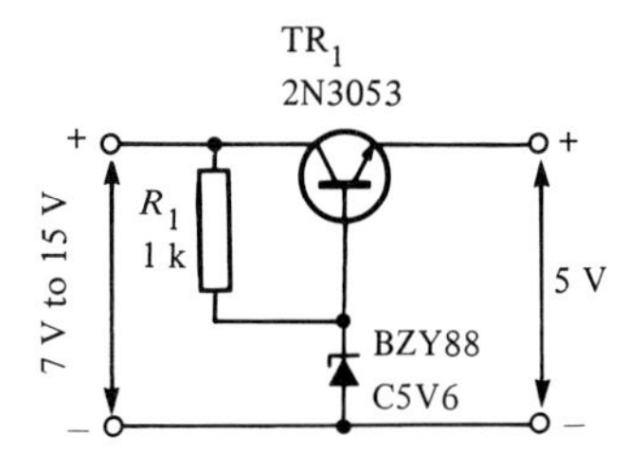

Figure 2.15 *+5 V supply*

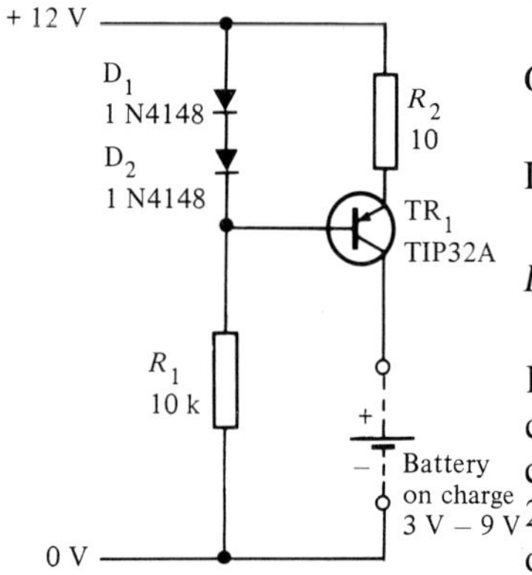

Figure 2.16 *Nicad charger*

Figure 2.12(b) shows how a bipolar transistor can be used in conjunction with a zener diode to form a constant voltage source. It should be noted that the output voltage produced by this arrangement will be approximately 600 mV less than the zener voltage. Hence:

Output voltage, $V_{out} = V_Z - V_{BE} \approx V_Z - 0.6\,V$

Where V_Z is the zener voltage and V_{BE} is the base-emitter voltage drop (approximately 0.6 V).

Figure 2.15 shows a practical application of this circuit in the shape of a simple 5 V supply. The circuit is capable of dealing with load currents of up to about 100 mA (this theme is developed at some length in Chapter 3).

Figure 2.12(c) shows how a bipolar transistor can be used as a constant current source. The base voltage is held constant (at approximately 1.2 V) using two forward biased silicon diodes. The output current will depend upon the value of resistance placed in the emitter such that:

$$\text{Output current, } I_{out} = \frac{V_B - V_{BE}}{R_E} \approx \frac{1.2 - 0.6}{R_E} = \frac{0.6}{R_E}\,A$$

It is usually more convenient to work in mA rather than A and hence:

$$I_{out} = \frac{600}{R_E}\,mA$$

Figure 2.16 shows how the constant current source can be used as a simple charger for AA-size nickel cadmium batteries. These will be fed with a current of 60 mA. In this application, the supply voltage must be at least 2 V greater than the nominal cell voltage but should not be so high that dissipation becomes excessive. In any event, it is advisable to fit the transistor with a small push-fit heatsink of 85 °C/W, or better.

Figure 2.12(d) shows a long-tailed pair which acts as a differential amplifier stage. The output voltage of this circuit is proportional to the difference between the two input voltages. Circuits of this type are ideal for use when signals are balanced (i.e. neither side of the signal is returned to 0 V). The signal is applied differentially (between the two bases) and extracted differentially (between the two collectors). To preserve symmetry, the two transistors should have near identical characteristics. The emitter resistor (or 'tail'), R_5, supplies current which is shared between the two transistors. This resistor should, ideally, be replaced by a constant current source.

Figure 2.12(e) shows a current mirror which can be used to form the basis of a constant current source. A current, I_1, flowing in TR_1 (connected as a diode) causes an equal collector current, I_2, to flow in TR_2.

Figure 2.17 shows how the long-tailed pair can be combined with a current mirror to form a practical differential amplifier stage in which the tail resistor can be replaced by an active current source. The value of collector current in TR_4 is set, by appropriate choice of R_5, at approximately 2 mA. This current is then shared between TR_1 and TR_2.

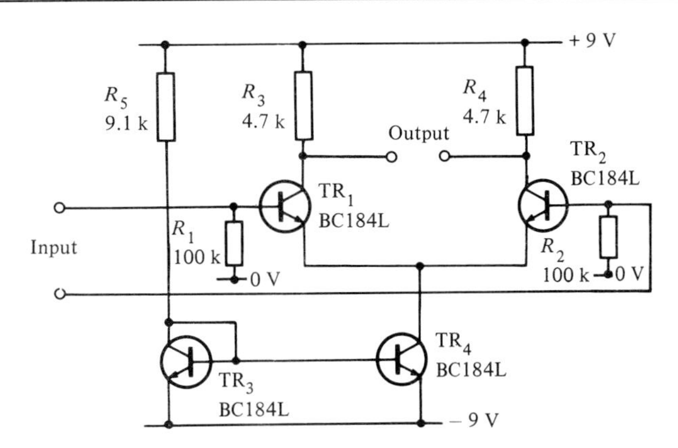

Figure 2.17 *Differential amplifier*

Integrated circuits

Integrated circuits free the equipment designer from the need to construct circuits with individual discrete components such as transistors, diodes, and resistors. The secret, of course, is that, with modern integrated circuit technology, it is possible to fabricate a large number of equivalent discrete components on a tiny slice of silicon. Not only is the resulting arrangement more compact than its discrete component equivalent but it is also considerably cheaper and very much more reliable.

At this stage we should, perhaps, make one further point for the benefit of the newcomer. It is not necessary to have a detailed understanding of the internal circuitry of an integrated circuit in order to be able to make effective use of it. Rather, we have to be aware of some basic ground rules concerning the supply voltage rails, and input and output requirements.

A relative measure of the number of individual semiconductor devices within the chip is given by referring to its 'scale of integration'. The following terminology is commonly applied:

Table 2.10

Scale of integration	*Abbreviation*	*Number of logic gates**
Small	SSI	1 to 10
Medium	MSI	10 to 100
Large	LSI	100 to 1000
Very large	VLSI	1000 to 10000
Super large	SLSI	10,000 to 100,000

*Or circuitry of equivalent complexity.

With the exception of a handful of very simple circuits, the availability of a host of low-cost integrated circuits have largely rendered discrete circuitry obsolete. Integrated circuits offer a number of significant advantages over discrete component circuitry designed to fulfil the same function. These may be summarized as follows:

(a) Higher reliability – despite their complexity, integrated circuits generally offer higher reliability than discrete component circuits.
(b) Lower cost – integrated circuits are often cheaper than the equivalent discrete circuitry.
(c) Smaller space requirements – integrated circuits require very little space. Typically this is around 10% of the space required by comparable discrete circuitry.
(d) Easier circuit layout – since integrated circuits are constrained to use a minimal number of external connections, circuit layout is greatly simplified.
(e) Matched performance – semiconductor devices fabricated using integrated circuit technology usually have very closely matched characteristics. Performance is thus more readily assured than would be possible with very significant variations in discrete component characteristics.

The disadvantages of integrated circuits can be summarized as follows:

(a) Lack of flexibility – it is not generally possible to modify the parameters within which an integrated circuit will operate as easily as it is to modify the performance of discrete circuitry.
(b) Performance limitations – integrated circuits may have severe limitations in applications where:

    (i) High currents are involved (e.g. high current regulators).
    (ii) High voltages are involved (e.g. high voltage regulators).
    (iii) High power levels are involved (e.g. high power audio amplifiers).
    (iv) High frequency signals are involved (e.g. RF amplifiers).
    (v) Very low level signals are involved (integrated circuits are inherently noisy).

Integrated circuits may be divided into two general classes; linear (analogue) or digital. Examples of linear integrated circuits are operational amplifiers (see Chapter 5) whereas examples of digital integrated circuits are logic gates (see Chapter 6).

Encapsulation

Various forms of encapsulation are used for integrated circuits. The most popular of these are the dual-in-line (DIL) types which are available in both plastic and ceramic case styles (the latter using a glass hermetic sealant). Common types have 8, 14, 16, 28 and 40 pins, the pin numbering conventions for which are shown in Figure 2.18.

Flat package construction (featuring both glass-metal and glass-ceramic seals and welded construction) are popular for planar mounting on flat circuit boards. No holes are required to accommodate the leads of such

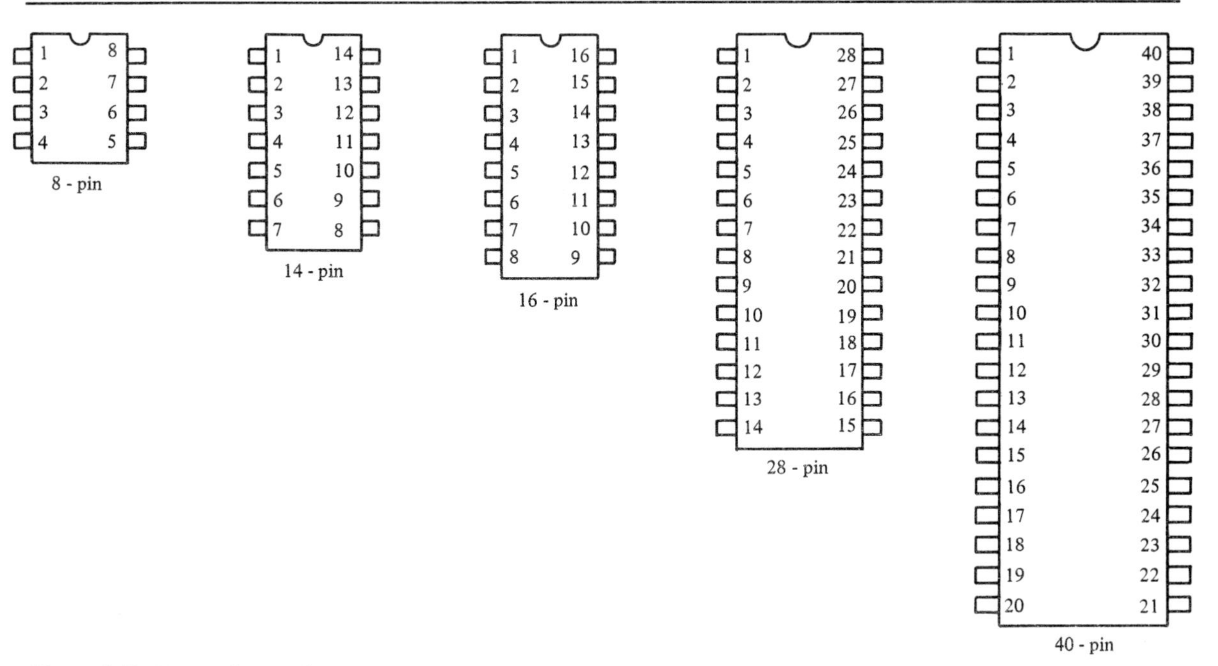

Figure 2.18 *Pin numbering for commonly used DIL integrated circuit packages (viewed from above)*

devices which are arranged on a 0.05 in pitch (i.e. half the pitch used with DIL devices). This form of encapsulation requires special handling and is not generally suitable for use with prototype or 'one-off' construction techniques.

3 Power supplies

This chapter deals with the unsung hero of most electronic circuits: the power supply. All too often the power supply is taken for granted and is not accorded the priority it really deserves; we glibly assume that it is there and is doing its job properly.

The correct operation of the power supply is, however, crucial to most electronic circuits. As an example, consider the case of a power supply used in conjunction with a high quality audio amplifier. Poor regulation within the power supply will result in significant distortion at high volume levels as the output voltage falls under load. A large amount of residual mains ripple present in the power supply output will result in a noticeable hum at low volume levels. The lesson from this, of course, is that there is little point in refining other areas of circuitry without first taking positive steps to improve the power supply.

Power supply characteristics and specifications

At first sight it may appear that there is little scope for variation in the design of such a mundane piece of equipment as a power supply. However, when one considers the widely differing supply requirements of electronic circuitry, one soon perceives the need for a range of different types of power supply. These may be broadly classified under the following six headings:

1 Classification according to the degree of smoothing

Unsmoothed: A significant proportion of mains ripple (this component is at twice the mains frequency in full-wave rectified circuits) is present as an a.c. component superimposed on the d.c. output.

Smoothed: Circuitry has been incorporated in order to substantially minimize the ripple component present.

2 Classification according to degree of regulation

Unregulated (unstabilized): The d.c. output is smoothed but no additional circuitry is provided in order to overcome the effect of line and load variations.

Regulated (stabilized): The d.c. output is smoothed and additional circuitry has been included in order to combat the effects of line and/or load variations.

3 Classification according to the degree of protection

Unprotected: The power supply is not protected against excessive values of load-current or short-circuiting of the output.

Protected: The power supply incorporates a means of protection against the effects of excessive load current or a short-circuit at the output.

Crow-bar protected: The load circuit is protected against catastrophic failure within the regulator which can be instrumental in placing an excessive voltage at the output.

4 Classification according to the operating conditions

Switched-mode: The series regulating element is operated under switching conditions (usually at frequencies in excess of 50 kHz).

5 Classification according to the type of circuitry employed

Discrete: Discrete devices (i.e. transistors) are used in the regulating or stabilizing stages.

Integrated: Integrated circuits are employed in the stabilizing or regulating stages.

Hybrid: A mixture of both discrete and integrated circuits are employed in the stabilizing or regulating stages.

6 Classification according to output voltage or current

High-current:
Low-current:
High-voltage:
Low-voltage:
} These terms are considered to be self-explanatory.

The specification of a power supply usually involves such obvious parameters as input and output voltage, maximum load current, etc. Specifications with which readers may be less familiar include:

(a) *Efficiency*. Ideally, all the power drawn from the a.c. mains supply would be usefully delivered to the load connected to the output of the power supply. In practice, however, some power is lost within the power supply itself. The efficiency of the power supply is given by:

$$\text{Efficiency} = \frac{\text{d.c. power output}}{\text{a.c. power input}} \times 100\%$$

(b) *Ripple.* The ripple present as an a.c. component superimposed on the d.c. output of a power supply may be specified in several ways including the r.m.s. or peak-peak value of the ripple voltage or by quoting a ripple factor where:

$$\text{Ripple factor} = \frac{V_r}{V_o}$$

Where V_r is the r.m.s. value of ripple voltage and V_o is the d.c. output voltage.

(c) *Ripple rejection.* This is a measure of the ability of a regulator or smoothing circuit to reduce the a.c. ripple component present. Ripple rejection is usually expressed in decibels where:

$$\text{Ripple rejection} = 20\log_{10}\left(\frac{V_{ri}}{V_{ro}}\right)$$

where V_{ri} and V_{ro} are the r.m.s. (or peak-peak) values of ripple voltage present at the input and output of the smoothing filter or regulator.

(d) *Load regulation.* This is the percentage rise in d.c. output voltage when the load is removed from the power supply. Load regulation is defined as:

$$\text{Load regulation} = \left(\frac{V_{onl} - V_{ofl}}{V_{ofl}}\right) \times 100\%$$

Where V_{ofl} is the d.c. output voltage under full-load conditions and V_{onl} is the d.c. output voltage under no-load conditions.

(e) *Line regulation.* This is the per-unit change in d.c. output voltage divided by the corresponding per-unit change in a.c. input voltage. Line regulation is hence defined as:

$$\text{Line regulation} = \frac{\left(\dfrac{V_{ih} - V_{ol}}{V_{ol}}\right)}{\left(\dfrac{V_{oh} - V_{il}}{V_{il}}\right)}$$

Where V_{ih} and V_{oh} are respectively the a.c. input and d.c. output voltages under the maximum a.c. output voltage state, while V_{il} and V_{ol} are respectively the a.c. input and d.c. output voltages under the minimum a.c. input voltage state.

(f) *Output impedance.* This is the ratio of the change in d.c. output voltage to the corresponding change in output current as the load on the power is varied. The output impedance of a power supply is thus given by:

$$\text{Output impedance} = \frac{V_{onl} - V_{ofl}}{I_{ofl}}$$

Where no-load and full-load conditions whilst I_{ofl} is the full-load output current.

A typical general purpose low-voltage d.c. power supply could be expected to perform according to the following specifications:

Output voltage: Variable from 2 V to 24 V.
Output current: 2 A maximum.
Output impedance: Less than 0.1 Ω.
Line regulation: Better than 0.2%.
Load regulation: Better than 0.5%.
Ripple: Less than 10 mV peak-peak at full-load.

Before describing some typical power supply circuitry, it is worth making a few points concerning safety. Not only is the power supply one of the few areas in which a potentially lethal shock hazard exists, but it is also a region in which strict attention must be placed on the rating of components.

The consequences of grossly exceeding the d.c. working voltage rating of a large electrolytic reservoir capacitor, for example, can be dire – the capacitor will overheat and may eventually explode! The moral is simple; always allow a generous margin for safety and never trust a component to operate continuously at, or near, its maximum ratings. To do so would be to ignore the possibility of such abnormal, but perfectly plausible, events as momentary surges in mains voltage, transient spikes, and short circuit failures of the load.

Mains input arrangements

The input of a mains operated item of equipment should normally be fitted with a suitably rated mains switch and fuses. It should go without saying that the switch MUST be specifically designed for a.c. mains operation. Furthermore, it should preferably be a double-pole single-throw (DPST) type. The switch may be of the toggle, latching push-button, rocker or rotary type. In the latter case it may be mechanically ganged with another component such as a variable resistor or multipole rotary switch. For heavy current applications a rocker switch is usually preferred. Such devices are also available with integral mains indicators though these components tend to be somewhat larger than their miniature counterparts.

Fuses should be fitted in both mains supply leads rather than just one. While one fuse will provide adequate protection in the event of a component failure in the secondary circuit, it may not protect against all eventualities (e.g. a failure between the primary winding and metal core or earthed screen). Two fuses will provide ample protection and these should normally be standard quick-blow types rated at approximately 150% to 200% of the normal working primary current.

With high current supplies which utilize very large values of reservoir capacitor (e.g. 4700 μF and above) problems sometimes arise due to the large inrush of current when the mains is first applied. In such applications, slow-blow buses should be substituted for standard types. Nominal current ratings should, however, be the same.

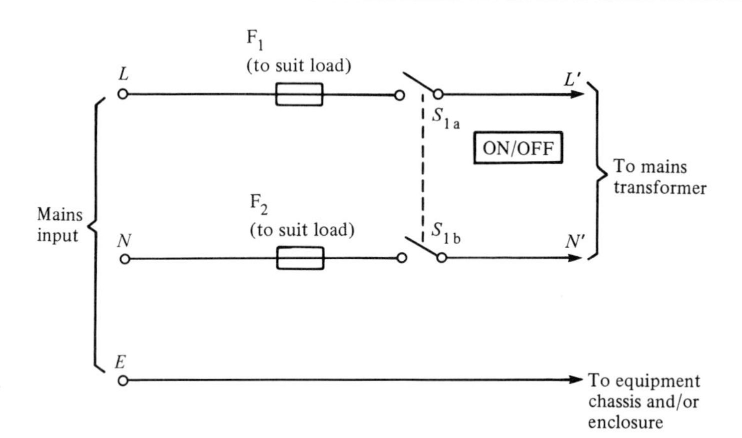

Figure 3.1 *Typical mains input arrangement*

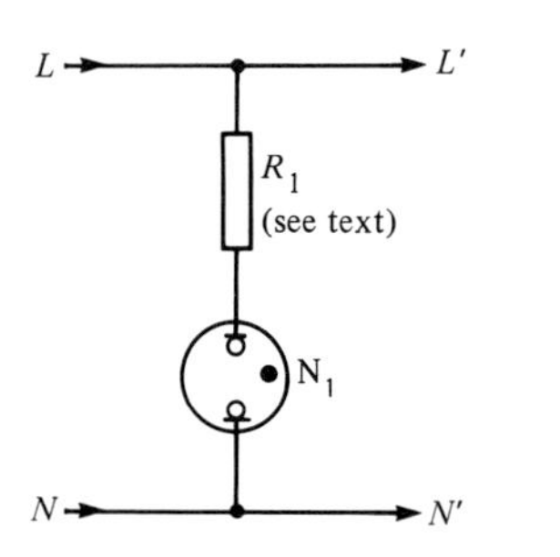

Figure 3.2 *Neon mains indicator*

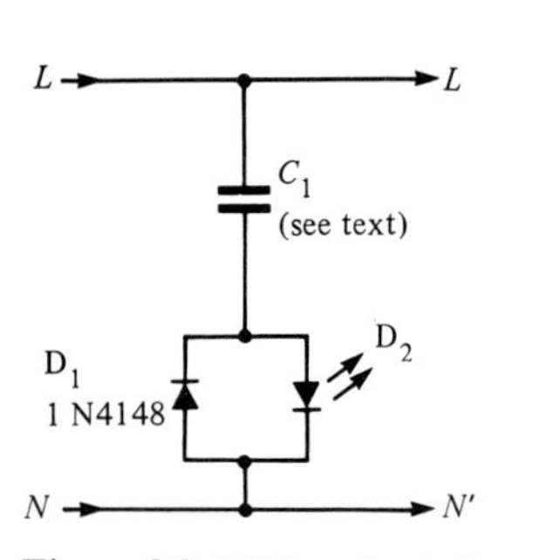

Figure 3.3 *LED mains indicator*

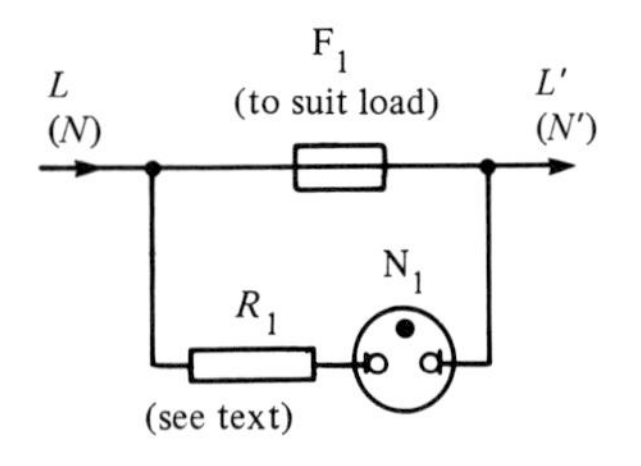

Figure 3.4 *Neon 'blown-fuse' indicator*

The circuit shown in Figure 3.1 is suitable for most applications involving mains input. Where a mains indicator is required, this can be very easily provided with the aid of a panel mounting neon indicator as shown in Figure 3.2. The value of series current limiting resistor, R_1, should be 100 kΩ 0.5 W for 100 V to 125 V a.c. mains supplies or 270 kΩ 0.5 W for 200 V to 250 V a.c. mains supplies. Neon mains indicators are also available with integral series current limiting resistors. When using an indicator of this type, no external resistor will be required and the indicator may simply be wired directly across the incoming mains supply.

In some cases (e.g. to match other panel mounting indicators) an LED mains indicator will be preferred. The arrangement shown in Figure 3.3 may then be used. It is, however, important to ensure that the capacitor is suitably rated for mains operation. For operation on 100 V to 125 V a.c. mains supplies C_1 should have a value of 220 nF and be rated at 125 V a.c. working. Where the mains supply is in the range 200 V to 250 V a.c., the value should be reduced to 100 nF, 250 V a.c. working.

In some equipment, a 'blown fuse' indicator may be useful and Figures 3.4 and 3.5 show arrangements based on neon and LED indicators. The values of R_1 and C_1 in these circuits are the same as those employed in the circuits of Figures 3.2 and 3.3.

Modern transformers often have two primary windings (to permit operation on both 110 V and 220 V a.c. supplies) and Figure 3.6 shows a typical dual input voltage arrangement based on the use of two links, LK_1 and LK_2, while Figure 3.7 shows a similar arrangement based on the use of a double-pole double-throw (DPDT) mains selector switch, S_1.

In some applications the use of a mains filter is desirable. Such a device can be instrumental in reducing the effects of mains borne noise and interference. Mains interference generally falls into one of three categories:

(a) Supply borne common mode noise. The noise signal is conveyed equally on the live and neutral conductors while the earth acts as a return path.
(b) Supply borne differential mode noise. The noise signal travels along the live conductor and returns along the neutral conductor.

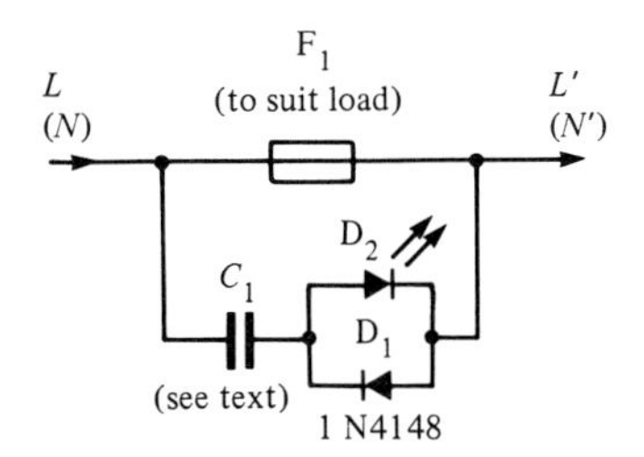

Figure 3.5 *LED 'blown-fuse' indicator*

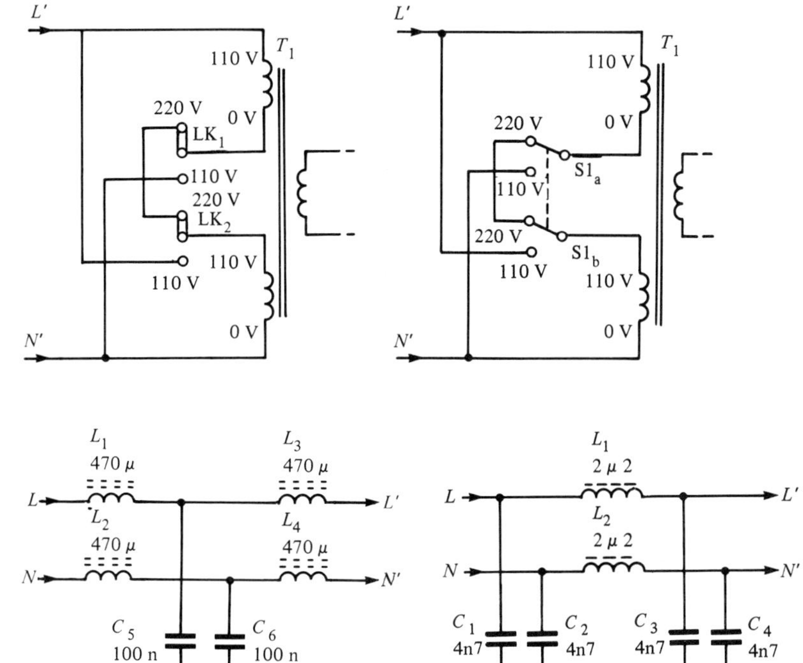

Figure 3.6 *Dual mains input voltage selector based on links*

Figure 3.7 *Dual mains input voltage selector based on a suitably rated DPDT switch*

Figure 3.8 *Mains input filter for use up to 30 MHz*

Figure 3.9 *Mains input filter for use above 30 MHz*

(c) Noise radiated by mains wiring and coupled electromagnetically and/or electrostatically into the equipment.

Where equipment is likely to be susceptible to the effects of (a) or (b) a filter should be fitted to the incoming supply at the point at which it enters the equipment. The filter itself should be a low-pass type and will be invariably based upon one, or more, double π-sections. The filter should be enclosed in a screened and grounded metal case.

Figure 3.8 shows the circuit of a filter designed for noise suppression up to 30 MHz while Figure 3.9 shows a similar unit designed for operation above 30 MHz. In severe cases, it is quite permissible to cascade the two filters in order to provide effective wideband noise suppression.

It should be noted that all filter components must be adequately rated. Capacitors should be suitable for continuous 240 V a.c. operation and should possess negligible series inductance. Inductors must be capable of carrying the full r.m.s. supply current indefinitely without appreciable temperature rise due to I^2R losses and should have inherently low shunt capacitance. The inductor must also not be prone to saturation at the peak supply current (this will be well in excess of the r.m.s. supply current).

In common with most mains filters, the filter circuits shown in Figures 3.8 and 3.9 produce a modest value of earth current which may pose something of a problem when a sensitive earth leakage trip is fitted

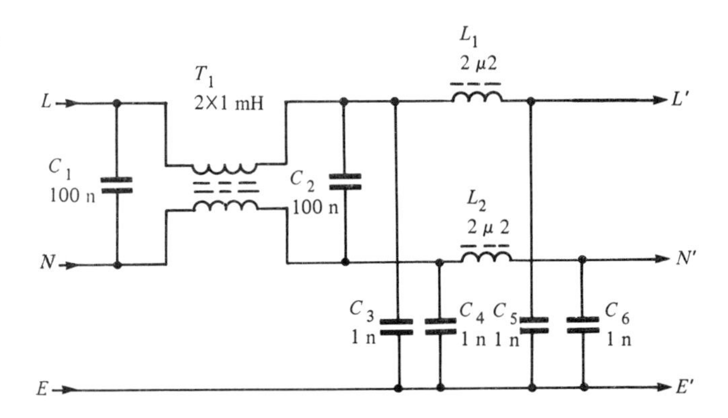

Figure 3.10 *Wideband low earth current mains filter*

to a particular mains circuit (a single 100 nF capacitor connected between live and neutral will pass a current of approximately 8 mA). Figure 3.10 shows the circuit diagram of a filter which has a negligible earth current (approximately 200 μA r.m.s.).

The effects of radiated mains interference can be minimized by adequate screening and earthing of the equipment. In severe cases it may be desirable to use screened mains cables and distribute mains supplies within an earthed metal conduit.

As well as providing two separate primary windings, most general purpose transformers provide two secondary windings which can be used individually or may be wired in either series or parallel. Figure 3.11 shows the various permutations possible with such a component.

Parallel connection of transformer secondaries is permissible provided the secondaries are accurately matched as regards output voltage. Parallel connection should be avoided unless the manufacturer's specification explicitly states that this mode can be employed. The penalty for not observing such a precaution is that very large circulating currents may flow due to an imbalance in secondary voltages. This may result in excessive dissipation within the transformer, poor regulation, and low efficiency.

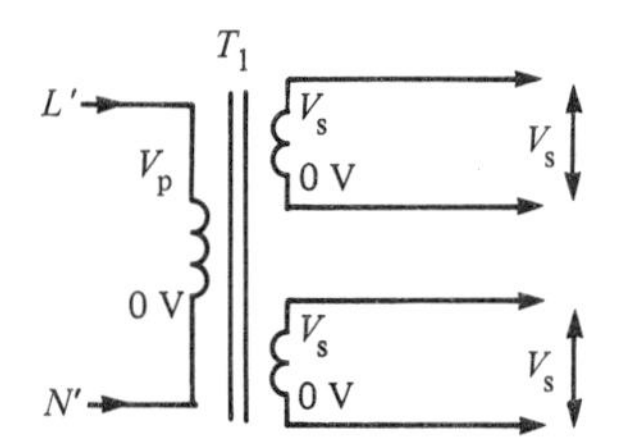

Two individual secondary outputs each rated at I_s

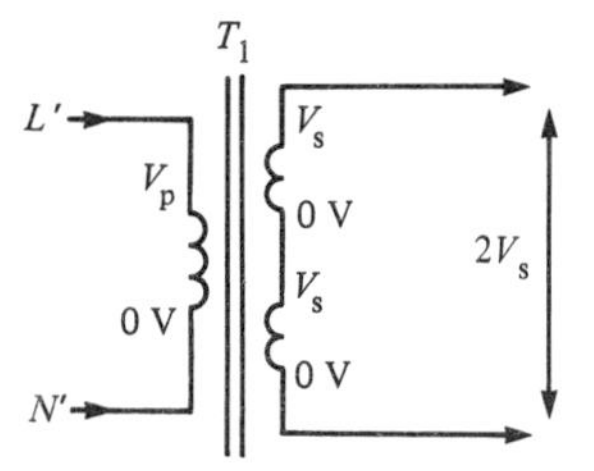

Series connected secondaries
Maximum output current = I_s

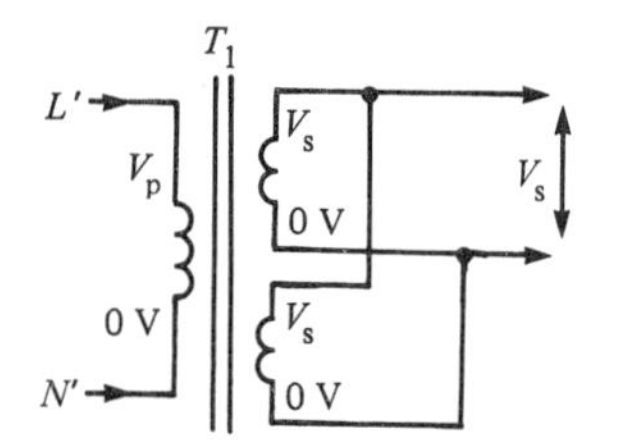

Parallel connected secondaries.
Maximum output current = $2I_s$

Figure 3.11 *Three ways of connecting a mains transformer having dual secondary windings*

Rectifier arrangements

Figure 3.12 shows a simple full-wave bridge rectifier arrangement which forms the basis of the vast majority of power supplies. The rectifier may take the form of either an encapsulated bridge (in which all four diodes are moulded or potted in epoxy) or four individual diodes.

Four-pin DIL types are suitable for use in very low current applications (900 mA or less). 'In-line' types are commonly used for currents of up to 4 A while square horizontally mounted block types are commonly available for currents of between 2 A and 35 A maximum. High power bridge-epoxy moulded or epoxy-potted bridge rectifiers normally have an isolated metal case which permits good thermal contact with a heatsink.

Typical values of no-load output voltage for the circuit of Figure 3.12 are as follows:

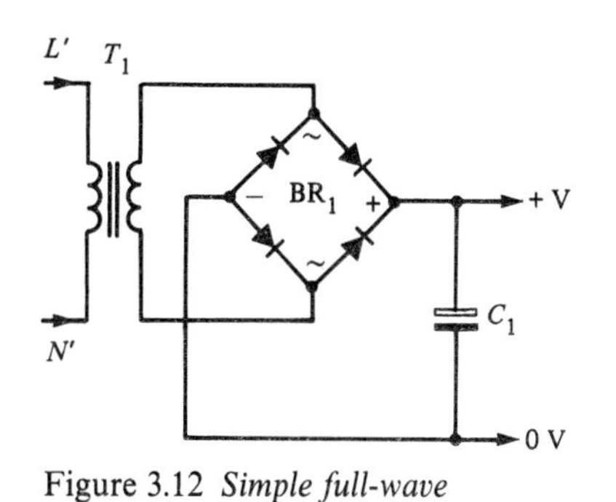

Figure 3.12 *Simple full-wave bridge rectifier arrangement*

Table 3.1

R.m.s. secondary voltage	*No-load d.c. output voltage*
4.5 V	6.3 V
6 V	8.5 V
9 V	12.7 V
12 V	16.9 V
15 V	21.2 V
18 V	25.4 V
20 V	28.2 V
24 V	33.9 V

While the no-load output voltage produced by the arrangement shown in Figure 3.12 will be approximately 1.4 times the r.m.s. secondary voltage, design calculations should allow for a 'worst-case' full-load forward voltage drop of 2 V within the bridge rectifier.

When using individual rectifiers, and to allow some margin of safety, each diode should be rated for a continuous forward current equivalent to the d.c. load current. In the case of an encapsulated bridge, the maximum continuous forward current will be the same as the maximum d.c. load current. For a capacitive input filter, the peak current should be at least twenty times the maximum continuous rating. Furthermore, when the rectifier case temperature exceeds 50 °C it should be derated proportionately to zero forward current at 150 °C. The maximum repetitive reverse voltage rating (V_{rrm}) should be at least equal to the off-load output voltage (i.e. approximately 1.4 times the r.m.s. secondary voltage).

The value of reservoir capacitor, C_1, will depend on the load current required and the amount of ripple that can be tolerated. The values shown in Table 3.2 should be adequate for most applications:

Table 3.2

Nominal load current	*Reservoir capacitance*
125 mA	470 μF
250 mA	1,000 μF
500 mA	2,200 μF
1 A	4,700 μF
2 A	6,800 μF
4 A	10,000 μF
8 A	22,000 μF

Readers should note that the reservoir capacitor must have an adequate voltage rating. To provide some margin of safety, this should be approximately 1.4 times the off-load output voltage (i.e. twice the r.m.s. secondary voltage).

When dual outputs are required, a bridge rectifier can be used in

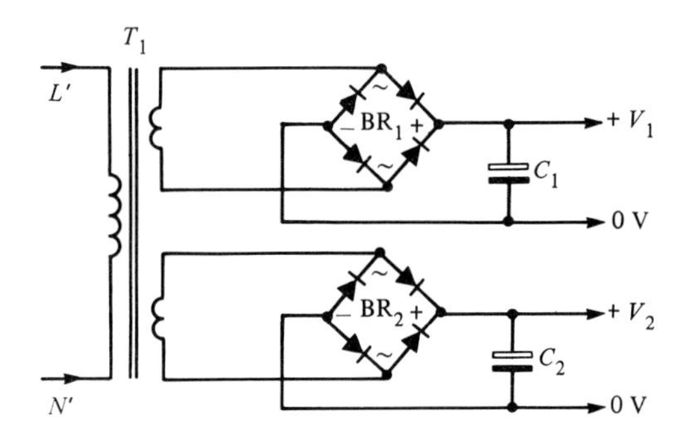

Figure 3.13 *Dual output arrangement using two separate bridge rectifiers*

Figure 3.14 *Dual output arrangement with a common 0 V connection*

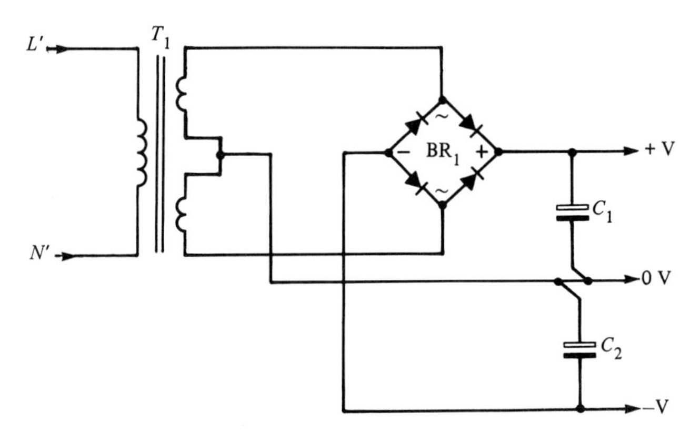

Figure 3.15 *Dual output arrangement with a common 0 V using a single bridge rectifier*

Figure 3.16 *Half-wave voltage doubler arrangement*

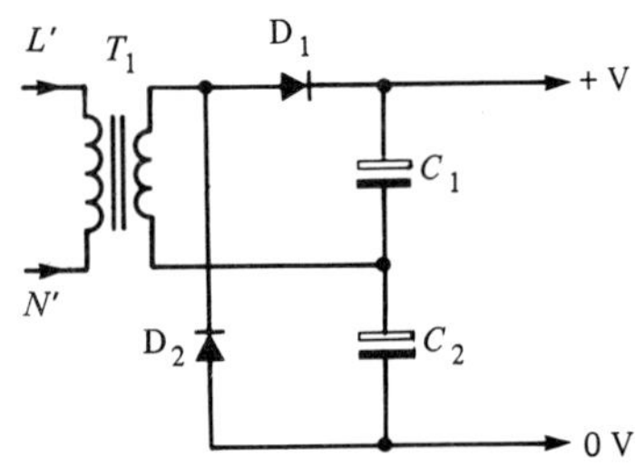

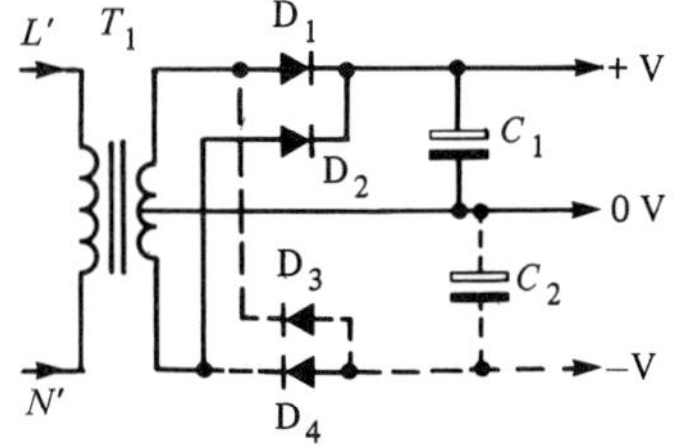

Figure 3.17 *Full-wave bi-phase rectifier arrangement using a centre tapped secondary*

conjunction with a transformer having two identical secondaries, as shown in Figure 3.13. A symmetrically split d.c. supply can be obtained by linking the negative output of one supply to the positive output of the other and then treating this as the common 0 V rail, as shown in Figure 3.14. Alternatively, a single bridge rectifier may be used in the arrangement of Figure 3.15. Note, however, that the maximum reverse repetitive voltage rating (V_{rrm}) for the diodes in this circuit should be equal to 2.8 times the r.m.s. voltage produced by each individual secondary winding.

A half-wave voltage doubler arrangement is shown in Figure 3.16. This circuit produces an off-load d.c. output which is 2.8 times the r.m.s. secondary voltage. The circuit is, however, generally only suitable for relatively small output currents (typically 250 mA, or less).

Figure 3.17 shows a full-wave biphase rectifier arrangement based on the use of a centre-tapped secondary winding. The circuit may be readily modified for dual outputs by adding two extra diodes, D_3 and D_4, together with an additional reservoir capacitor, C_2. The circuit of Figure 3.17 may also be realized using a transformer with two individual secondaries (appropriately rated) by merely series linking the two windings as shown in Figure 3.18.

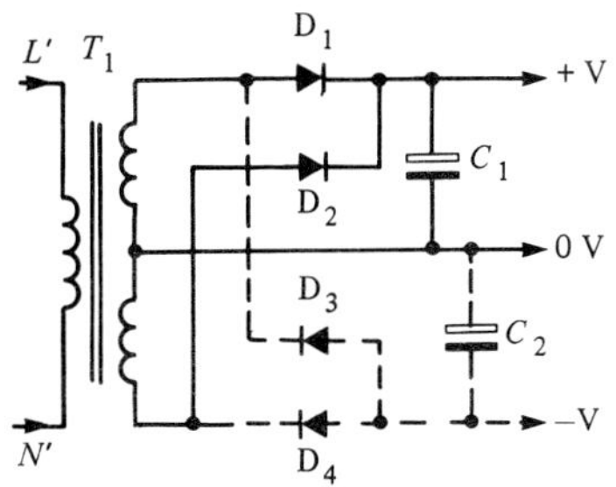

Figure 3.18 *Full-wave bi-phase rectifier arrangement using dual secondary windings*

Discrete component regulators

Regulation of the output voltage of a power supply is important in nearly every application. Happily, for the vast majority of applications, this can be very easily achieved using a handful of components.

The simple shunt zener voltage regulator shown in Figure 3.19 is only suitable for low current applications (50 mA or less). In practice a circuit of this type can be expected to produce a regulation figure of around 10%, or better. The output impedance of such an arrangement will normally be of the order of 15 Ω, or less.

The performance of the single-stage shunt zener diode regulator can be improved by cascading two similar sections, as shown in Figure 3.20. This circuit can be expected to provide a regulation of better than 5% with an output impedance of less than 5 Ω).

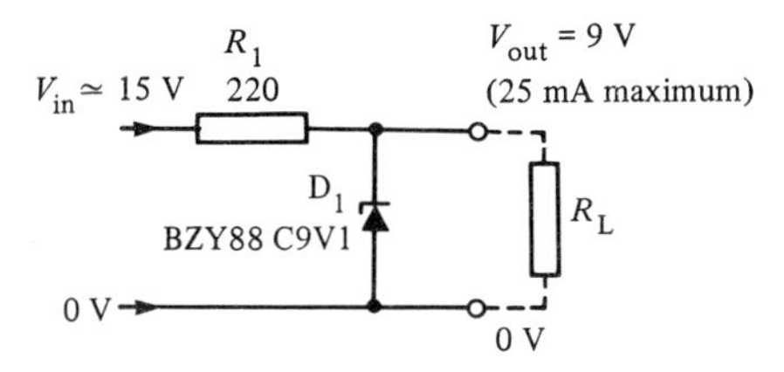

Figure 3.19 *Simple zener diode shunt voltage regulator*

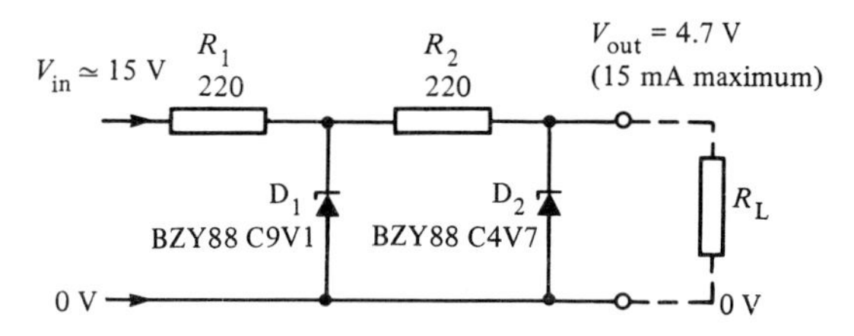

Figure 3.20 *Cascaded zener diode shunt voltage regulator*

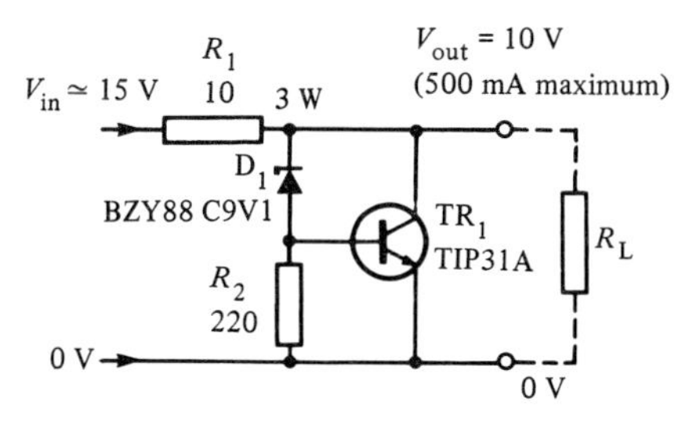

Figure 3.21 *Amplified zener diode shunt voltage regulator*

A further improvement can be obtained by diverting current from the load using a shunt connected transistor, as shown in Figure 3.21. This 'amplified zener' arrangement is suitable for applications in which the load current is very much greater than the maximum rated current of the zener (found by dividing its maximum permissible power dissipation by its nominal zener voltage). The compound arrangement of zener and transistor behaves like a superior zener diode having a much higher power rating. The circuit shown in Figure 3.21 has an output impedance of

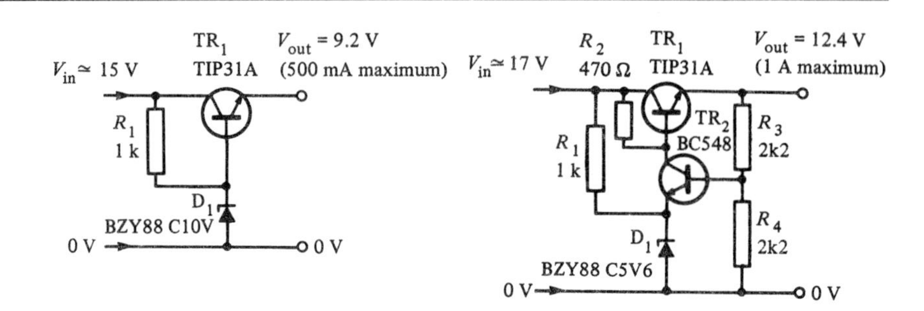

Figure 3.22 *Basic transistor series voltage regulator*

Figure 3.23 *Improved series voltage regulator*

approximately 0.3 Ω. The output voltage will be approximately 0.8 V greater than the nominal zener voltage.

Unfortunately, high current shunt regulators have several obvious disadvantages not least of which is associated with the power dissipation in the series dropping resistor. This must be adequately rated (a wirewound resistor will usually be required) and mounted so that convection cooling is possible. Another problem is that maximum dissipation in the shunt element occurs under no-load conditions. It, too, will require heatsinking.

At this point it is worth mentioning that shunt regulators do have one advantage over their series connected counterparts; a collector-emitter short circuit failure within the transistor will make the output fall to zero. Unless special 'crowbar' protection (see page 73) is incorporated, a similar failure within a comparable series regulator will cause the output voltage to rise to the full unregulated d.c. output from the bridge rectifier.

A basic series regulator is shown in Figure 3.22. The transistor is effectively connected as an emitter follower and thus the output voltage will be approximately 0.8 V less than the nominal zener voltage. Dissipation in the series element rises as the load current increases. Under 'worst-case' conditions (i.e. under full-load and with the minimum expected value of the a.c. mains supply), the input voltage should be at least 2 V greater than the desired output. In order to minimize dissipation in the transistor (and thus minimize heatsink ratings) the d.c. input should not, however, be very much greater than this value. A recommended range for the input would be between 2 V and 7 V greater than the desired output voltage. Thus, for a nominal 12 V output, the rectifier output should be within the range 14 V (minimum) to 19 V (maximum).

An improved series regulator is shown in Figure 3.23. Here the zener voltage is used as a 'reference' against which the output voltage is compared. TR_2 senses the difference between the two voltages and makes appropriate changes to the bias applied to the series transistor, TR_1. While the values shown in Figure 3.23 are for 12 V operation, the circuit is adaptable for outputs in the range 5 V to 10 V with currents up to 1 A. The output voltage of the circuit is given by:

$$V_{out} = \left(\frac{R_3 + R_4}{R_4}\right)(V_z + 0.6)\,\text{V}$$

Where V_z is the nominal zener voltage.

A further improvement can be obtained by replacing the transistor with an operational amplifier comparator, as shown in Figure 3.24. Almost any

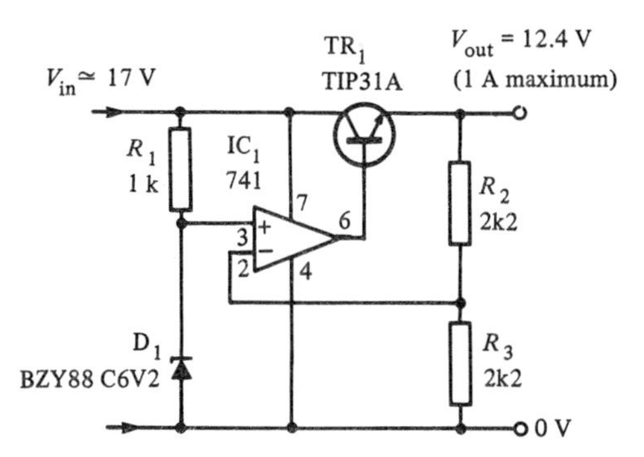

Figure 3.24 *Series voltage regulator based on an operational amplifier comparator*

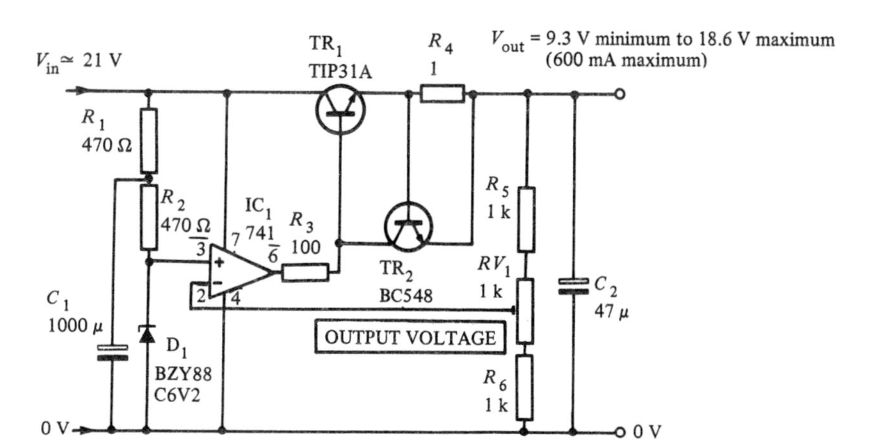

Figure 3.25 *Adjustable and current-limited voltage regulator*

bipolar, BIFET, or JFET operational amplifier can be used in this arrangement provided the input voltage does not exceed the maximum rated supply voltage. The regulation of the arrangement shown in Figure 3.24 is better than 1% and the output resistance typically less than 0.5 Ω.

Figure 3.25 shows several more refinements. R_1 and C_1 provide additional smoothing prior to the zener reference (C_1 is less effective if placed directly across D_1), the output is made adjustable by means of RV_1, and R_4/TR_2 provide current limiting.

The output current limit operates at a value of load current given by:

$$I = \frac{0.6}{R_4} \text{ A}$$

Fixed integrated circuit voltage regulators

Integrated circuit voltage regulators are available in a range of output

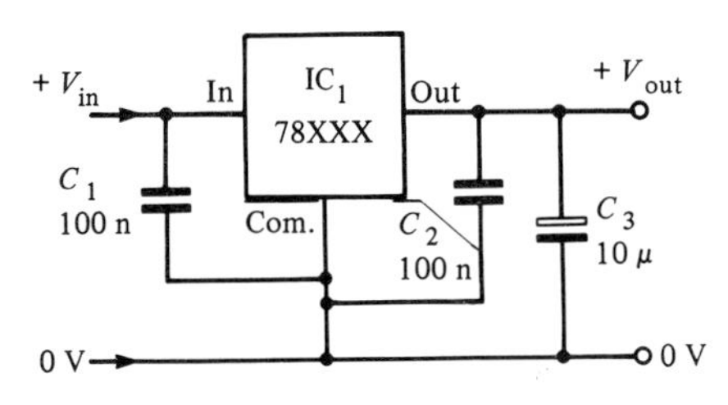

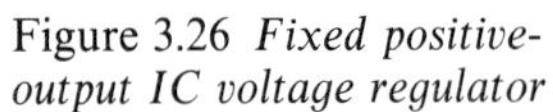
Figure 3.26 *Fixed positive-output IC voltage regulator*

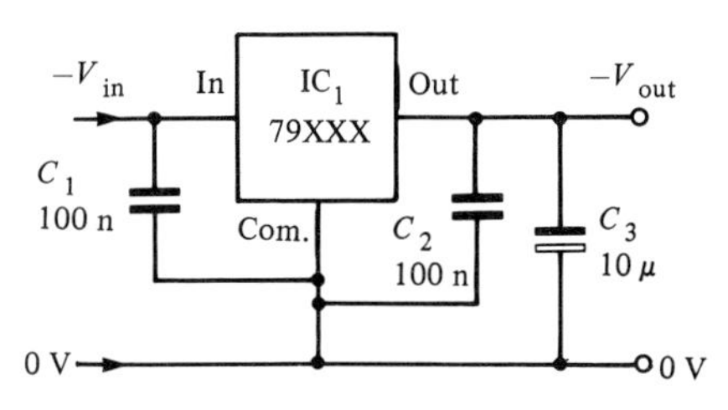

Figure 3.27 *Fixed negative-output IC voltage regulator*

voltages and usually feature internal fold-back current limiting as well as thermal shut-down.

The most popular series of fixed voltage regulators employ a TO220 plastic package and is prefixed with 78 (positive input and output) and 79 (negative input and output). These devices are available in a range of voltages (i.e. 5 V, 9 V, 12 V, 15 V, 18 V and 24 V) and are rated for a maximum load current of 1 A.

Basic fixed voltage regulator circuits using 78 and 79 series devices are shown in Figures 3.26 and 3.27. The circuit of Figure 3.26 is designed for positive input and positive output whilst that a Figure 3.27 is for negative input and negative output.

It is important to note that the worst-case unregulated d.c. input voltage should normally be at least 3 V greater than the nominal regulated output voltage. The penalty for not observing this precaution will be poor regulation and an unacceptable level of residual mains hum present at the output. Furthermore, the worst-case unregulated d.c. input voltage should not be allowed to exceed the nominal regulated output voltage by more than about 15 V otherwise regulator dissipation will be excessive. This, in turn, can lead to premature thermal shut-down.

Typical characteristics of 100 mA, 1 A and 2 A plastic fixed voltage regulators are shown in Tables 3.3, 3.4 and 3.5:

Table 3.3 ***100 mA series***

Type Positive output Negative output	*78L05* *79L05*	*78L12* *79L12*	*78L15* *79L15*	*78L24* *79L24*
Input voltage range (V)	7–30	14.5–35	17.5–35	27–35
Load regulation (%)	0.2	0.2	0.3	0.4
Line regulation (%)	1	1	1.5	1.5
Ripple rejection (dB)	60	55	52	49
Output resistance (mΩ)	200	400	500	850
Output noise voltage (10 Hz to 100 kHz) (μV)	40	70	90	200
Short-circuit current (mA)	75	35	25	20

Table 3.4 *1 A series*

Type	*Positive output* *Negative output*	*7805* *7905*	*7812* *7912*	*7815* *7915*	*7824* *7924*
Input voltage range (V)		7–25	14.5–30	17.5–30	27–38
Load regulation (%)		0.2	0.4	0.5	0.6
Line regulation (%)		0.2	0.2	0.3	0.3
Ripple rejection (dB)		71	61	60	56
Output resistance (mΩ)		30	75	95	150
Output noise voltage (10 Hz to 100 kHz) (μV)		40	80	90	170
Short-circuit current (mA)		750	350	230	150

Table 3.5 *2 A series*

Type	*Positive output*	*78S05*	*78S12*	*78S15*	*78S24*
Input voltage range (V)		8–35	15–35	18–35	27–40
Load regulation (mV)		100	160	180	250
Line regulation (mV)		100	240	300	480
Ripple rejection (dB)		60	53	52	48
Output resistance (mΩ)		17	18	19	28
Output noise voltage (10 Hz to 100 kHz) (μV)		40	75	90	170
Short-circuit current (mA)		500	500	500	500

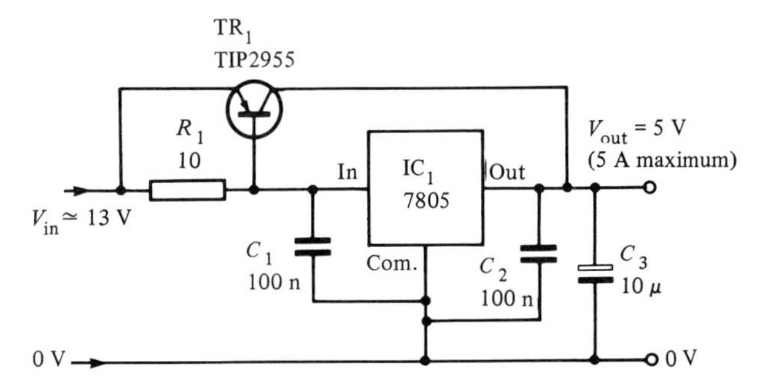

Figure 3.28 *Current boosted positive-output IC voltage regulator*

Current boosted regulators

The output current of a fixed voltage three-terminal regulator can be very easily increased using an additional power transistor. Figures 3.28 and 3.29 respectively show arrangements based on positive and negative voltage regulators. It should be noted that the positive rail version requires the use of a p-n-p transistor while the negative rail version uses an n-p-n transistor.

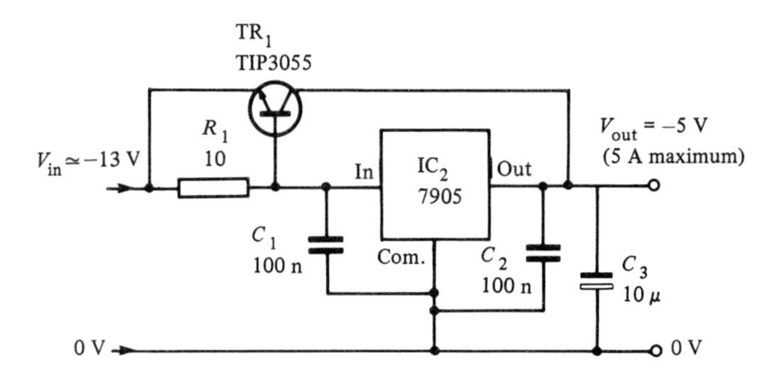

Figure 3.29 *Current boosted negative-output IC voltage regulator*

The power transistors specified in Figures 3.28 and 3.29 are plastic TAB versions of the ever-popular metal cased TO3 encapsulated PNP3055 and 2N3055. The TIP2955 and TIP3055 are, however, easier to mount and share the same 15 A maximum collector current rating as their metal cased counterparts.

The regulator operates normally (i.e. the transistor remains idle) for current of less than approximately 60 mA but, beyond this figure, the transistor begins to conduct, passing collector current into the load.

The circuit is capable of delivering load currents well in excess of 5 A and the output resistance will typically be 0.1 Ω, or less. Care must be taken to ensure that the heatsink arrangements for TR_1 are adequate for continuous dissipation at full-load. When operated from an unregulated d.c. input of 12 V, for example, the heatsink should be rated at 2 °C/W (or less) for load currents of 2 A. For a 5 A load the heatsink should be rated at 0.75 °C/W (or less).

The circuit may be used with other fixed voltage regulators 9 V, 12 V etc.) provided the unregulated d.c. input voltage is increased so that it exceeds the rated output voltage by between 4 V and 7 V. A 12 V version would, for example, require a d.c. input of approximately 17 V.

It is important to note that the usual current limiting associated with a three-terminal regulator is no longer provided by this circuit and the short-circuit output current may be excessive (limited by the d.c. resistance of transformer secondary and component wiring). Provided the transformer and rectifier are both adequately rated, over-current protection can be provided by means of a 5 A quick-blow fuse.

Voltage boosted regulators

The output voltage from a regulator can be very easily increased in either fixed increments (as shown in Figures 3.30 and 3.31) or be made continuously variable (as shown in Figure 3.32).

A forward biased diode placed between the regulator common connection and 0 V will increase the regulator output by 0.6 V (i.e. the same as the forward voltage of the diode). Two series connected forward biased diodes will increase the regulator output by 1.2 V, and so on.

The arrangement of Figure 3.30 can be cumbersome when the required

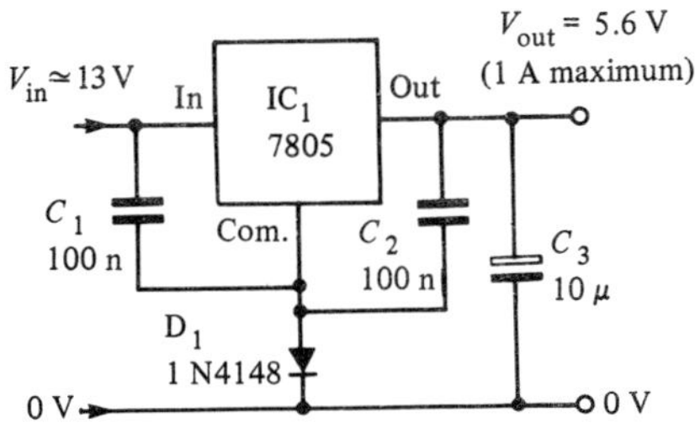

Figure 3.30 *Simple voltage boosted IC voltage regulator using a diode pedestal*

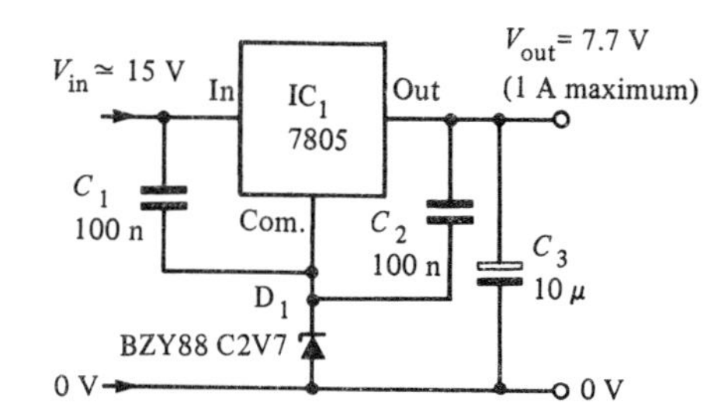

Figure 3.31 *Simple voltage boosted IC voltage regulator using a zener diode pedestal*

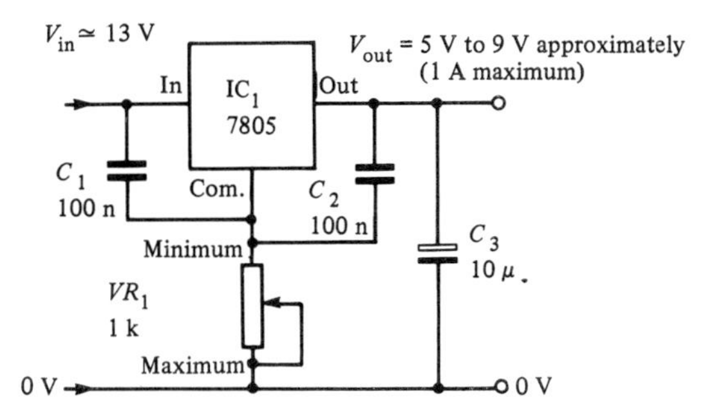

Figure 3.32 *Method of obtaining a variable output voltage from a fixed IC voltage regulator*

output voltage is well in excess of the regulator rating and an alternative method using a zener diode is shown in Figure 3.31. In this case, the output voltage is increased by an amount equal to the zener voltage of the diode 'pedestal'.

The circuits shown in Figures 3.30 and 3.31 both provide output resistances and regulation which, for most purposes, is not significantly worse than that offered by the regulator itself. Where a fully variable output voltage is required, and provided a small increase in output resistance can be tolerated, the circuit shown in Figure 3.32 may be employed. The variable resistor used in this arrangement should be a good quality wirewound type. With the values specified, the output voltage will be variable from 5 V to approximately 9.1 V. If a 12 V regulator is used the output voltage range will be variable from 12 V to approximately 17.5 V. In this case, however, the unregulated input voltage should be increased to 24 V.

Constant current power supply

Occasionally, a constant current rather than constant voltage power supply will be required. Figure 3.33 shows how a constant current power supply can be very easily realized using a standard fixed voltage regulator.

The output current is governed by the voltage of *R* and it obeys the following relationship:

$$I = \left(\frac{5000}{R} + 4 \right) \text{mA}$$

Where R is expressed in Ω.

Typical values of R and output current are given in Table 3.6:

Table 3.6

R(Ω)	*Output current (mA)*
10	504
22	234
47	109
100	54
220	27
470	15

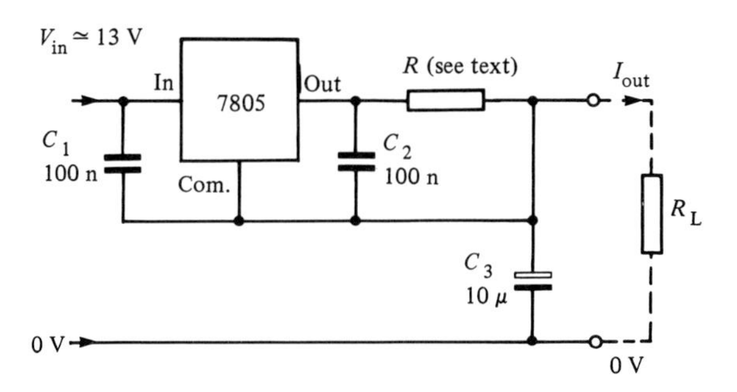

Figure 3.33 *Fixed constant current supply based on an IC voltage regulator*

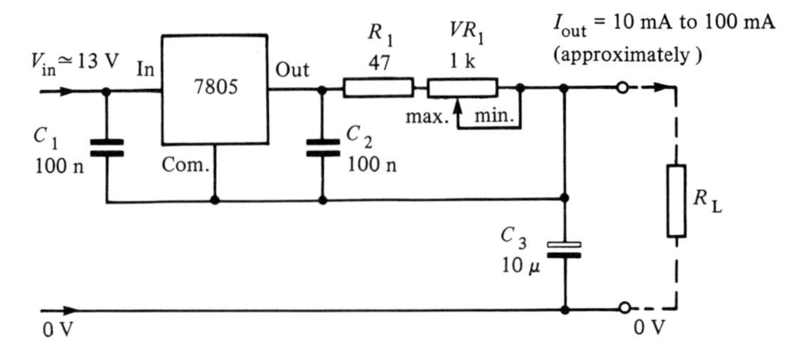

Figure 3.34 *Variable constant current supply based on an IC voltage regulator*

If a variable, rather than fixed, constant current output is required, R may be made variable (or preset), as shown in Figure 3.34. The fixed resistor determines the maximum current produced and the values shown provide a range of 10 mA to 100 mA.

Variable integrated circuit voltage regulators

Where a variable regulated output voltage is required it is expedient to make use of a variable integrated circuit voltage regulator such as the

LM317 or LM338. These versatile devices provide an adjustable output voltage range of 1.2 V to over 30 V (depending upon the upper limit of the unregulated d.c. input voltage) and incorporate the usual internal current limiting, thermal and safe operating area protection.

The characteristics of the LM317/LM338 family are listed in Table 3.7:

Table 3.7

Type	*LM317L*	*LM317M*	*LM317T*	*LM317K*	*LM338K*
Maximum load current	100 mA	500 mA	1.5 A	1.5 A	5 A
Output voltage range (V)	1.2–37	1.2–37	1.2–37	1.2–37	1.2–32
Input voltage range (V)	4–40	4–40	4–40	4–40	4–35
Load regulation (%)	0.1	0.1	0.3	0.1	0.1
Line regulation (%/V_{out})	0.01	0.01	0.02	0.01	0.005
Ripple rejection (dB)	65	65	65	65	60
Output impedance (mΩ)	10	10	10	10	3
Adjustment pin current (μA)	50	50	50	50	45
Thermal resistance (junction-case) (°C/W)	160	12	4	2.3	1
Maximum junction temperature (°C)	125	125	125	125	125
Maximum dissipation (W)	0.625	7.5	15	20	50
Package	TO92	TO202	TO220	TO3	TO3

Representative circuits using variable voltage regulators are shown in Figures 3.35 and 3.36. The circuit of Figure 3.35 uses an LM317K (which should be mounted on a heatsink rated at 2 °C/W, or better) and provides an output voltage which is fully variable from 1.5 V to 13 V at load currents of up to 1.5 A. Figure 3.36 shows the circuit of a 5 A power supply which has an adjustment voltage range of 9.5 V to 13.8 V (the LM338K should be mounted on a heatsink rated at 1 °C/W, or better). This power supply makes an ideal replacement for a 12 V lead-acid battery.

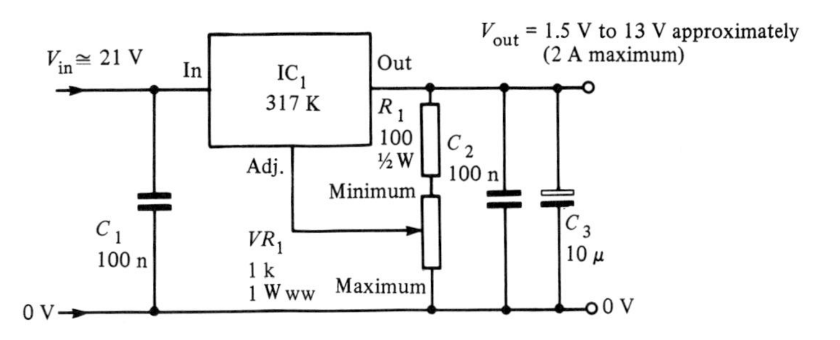

Figure 3.35 *Representative variable voltage supply based on an LM317 family device*

Where it is necessary to provide a means of adjusting the output current as well as the output voltage provided by a monolithic regulator, the L200

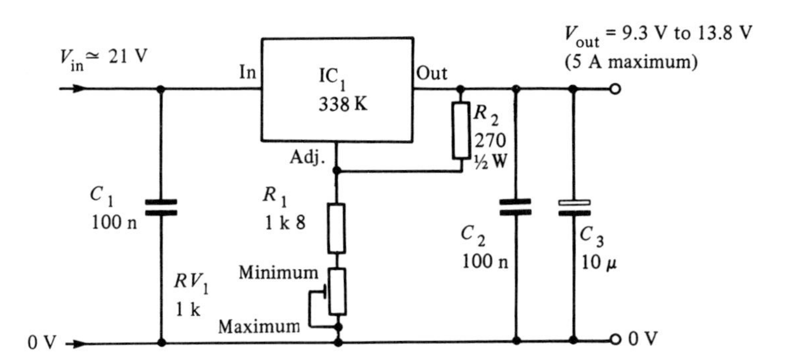

Figure 3.36 *Adjustable 5 A supply based on an LM338K*

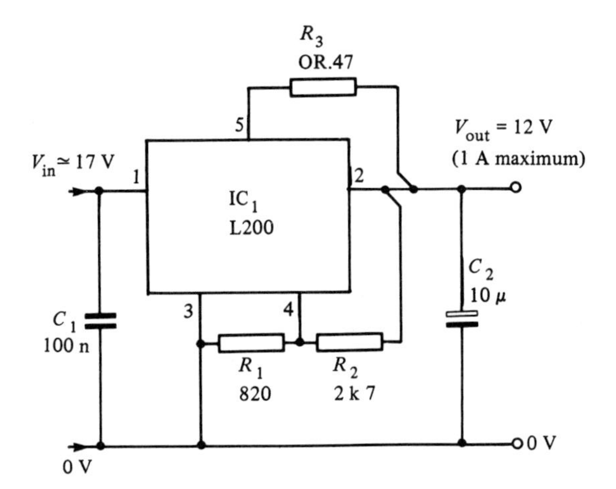

Figure 3.37 *Fixed 12 V, 1 A power supply based on an L200*

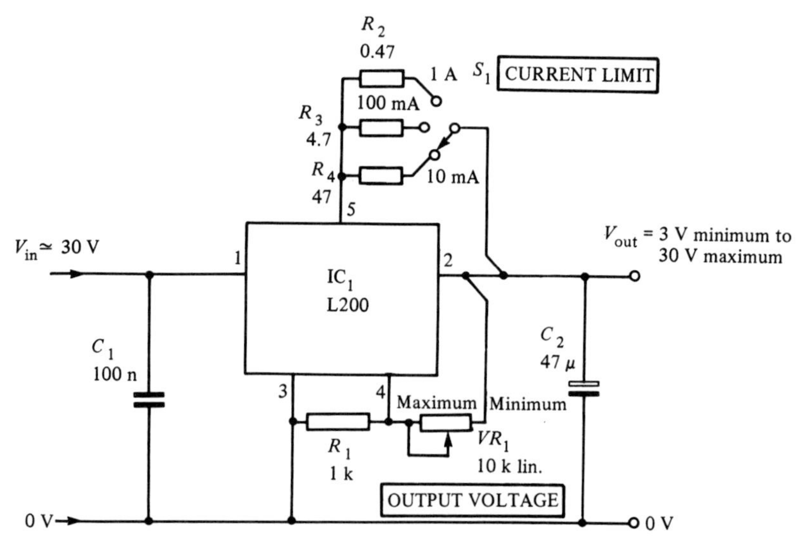

Figure 3.38 *Versatile L200 supply with variable output voltage and switch selected current limit*

provides an adjustable output voltage range of 2.8 V to 36 V together with an adjustable output current range with a 2 A upper limit. The device is housed in a 5-pin plastic package. Load regulation is typically 1.5% and the device incorporates safe operating area protection as well as input overvoltage protection (up to 60 V, 10 ms) and short-circuit protection.

The output voltage and output current of the L200 may be readily programmed by appropriate selection of resistors. Figure 3.37 shows a fixed voltage/fixed current regulator arrangement which provides an output of approximately 12 V at 1 A maximum. For different outputs, the following formulae may be used:

$$\text{Output voltage, } V_{out} = 2.77\left(1 + \frac{R_2}{R_1}\right)\text{V}$$

$$\text{Current limit, } I_{out\,max.} = \frac{0.45}{R_3}\text{A}$$

A typical application of the L200 (in the form of a variable voltage/variable current bench power supply unit) is shown in Figure 3.38. It should be noted that regulator dissipation will be large when operating near the maximum rated current (2 A) with low output voltages. An adequate heatsink (2 °C/W, or better) is therefore essential in order to prevent premature thermal limiting of the output current.

Dual rail power supplies

One often realizes the need for power supplies which provide dual positive and negative output rails. This is particularly true with circuits which contain operational amplifiers. Such devices require closely regulated supply rails of typically ±5 V, ±9 V, ±12 V or ±15 V.

A simple method of meeting such a requirement is shown in Figure 3.39.

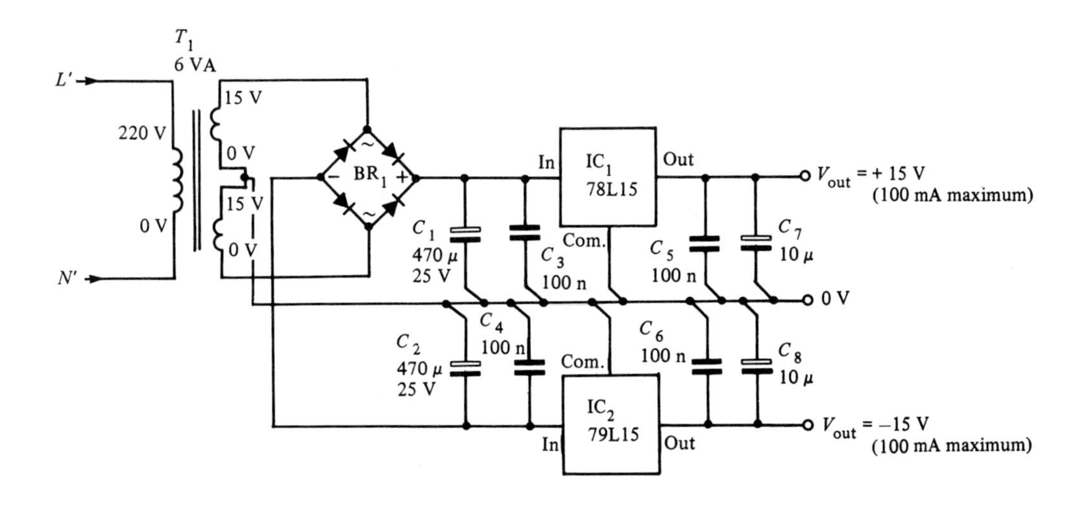

Figure 3.39 *Typical regulated supply with dual outputs suitable for operational amplifiers*

Here two fixed voltage regulators, of opposite polarity, are used in conjunction with a dual secondary transformer and single bridge rectifier. The current rating of the regulators should be greater than, or equal to,

Figure 3.40 *Operational amplifier supply based on an RC4195*

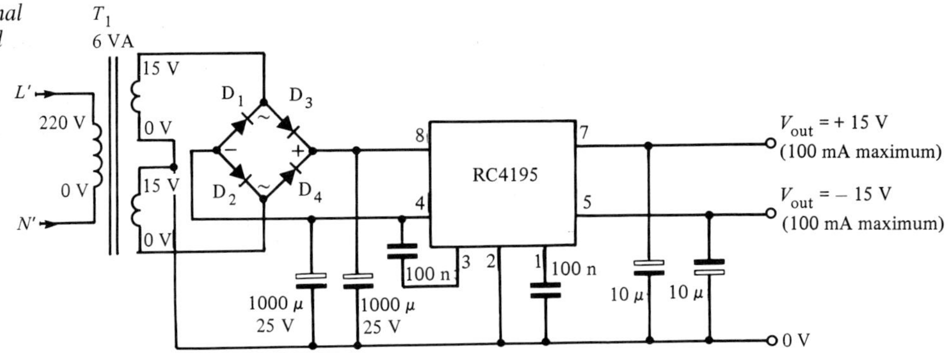

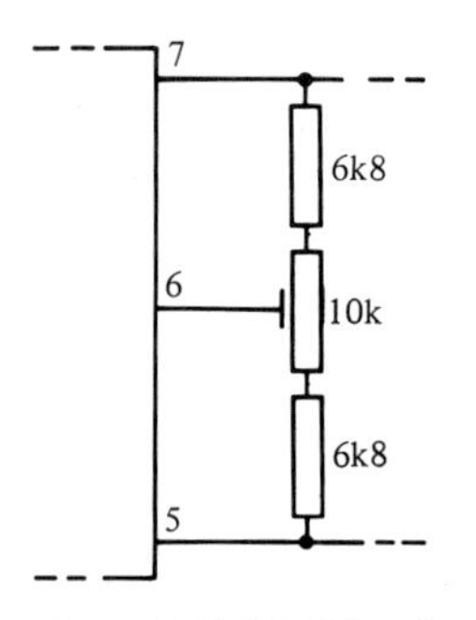

Figure 3.41 *Modifications to the circuit of Figure 3.40 for use when the output voltages are to be very accurately balanced*

the maximum load current. Circuits which are solely based on operational amplifiers rarely require supply currents in excess of 100 mA and thus 78L/79L series regulators will usually be adequate.

Provided the output load current does not exceed 100 mA on either rail, the circuit shown in Figure 3.40 makes an arguably more elegant alternative to that shown in Figure 3.39. This circuit employs an RC4195 dual fixed voltage regulator which incorporates internal current limiting and thermal shutdown. The additional circuitry shown in Figure 3.41 can be incorporated for applications in which it is desirable to accurately balance the output voltages.

The circuit of Figure 3.39 may be modified for applications in which

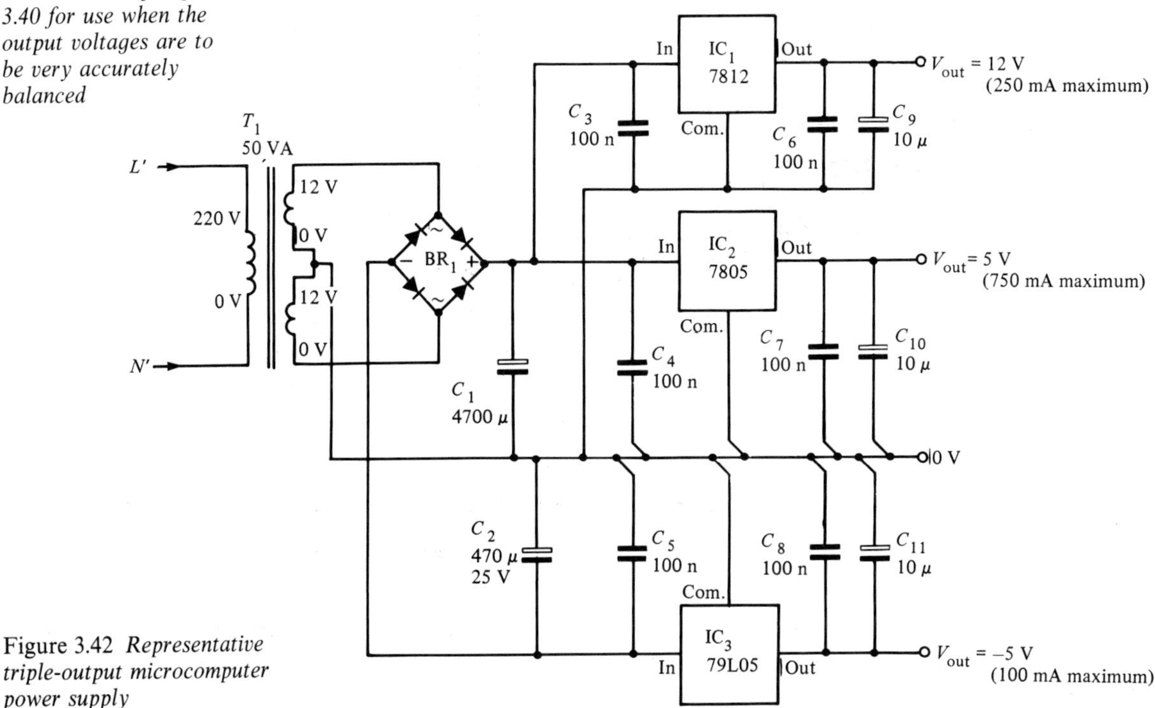

Figure 3.42 *Representative triple-output microcomputer power supply*

the dual output rails are required to support widely different load currents. In such a case the two regulators are selected with appropriate current ratings. Figure 3.42 shows the circuit of a microcomputer power supply having the following output rails:

(a) +12 V at 250 mA.
(b) + 5 V at 750 mA.
(c) − 5 V at 100 mA.

Since their mounting tabs are connected to the 'common' connection, IC_1 and IC_2 can be mounted on a common heatsink without the need for insulating kits. The heatsink should be rated at 3 °C/W, or better.

Crowbar protection

Short circuit failure within the series pass element of a regulator can have catastrophic consequences since the full unregulated d.c. voltage will be transferred directly to the output. In the case of TTL circuits, for example, a regulator may be fed with an unregulated d.c. input in excess of 10 V. If the output should ever rise much above 7 V, however, most TTL devices connected to the nominal 5 V rail will self-destruct virtually instantaneously. The dire consequences of such a failure can be avoided by the incorporation of a 'crowbar' over-voltage protection circuit of the type shown in Figure 3.43. This circuit places a virtual short-circuit across the supply whenever the rail voltage exceeds approximately 6.1 V. The circuit can be readily adapted for operation with other voltages using the formula:

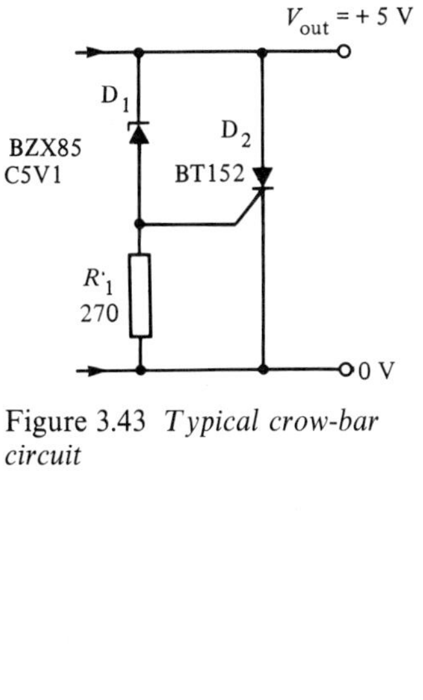

Figure 3.43 *Typical crow-bar circuit*

Crowbar voltage, $V_{th} = V_z + V_{gt}$

Where V_z is the zener voltage and V_{gt} is the thyristor gate trigger voltage (for the BT152, $V_{gt} = 1$ V approximately).

Once triggered, the thyristor remains in the conducting state until the supply is disconnected or a mains fuse ruptures. The action of the circuit may appear to be somewhat crude but it does make a useful 'last-ditch' protection.

Switched-mode power regulators

Switched-mode regulators are capable of operating with efficiencies of 80% to 90% as compared with generally less than 50% for conventional regulators. Since heat dissipation is minimized, switched-mode regulators can be made extremely compact. Unfortunately, the circuitry employed is rather more complex than that associated with more conventional linear regulators.

A simple switched-mode regulator based on discrete circuitry and providing 5 V at up to 1 A, is shown in Figure 3.44. This circuit operates at a switching frequency of approximately 125 kHz, has an output resistance of less than 0.2 Ω, and an output voltage adjustment range from 4.6 V to 5.8 V (approximately). The FET, TR_4, is a V-MOS power type with a maximum rated drain current of 2 A and the inductor, L_1 is wound

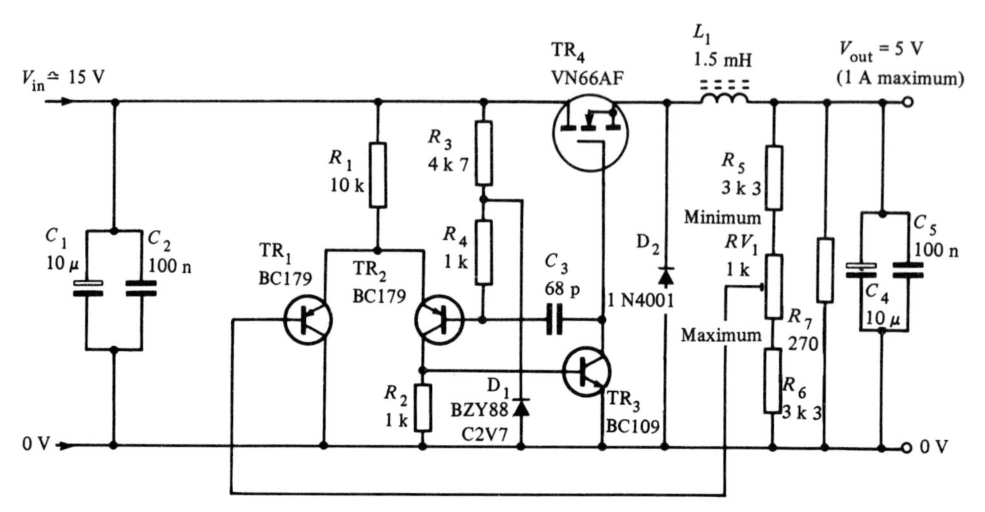

Figure 3.44 *Simple discrete switched mode regulator*

on a high-permeability ferrite pot core (see Chapter 1 for constructional details).

The d.c. output of a switched mode regulator can contain a significant proportion of high frequency noise and thus careful attention should be given to adequate decoupling and, where necessary, screening of the entire power supply module. Decoupling should involve the use of at least two capacitors of widely differing value and appropriate construction (e.g. 10 μF radial lead PCB electrolytic and 10 nF disc ceramic). Since the main reservoir capacitor will be ineffective for decoupling at the relatively high switching frequencies concerned, the input to the regulator should be similarly decoupled to prevent radiation of noise from the rectifier and transformer wiring.

4 Amplifiers

A common theme in many electronic circuits is the need to increase the level of voltage, current, or power present. This need is satisfied by some form of amplifier. While this chapter is devoted to amplifier circuits generally, it excludes all forms of circuit based on general purpose integrated circuit operational amplifiers. These circuits are discussed separately in Chapter 5.

Amplifier characteristics and specifications

Amplifiers may be categorized under a number of general headings including such characteristics as frequency response, class of operation, and the type of circuitry employed. The following six general categories are commonly used:

1 Classification according to frequency response

Audio frequency: A low frequency amplifier which typically operates over the frequency range 20 Hz to 20 kHz.

Radio frequency: A high frequency amplifier which is invariably frequency selective.

Wideband: An amplifier with an extended frequency response (typically from below 10 Hz to beyond 1 MHz). Typical frequency response characteristics for these three types of amplifier are shown in Figure 4.1.

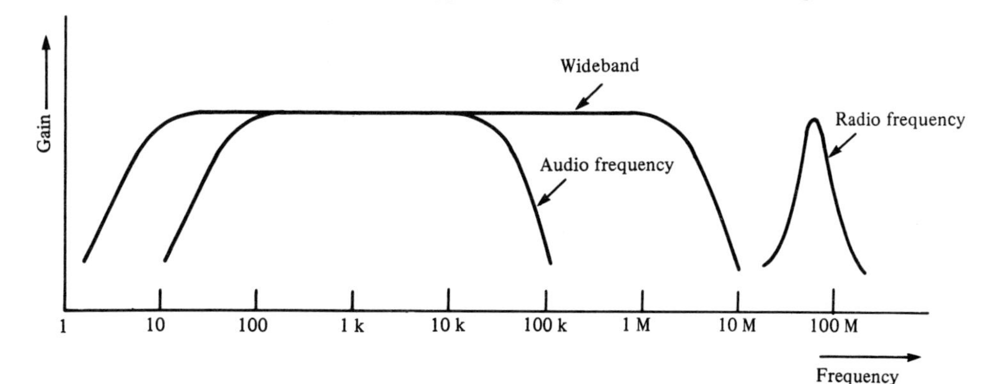

Figure 4.1 *Typical frequency response characteristics for audio frequency, radio frequency, and wideband amplifiers*

2 Classification according to operating point

Linear: Operation is confined to the linear part of the transfer characteristic.

Non-linear: Operation extends into the non-linear part of the transfer characteristic. (See also (5) below.)

3 Classification according to signal amplitude

Small-signal: Signals are of sufficiently small amplitude for transistor parameters to be considered constant, hence generalized small-signal equivalent circuits may be used in circuit analysis.

Large-signal: Signals are of relatively large amplitude and transistor parameters can not be relied upon to remain constant. Small-signal equivalent circuits are then not particularly relevant.

Pre-amplifier: A low level small-signal amplifier which precedes a large-signal amplifier or power output stage.

Power amplifier: A large-signal amplifier designed to develop an appreciable level of power in a specified load.

4 Classification according to noise and/or distortion

Low-noise: An amplifier which, by virtue of its design and selection of active devices, contributes a negligible amount of noise to the signal undergoing amplification.

Low-distortion: An amplifier which, by virtue of its design and appropriate selection of active devices, contributes a negligible amount of distortion to the signal undergoing amplification.

Figure 4.2 *Representative transfer characteristic for a class A amplifier*

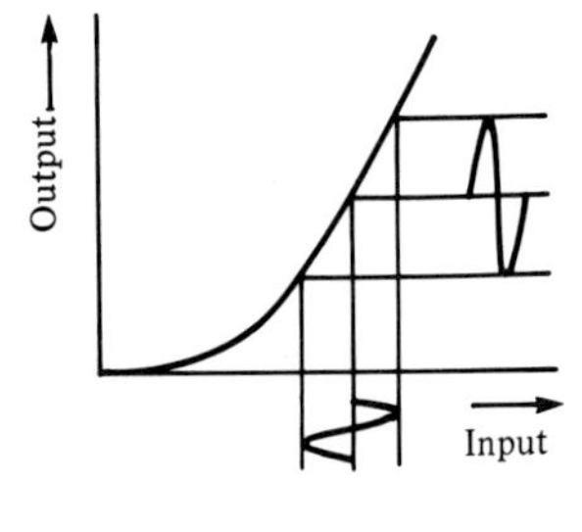

5 Classification according to class of operation

Class A: A linear amplifier in which bias is applied such that active devices are operated at the mid-point of their transfer characteristics (see Figure 4.2). In practice, this means that an appreciable value of collector or drain current flows under quiescent conditions. Furthermore, collector or drain current flows throughout the complete cycle of an input signal (i.e. conduction takes place over an angle of 360°). The maximum theoretical efficiency of a class A transistor output stage is only 50%.

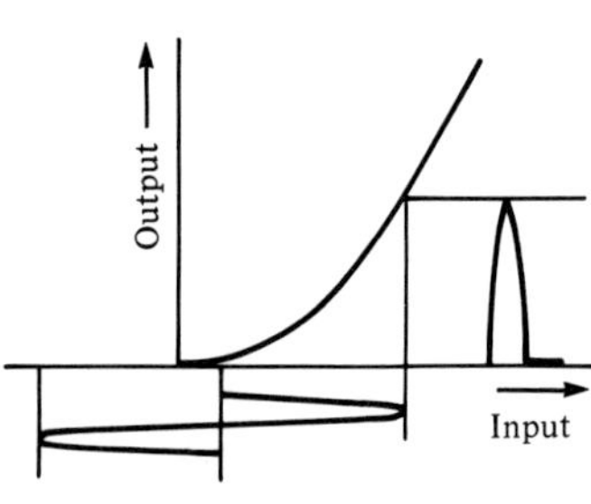

Figure 4.3 *Representative transfer characteristic for a class B amplifier*

Class B: An essentially non-linear amplifier in which the bias is adjusted for operation at cut-off (see Figure 4.3). In the absence of a signal, and during negative going half-cycles of the input signal, no collector current flows. The conduction angle is thus 180° and hence, for audio frequency applications, class B stages must employ push-pull output stages (the output transistors are alternately driven into conduction during successive half-cycles of the signal). The maximum theoretical efficiency of a class B output stage is 78.5%.

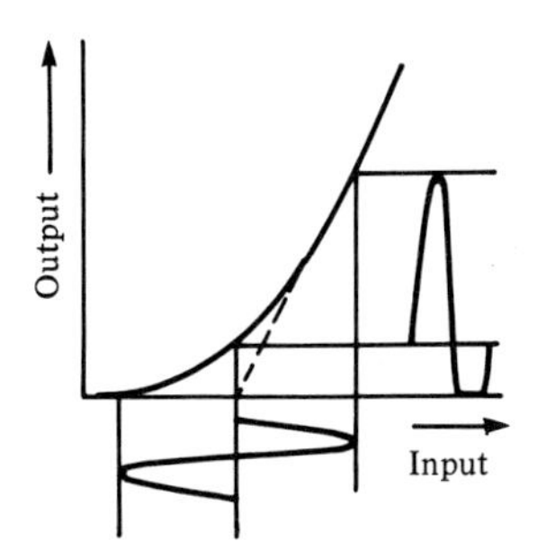

Figure 4.4 *Representative transfer characteristic for a class AB amplifier*

Class AB: The bias point is adjusted so that the stage is operated at projected cut-off (as shown in Figure 4.4). Distortion due to the non-linearity of the transfer characteristic (cross-over distortion) is thus minimized. A small value of quiescent collector or drain current will flow (typically this is between 5% and 15% of the peak value of collector or drain current). The conduction angle is typically between 200° and 220° and the efficiency is typically around 60%.

Class C: The bias is selected so that the stage is operated at beyond the cut-off point (see Figure 4.5). The conduction angle is thus less than 180° (typical values are between 90° and 150°). Distortion is too severe for audio frequency applications (it is impossible to reconstitute the signal in a push-pull output stage) and thus this mode is reserved for RF power applications in which the flywheel action of the tuned circuit can be employed to regenerate an acceptably sinusoidal output. The efficiency of a class C output stage is typically 80% or more.

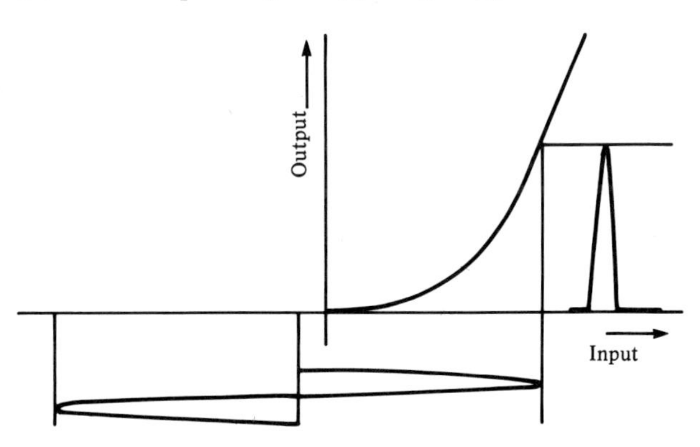

Figure 4.5 *Representative transfer characteristic for a class C amplifier*

Class D: The bias point is adjusted as for class C operation. However, the transistors are operated under switching conditions using a rectangular pulse signal rather than a sinewave. Efficiency is exceptionally high but, since class D stages can only cope with pulses, information has to be modulated on to the pulse train (e.g. using pulse width modulation) and the output signal recovered using a low-pass filter.

6 *Classification according to circuitry employed*

Single-stage: Only one active (amplifying) device is employed.

Multi-stage: More than one active (amplifying) device is employed.

a.c. coupled: Stages are coupled together in such a way that d.c. levels are isolated and only the a.c. components of a signal can be transferred from stage to stage (see Figure 4.6).

d.c. coupled: Stages are coupled together in such a way that d.c. levels are not isolated and both a.c. and d.c. signal components can be transferred from stage to stage (see Figure 4.6).

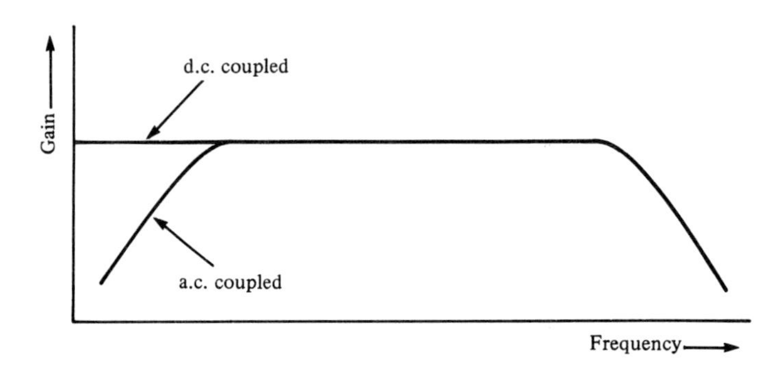

Figure 4.6 *Comparison of a.c. and d.c. coupled amplifier frequency response*

Discrete: Discrete devices (i.e. transistors) are employed in the amplifying stages.

Integrated: Integrated circuits are employed in the amplifying stages.

Hybrid: A mixture of both discrete and integrated circuits are employed in the amplifying stages.

Having dealt with classification of amplifiers, and before introducing a number of representative circuits, it is worth summarizing some of the more important performance characteristics associated with amplifiers.

(a) *Gain.* This is the ratio of output voltage to input voltage (voltage gain), output current to input current (current gain), or output power to input power (power gain). Gain is often expressed in decibels (dB) where:

$$\text{Voltage gain in dB} = 20\log_{10}\left(\frac{V_{out}}{V_{in}}\right)$$

$$\text{Current gain in dB} = 20\log_{10}\left(\frac{I_{out}}{I_{in}}\right)$$

$$\text{and power gain in dB} = 10\log_{10}\left(\frac{P_{out}}{P_{in}}\right)$$

Note that in the two former cases, the specification will only be meaningful where the input and output impedances are identical.

The Table 4.1 gives the equivalent power, voltage and current ratios for some common decibel values.

(b) *Input impedance.* This is the ratio of input voltage to input current and it is expressed in ohms. The input of an amplifier is normally purely resistive (i.e. the reactive component is negligible) in the middle of its working frequency range (i.e. the mid-band) and hence, in such cases, input impedance is synonymous with input resistance.

(c) *Output impedance.* This is the ratio of open-circuit output voltage to short-circuit output current and is measured in ohms. Note that this impedance is internal to the amplifier and should not be confused with the impedance of the load.

Table 4.1

Power gain (ratio)	*Decibels (dB)*	*Voltage or current gain (ratio)*
1	0	1
1.26	1	1.12
1.58	2	1.26
2	3	1.41
2.51	4	1.58
3.16	5	1.78
3.98	6	2
5.01	7	2.24
6.31	8	2.51
7.94	9	2.82
10	10	3.16
19.95	13	3.98
39.81	16	6.31
100	20	10
1000	30	31.62
10000	40	100
100000	50	316.23
1000000	60	1000

(d) *Frequency response.* Frequency response is usually specified in terms of the upper and lower cut-off frequencies of the amplifier. These frequencies are those at which the output power has dropped to 50% (otherwise known as the −3 dB points) or where the voltage gain has dropped to 70.7% of its mid-band value.

(e) *Bandwidth.* The bandwidth of an amplifier is usually taken as the difference between the two cut-off frequencies (see Figure 4.7).

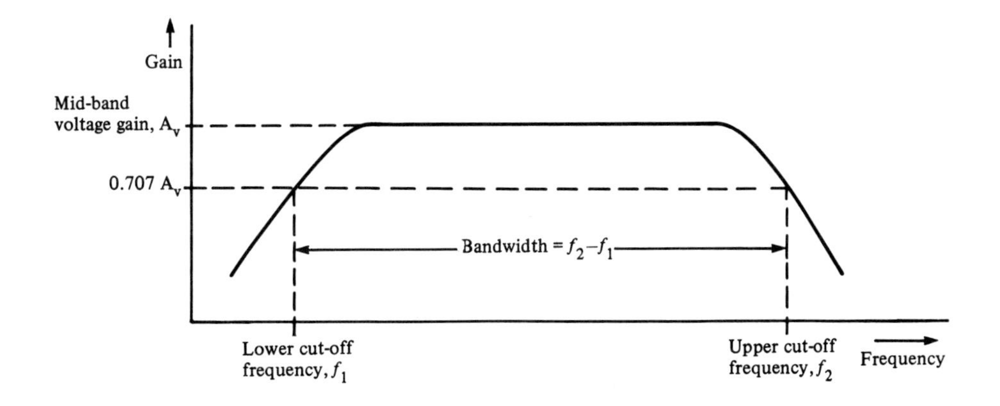

Figure 4.7 *Relationship between frequency response and bandwidth*

(f) *Distortion.* Distortion can take several forms, all of which are present to some extent in any amplifier.

Distortion is often expressed in terms of the percentage of r.m.s. harmonic voltage present in the output. This total harmonic distortion (THD) is given by:

$$\text{THD} = \frac{\text{harmonic output voltage (r.m.s.)}}{\text{total output voltage (r.m.s.)}} \times 100\%$$

When an amplifier is over-driven (or where the peak-peak output voltage is limited by the supply) the THD rises dramatically due to the clipping of the waveform (see Figure 4.8).

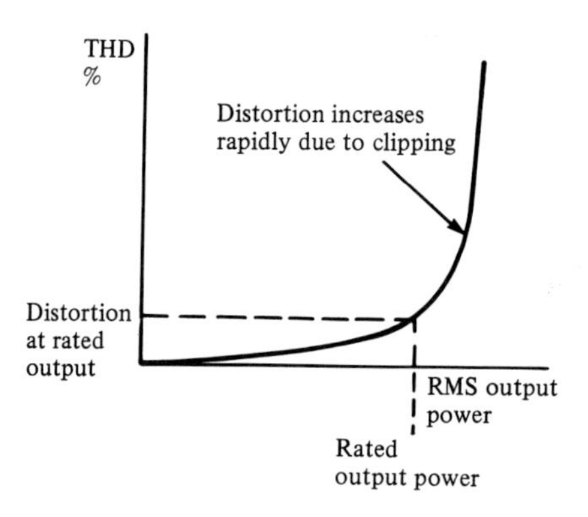

Figure 4.8 *Relationship between THD and output power*

(g) *Phase shift.* This is the phase angle between the input and output voltages measured in degrees. The measurement is usually carried out in the mid-band.

(h) *Hum and noise.* When no input signal is present, the residual output of an amplifier is attributable to several distinct sources including supply borne mains ripple (hum) and noise resulting from thermal agitation within components.

(i) *Supply.* The supply voltage and current requirements (note that some amplifiers need more than one supply rail).

(j) *Output voltage.* In certain applications the maximum undistorted output voltage swing (specified in either peak or peak-peak voltage) is an important parameter.

(k) *Output power.* This is the r.m.s. output power delivered to a load of specified impedance at a particular level of distortion and at a specified frequency. This parameter is obviously only relevant to power amplifiers.

(l) *Efficiency.* Not all the power drain from an amplifier's supply rails is developed in its load. The efficiency of an amplifier is thus defined as:

$$\text{Efficiency} = \frac{\text{signal power output}}{\text{d.c. power input}} \times 100\%$$

Efficiency is usually quoted at the rated output power and varies according to the class of operation employed.

(m) *Noise figure.* Noise figure is a measure of the deterioration of signal to noise ratio resulting from passing a signal through an amplifier. Noise figure is thus defined as:

$$\text{Noise figure} = \frac{\text{signal to noise ratio at the input}}{\text{signal to noise ratio at the output}}$$

$$= \frac{\left(\dfrac{P_{si}}{P_{ni}}\right)}{\left(\dfrac{P_{so}}{P_{no}}\right)} = \frac{P_{no}}{P_{ni}} \times \frac{1}{A_p}$$

Where P_{si} is the signal power at the input, P_{ni} is the noise power at the input, P_{so} is the signal power at the output, P_{no} is the noise power at the output, and A_p is the amplifier power gain. Noise figure is usually quoted in decibels. For optimum low-noise performance transistor amplifiers should be operated with collector currents of approximately 100 μA.

Not all of the above characteristics will be important in any given application and an essential first stage in designing an amplifier will be to select those characteristics which are important and then specify the parameters required.

As an example, consider the case of a simple baby alarm based upon the use of two miniature 40 Ω loudspeakers (one to be used as a microphone and one to be used as a loudspeaker). Let us assume that an initial experiment on the loudspeaker suggests that a voltage of 10 mV r.m.s. will be generated by the loudspeaker when a baby is demanding attention and that a power of 100 mW r.m.s. would be sufficient to alert us to the fact that all is not well in the nursery.

Since $P = V^2/R$, a power of 100 mW r.m.s. into 40 Ω would be produced by an amplifier output voltage of 2 V r.m.s. We have established that a baby produces a 10 mV input signal and thus we would be looking for a voltage gain of 2 V/10 mV or 200.

Items (g), (h), (j), (l), and (m) may be considered relatively unimportant in the context of this example and hence we might arrive at a specification which takes the following form:

Voltage gain:	200 minimum.
Input impedance:	Medium, say 10 kΩ.
Output impedance:	40 Ω load.
Frequency response:	200 Hz to 4 kHz.
Supply:	12 V (at 50 mA nominal).
Output power:	100 mW minimum for 100 mV input.

Note that, in this application, it is not necessary (nor is it desirable) to match the input and output impedances. The reason is simply that to attempt to do so would result in a loss of signal (we are not, in this case, concerned with maximum power transfer).

The input impedance of the amplifier should be high in comparison to that of the input transducer (but not so high that excessive noise and hum is produced). Conversely, the output impedance of the amplifier should be low in comparison to the output transducer.

For this particular application we would not wish to extend the amplifier's frequency response to cover the range normally associated with high-quality audio applications (we will assume that a baby is not a particularly musical creature!). Furthermore, the low cost loudspeakers used for input and output will be incapable of coping with the complete audio bandwidth.

Before delving into some representative amplifier circuits it is worth making just one final point. Whereas the use of discrete circuitry can be expedient for simple single-stage amplifiers and matching circuits (or where high voltages are encountered or low-noise performance is required), integrated circuits invariably provided solutions to design problems which are nowadays both cost and performance effective. The moral is simple: before starting work on traditional discrete component circuitry, designers would be wise to first acquaint themselves with the ever increasing range of available integrated circuits.

Single-stage audio amplifiers

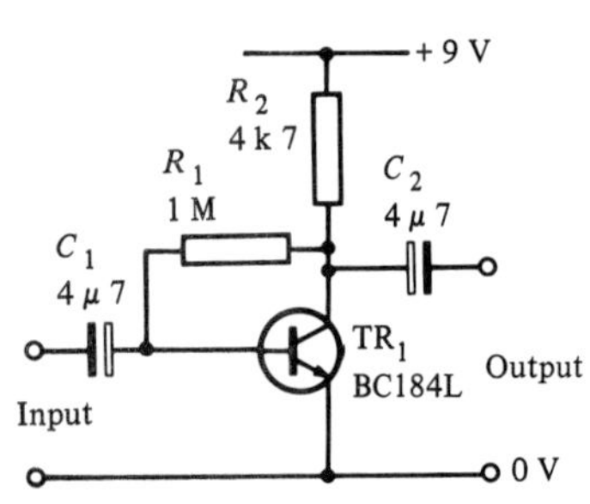

Figure 4.9 *Simple single-stage common-emitter amplifier*

The simplest conceivable small-signal amplifier stage is shown in Figure 4.9. The transistor (an n-p-n type having a current gain of 150, or more) is connected in common-emitter mode and operated in class A with base bias current fed back from the collector, via R_1.

Signals are coupled, into and out of the stage, by means of C_1 and C_2 respectively. Since these components determine the low frequency response of the stage, their values should be sufficiently large to have negligible reactance at the lowest signal frequency. The upper cut-off frequency is largely determined by shunt stray capacitance across the load and the internal collector-base capacitance of the transistor. In practice, the upper cut-off frequency lies well outside the audio bandwidth and hence this does not usually represent a restriction on performance. Where it is desirable to limit the upper frequency response, this can be achieved by adding an appropriate value of shunt capacitance in parallel with R_1.

A collector current of approximately 1 mA is typical for the circuit of Figure 4.9 and hence the quiescent voltage appearing at the collector will be approximately 4.3 V.

The optimum value of bias resistor, R_1, is that which, for a transistor having a 'typical' value of current gain, fixes the quiescent value of collector-emitter voltage (V_{CE}) at half the d.c. supply voltage. This value ensures the largest possible output voltage swing with symmetrical clipping (the maximum possible peak-peak output voltage swing will be roughly equal to the d.c. supply voltage).

The voltage gain provided by the arrangement in 4.9 depends upon the value of transistor current gain (h_{fe}) and effective load resistance. (Remember that the a.c. collector load comprises R_2 acting in shunt with whatever load impedance is presented by the following stage.) In practice, voltage gains of between 50 and 150 are typical for this arrangement.

The input impedance is usually in the region of 3 kΩ to 8 kΩ and the output impedance approximately 4 kΩ. With the values quoted in Figure 4.9, frequency response extends from approximately 10 Hz to over 200 kHz.

The circuit of Figure 4.9 provides a.c. as well as d.c. negative feedback and this results in less voltage gain than could otherwise have been achieved. This limitation can be overcome by splitting the bias resistor and bypassing (using a suitably large value of capacitor) the a.c. feedback component, as shown in Figure 4.2.

While the circuits of Figures 4.9 and 4.10 both incorporate a measure of d.c. stabilization, the circuit of Figure 4.11 is preferred whenever there is likely to be a very large variation in ambient temperature or transistor current gain. The improvement arises from the use of series-current (rather than shunt-voltage) negative feedback.

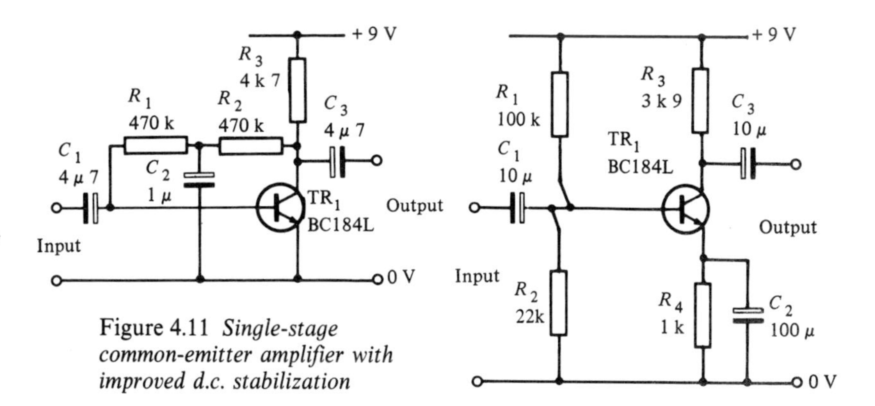

Figure 4.10 *Improved version of Figure 4.9 offering high gain*

Figure 4.11 *Single-stage common-emitter amplifier with improved d.c. stabilization*

A good rule of thumb is that the emitter voltage should be approximately 20% of the d.c. supply (i.e. around 2.5 V for a 12 V supply) and that the quiescent value of collector-emitter voltage should be equal to the quiescent voltage drop across the load. The maximum possible peak-peak output voltage swing is therefore somewhat less than that of the previous circuit when operated from identical supply rails. The frequency response of Figure 4.10 extends from approximately 30 Hz to well over 250 kHz; the upper cut-off frequency being largely dependent upon stray capacitive reactance.

In order to significantly extend the upper frequency response of Figure 4.11, an appropriate value inductor may be incorporated in series with the load resistor, as shown in Figure 4.12. The value of the inductor is something of a compromise. In order maintain a reasonably flat

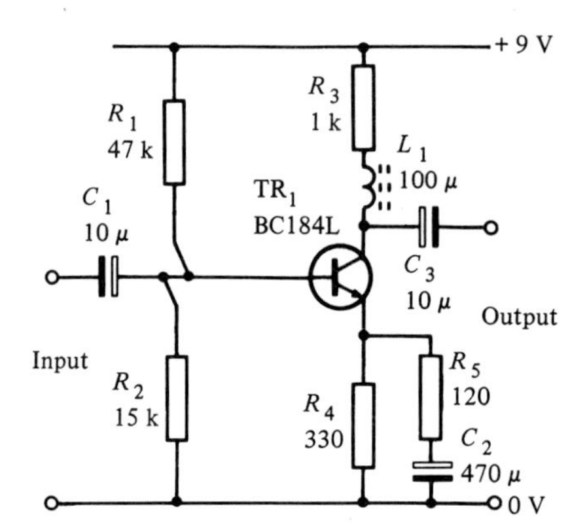

Figure 4.12 *Wideband single-stage common-emitter amplifier*

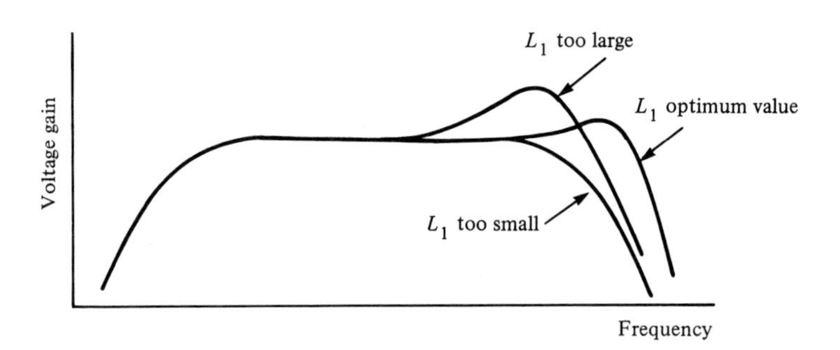

Figure 4.13 *Frequency response of the amplifier in Figure 4.12*

frequency response it must not be too large. It must equally well not be made too small or it will be ineffective in extending the frequency response. The effect of the peaking inductor is shown in Figure 4.13.

The value of the inductor can only be calculated precisely where the equivalent value of shunt capacitive reactance is known. In practice, however, values of between 22 μH and 220 μH make a good starting point for experimentation.

The best way of assessing the effect of the peaking inductor is to apply a square wave (at, say, 50 kHz) to the input and selecting the value of inductor that produces the most rectangular square wave at the output as observed using an oscilloscope. If this procedure is adopted it is, however, absolutely essential to use compensated oscilloscope probes and leads otherwise the results will be meaningless.

Single-stage RF amplifiers

As an alternative to extending the bandwidth of an amplifier, there are often times when the response needs to be restricted. A typical example is that of the selective amplifier stages associated with a radio receiver. In such applications a tuned load is normally employed in order to achieve the desired degree of selectivity. Figure 4.14 shows one such arrangement. Here the collector load resistor is replaced with an L-C tuned circuit resonant at the desired frequency of operation.

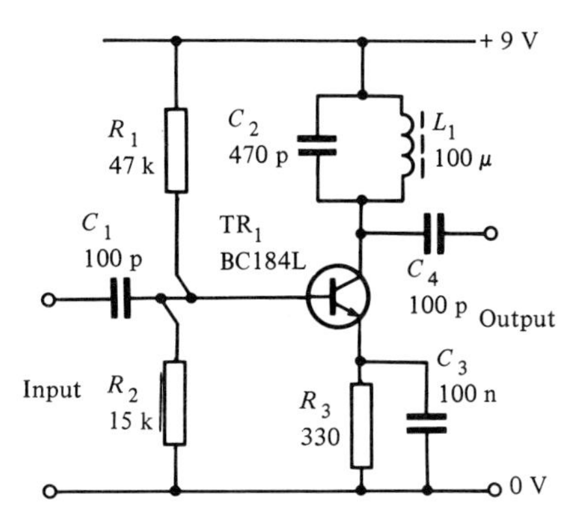

Figure 4.14 *Single-stage common-emitter radio frequency amplifier*

The bandwidth and voltage gain at resonance will be dependent upon the Q-factor and dynamic resistance of the tuned circuit. The following general relationships apply:

$$Q\text{-factor} = \frac{2\pi f_o L}{r_s}$$

$$\text{Bandwidth} = \frac{f_o}{Q}$$

and voltage gain α dynamic resistance $= \dfrac{L}{Cr_s}$

Where r_s is the equivalent series loss resistance.

It should be noted that the transistor and load both effectively appear in parallel with the tuned circuit. This results in an increase in series loss resistance and a reduction in Q. With the arrangement shown, and using a tuned circuit with a Q-factor of 100, the voltage gain at resonance will be in the region of 50 to 100, and the bandwidth will be between 10% and 20% of the nominal centre frequency.

Where a very selective amplifier stage is required (as in the intermediate frequency amplifier stages of an AM receiver) we must take steps to minimize the damping of the tuned circuit and, at the same time, improve the method of interstage coupling employed.

Figure 4.15 shows a typical arrangement of an intermediate frequency amplifier for operation at 455 kHz. In order to achieve optimum impedance matching, the collector is tapped into the tuned circuit transformer, IFT1. A medium/low impedance secondary winding is used to couple from the transformer to the base of the next stage. The circuit of Figure 4.15 provides a bandwidth of approximately 50 kHz at a nominal 455 kHz centre frequency. Where several similar stages are connected in tandem, the overall voltage gain is increased accordingly, while the overall bandwidth is reduced. A two-stage 455 kHz IF amplifier using three transformers can be expected to provide a bandwidth of between 10 kHz and 15 kHz and would thus be ideal for reception of AM signals.

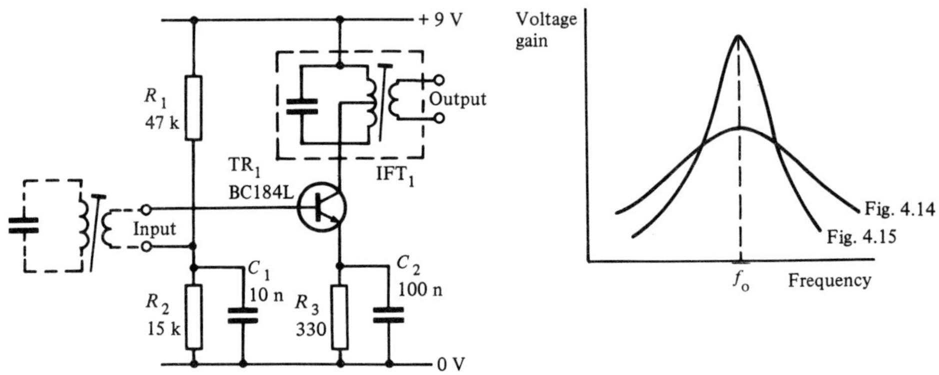

Figure 4.15 *Single-stage common-emitter intermediate frequency amplifier*

Figure 4.16 *Comparison of normalized frequency response characteristics of the circuits of Figures 4.14 and 4.15*

A comparison of the normalised frequency response of two arrangements depicted in Figures 4.14 and 4.15 is shown in Figure 4.16.

A common-base amplifier of the type shown in Figure 4.17 is ideal for applications in which a very low input impedance (20 Ω to 50 Ω) is required. The output impedance is very high and thus the collector may be connected directly across the tuned circuit load. In order to prevent loading of the tuned circuit (and consequent lack of selectivity) the next stage must present a high input impedance (ideally 50 kΩ or more). Alternatively, the collector inductor may form part of a tuned step-down transformer, as shown in Figure 4.18.

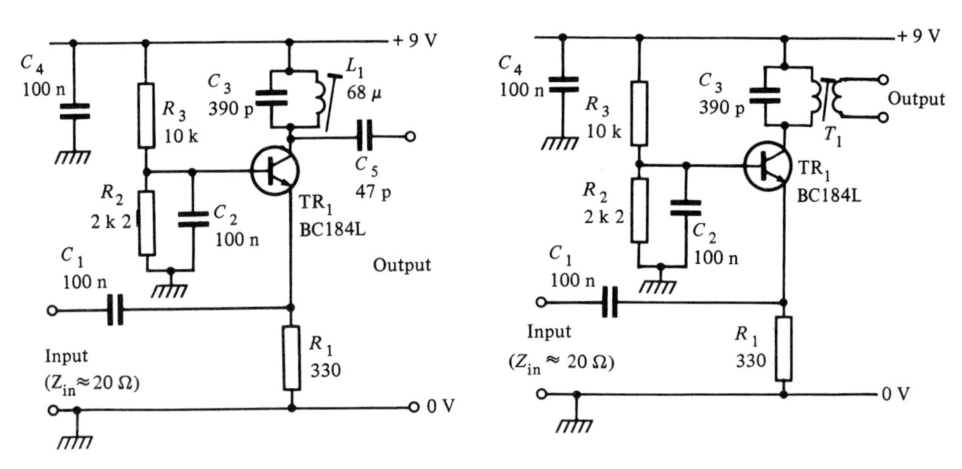

Figure 4.17 *Single-stage common-base radio frequency amplifier*

Figure 4.18 *Alternative version of Figure 4.17 using transformer coupling*

Both of the previous circuits are suitable for operation over a very wide frequency range (wider than that associated with a comparable common-emitter amplifier stage) and, provided appropriate values and component types are selected this may extend from around 10 kHz to well over 100 MHz. Above 30 MHz the following precautions should be observed:

(a) Use capacitors which have low values of self-inductance (e.g. miniature ceramic plate types).

(b) Use RF transistors.
(c) Pay careful attention to circuit layout (separating and screening input and output tuned circuits).
(d) Use distributed decoupling of the supply rail (i.e. incorporate several decoupling capacitors of appropriate value at strategic points along the supply rail).
(e) Provide an adequate common earth plane on the PCB – any vacant areas of PCB should be left with foil which is returned to the 0 V rail.

A high performance cascade RF amplifier using junction gate FETs is shown in Figure 4.19. This amplifier provides a typical gain of 28 dB over the range 30 MHz to 150 MHz together with a noise figure of approximately 2 dB. The amplifier will normally be used with 50 Ω input and output impedances. Construction should obey the general rules mentioned with the previous circuit.

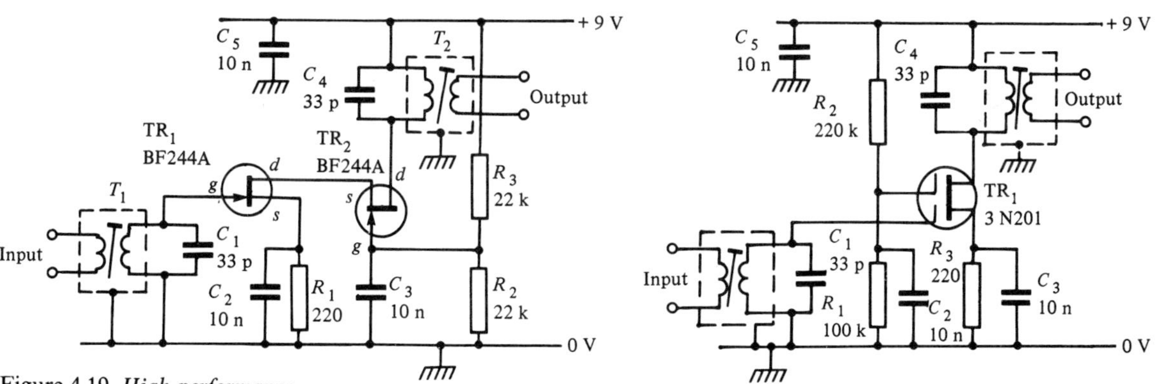

Figure 4.19 *High-performance radio frequency amplifier using a cascade FET arrangement*

Figure 4.20 *High-performance radio frequency amplifier using a dual-gate MOSFET*

Figure 4.20 shows another high performance RF amplifier. This stage is a close relative of the previous circuit but is based on a dual gate MOSFET rather than an individual junction gate device. While values are quoted for operation at 30 MHz, this circuit is usable to over 400 MHz. Performance is otherwise similar to the previous circuit.

A low-power class C RF amplifier is shown in Figure 4.21. Values quoted are for operation at 1 MHz, but the circuit is suitable for use at much higher frequencies. When operated from a 12 V supply rail, this circuit provides an output of 250 mW into a 50 Ω load at a frequency of 1 MHz. The output power increases to approximately 1 W into 50 Ω with a 24 V d.c. supply. The value of R_1 should be increased as the supply voltage increases. At 24 V, for example, R_1 should be increased to 22 kΩ in order to maintain class C operation. The circuit requires an input of approximately 500 mV r.m.s. and offers an input impedance of around 100 Ω.

The 2N2219A is adequate for operation at frequencies up to 30 MHz. Above this, a 2N3866 should be employed. Alignment involves first tuning L_2 and TC_1 for maximum collector current and then adjusting TC_2 for

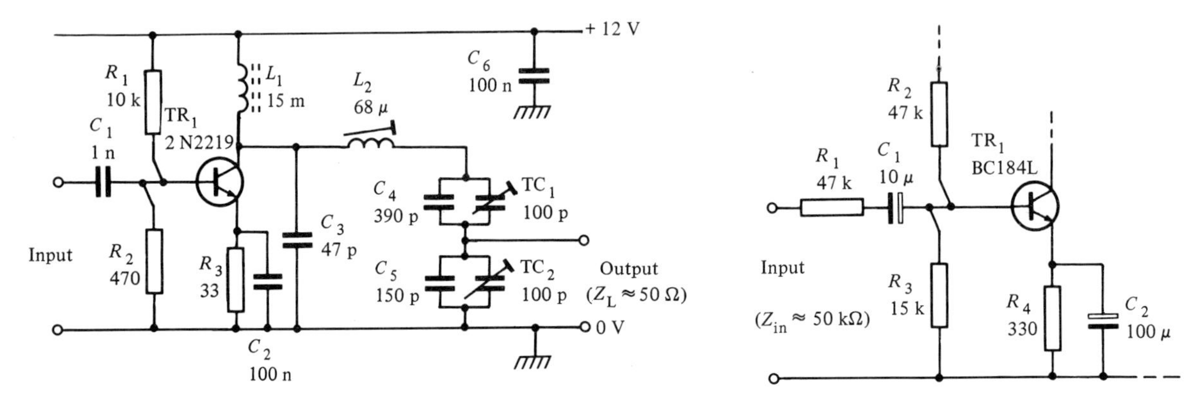

Figure 4.21 *Single-stage class C radio frequency power amplifier*

Figure 4.22 *Adding series resistance to the input of an amplifier in order to raise the effective input impedance*

maximum output. There is some interaction between the three adjustable components and thus the procedure should be repeated several times for optimum results. The output of the circuit is reasonably pure but, where the signal is to be radiated, it should be applied to a multistage low-pass or band-pass filter in order to remove unwanted harmonics which may otherwise cause interference to other services.

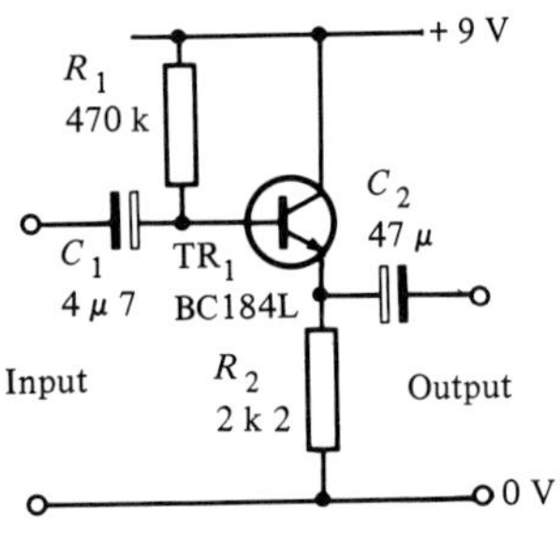

Figure 4.23 *Simple emitter follower stage*

Impedance matching circuits

All of the simple amplifier circuits described so far have had input impedances of less than 10 kΩ. While it is possible to raise the input impedance of a circuit by simply adding a resistor of suitable value in series with the input (see Figure 4.22) this also results in a loss of voltage gain. A better method of raising input impedance is with the aid of an emitter follower. Such a stage provides a voltage gain of just less than unity but the inherent current gain raises the input impedance by a factor roughly equivalent to the current gain (h_{fe}) of the transistor.

Figure 4.23 shows a simple emitter follower. Base bias current is supplied by R_1 while the load, R_2, is placed in the emitter. Voltage gain is very slightly less than 1 and the input impedance is approximately 200 kΩ. The quiescent voltages at the base and emitter are respectively 5.3 V and 4.7 V, approximately. When driven from a 600 Ω source, the output impedance is approximately 10 Ω.

Figure 4.24 shows an alternative form of emitter follower arrangement in which the base is biased at a potential equal to approximately half the supply voltage. The optimum value of base potential is, in fact, greater than half the supply voltage by an amount equal to the base-emitter voltage drop. In this condition the emitter voltage should be exactly half the supply and the maximum undistorted peak-peak output voltage (roughly equal to the supply voltage) will result. The input impedance of the stage is determined by the parallel combination of base bias resistors and transistor input impedance. The input impedance of the circuit shown in Figure 4.24 is approximately 50 kΩ.

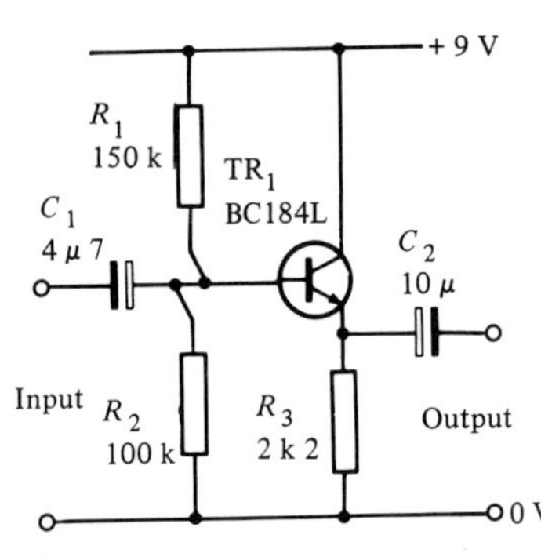

Figure 4.24 *Alternative emitter follower stage*

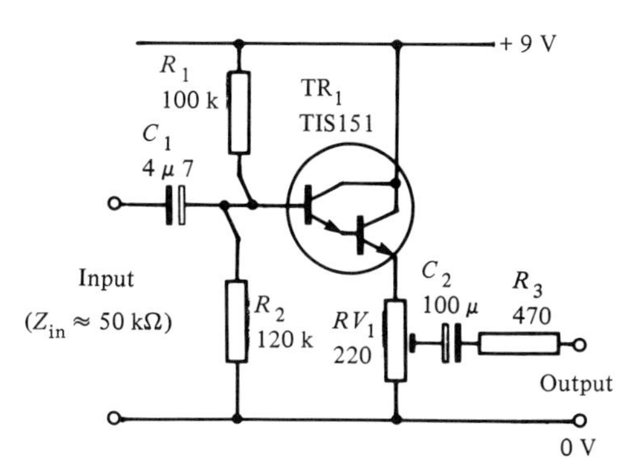

Figure 4.25 *Emitter follower based on a Darlington transistor*

The input impedance can be raised still further by using a compound Darlington transistor in the arrangement of Figure 4.24. Such a device has a very high current gain (equivalent to the product of the individual current gains of the two internal transistors) and hence the base input current is extremely small.

Figure 4.25 shows a typical low impedance line driver employing a Darlington transistor. The output level of this arrangement is adjustable and the unit has nominal input and output impedances of 50 kΩ and 600 Ω respectively.

Higher input impedances can be achieved by using a 'bootstrapped' emitter follower circuit like that shown in Figure 4.26 which has an input impedance of approximately 700 kΩ.

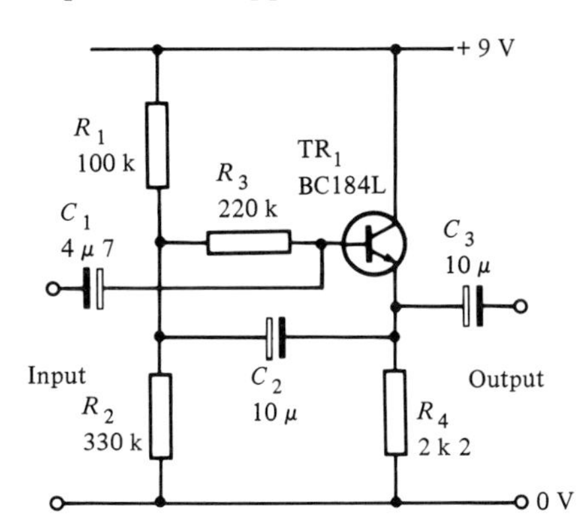

Figure 4.26 *Bootstrapped emitter follower*

Yet another alternative is that of using an FET source-follower along the lines of Figure 4.27. The FET requires no bias and, since its input resistance is exceptionally high, the effective input resistance of the circuit is simply equal to the resistance connected between the gate and the common 0 V rail (i.e. 1 MΩ). Like its emitter follower counterpart, the circuit has a voltage gain of slightly less than 1. The circuit is not, however, suitable for input voltages in excess of approximately 500 mV peak-peak. Where large amplitude input signals are expected, the circuit of Figure 4.29 should be employed.

Figure 4.28 shows how a simple switched input attenuator can be

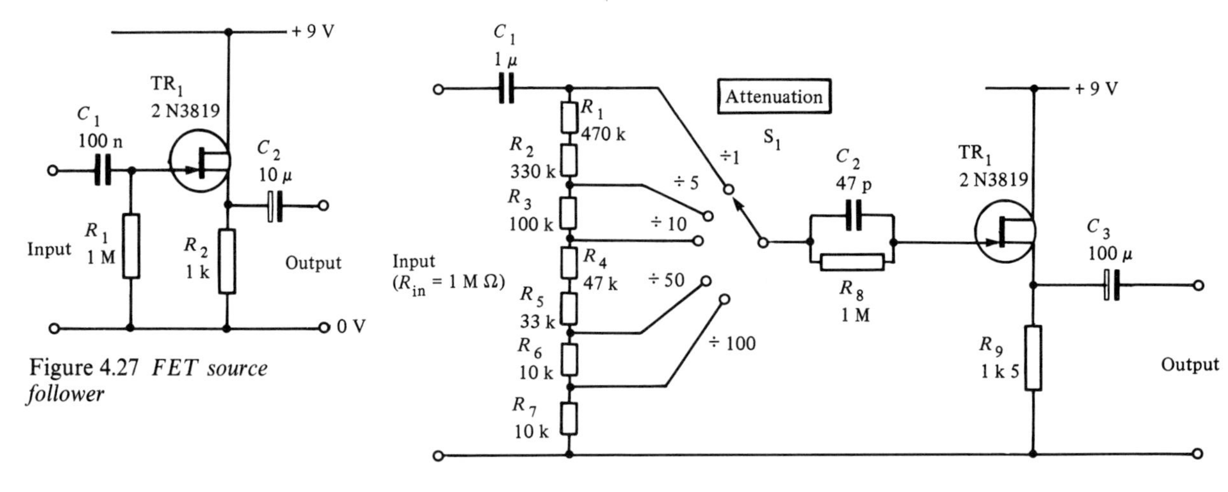

Figure 4.27 *FET source follower*

Figure 4.28 *Switched input attenuator based on a FET*

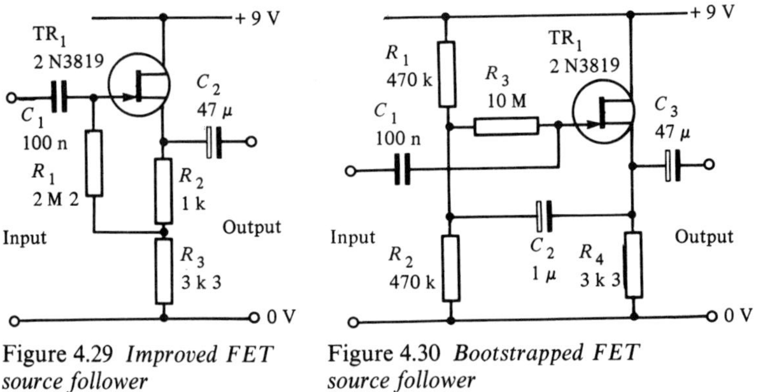

Figure 4.29 *Improved FET source follower*

Figure 4.30 *Bootstrapped FET source follower*

implemented using an FET source follower. The circuit offers a constant input impedance of 1 MΩ and divides the input voltage by 5, 10, 50, and 100. C_2 extends the upper frequency limit of the circuit and helps to compensate for stray reactance within the switch attenuator and the internal gate-source capacitance associated with TR_1.

The simple source follower of Figure 4.26 is only suitable for use with signals of relatively small amplitude. Figure 4.29 overcomes this problem while preserving correct operating conditions for the FET. The output voltage swing from this stage can approach 5 V peak-peak without appreciable distortion. Voltage gain is still unity but the input impedance is raised (by partial bootstrapping) to approximately 10 MΩ.

An ultra high impedance input stage is shown in Figure 4.30. This circuit is comparable to that shown in Figure 4.26 but has an input impedance of approximately 100 MΩ.

Multi-stage amplifiers

When designing multi-stage amplifiers, it is important to ensure that stages are stabilized adequately by providing a.c. and d.c. negative feedback over several stages. A simple d.c. coupled two-stage amplifier is shown in Figure 4.31. TR_1 operates in common-emitter mode while TR_2 operates as an emitter follower. Base bias for TR_1 is derived from the emitter of the second stage and this helps promote good stability. The voltage gain is proportional to the ratio R_3/R_2; the values shown producing a voltage gain of approximately 40. Where it is necessary to alter the gain it is possible to change R_2 (up to a maximum of about 1 kΩ) without greatly affecting the bias conditions.

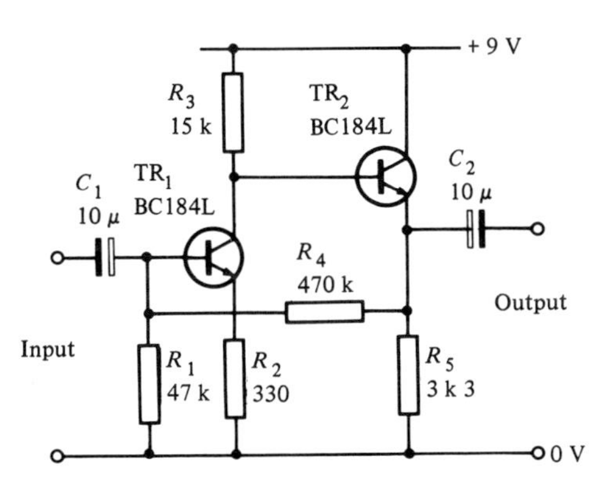

Figure 4.31 *Simple two-stage amplifier*

The circuit of Figure 4.31 provides an input impedance of approximately 20 kΩ (increasing as R_2 increases) and has a frequency response which extends from below 10 Hz to over 500 kHz. The upper frequency cut-off may be reduced by placing an appropriate value capacitor in parallel with R_3; e.g. a value of 100 pF will reduce the bandwidth to approximately 100 kHz.

With a nominal 15 V supply, the circuit of Figure 4.31 is able to deliver a maximum output of 10 V peak-peak into a load of 1 kΩ. The value of R_4 should be reduced for operation on lower supply voltages. The arrangement may be easily modified in order to produce a considerably greater voltage gain. This is achieved by connecting TR_2 as a common-emitter stage rather than as an emitter follower, as shown in Figure 4.32.

The circuit of Figure 4.32 provides a voltage gain which is variable from approximately 200 to over 1000. (Note, however, that the bandwidth will be reduced as the gain is increased.) For high quality audio and instrumentation applications, TR_1 should preferably be a low noise type (e.g. BC109). With RV_1 set to 150 Ω, the voltage gain is 375 and the bandwidth extends from 12 Hz to 180 kHz.

A notable disadvantage of simple single-stage emitter followers is that the input impedance varies with the impedance of the load. A useful refinement of the emitter follower which does not suffer from this problem

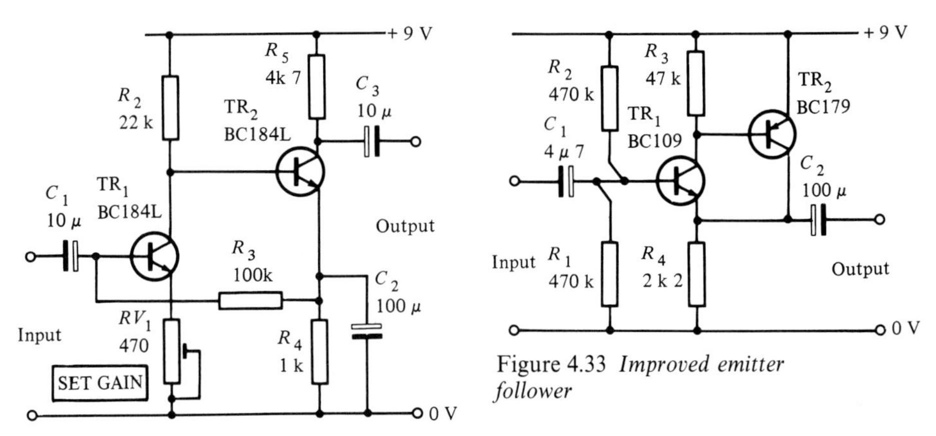

Figure 4.32 *High-gain two-stage amplifier*

Figure 4.33 *Improved emitter follower*

is shown in Figure 4.33. This circuit still provides a voltage gain of unity and zero phase shift. The transistors have been specified for low-noise operation and this circuit is ideal for more critical applications such as buffering a non-constant impedance filter from a high impedance input source.

Power amplifiers

Whereas all of the amplifiers described thus far provide power gain as well as voltage and/or current gain, the term 'power amplifier' is often used to describe a type of amplifier which is capable of delivering an appreciable power (typically 100 mW, or more) into a load which is invariably of low impedance (600 Ω, or less).

Linear audio frequency power amplifiers may be operated in class A (in which case they tend to be somewhat inefficient), class AB, or class B. Due to their limited efficiency (typically no more than 25% at full output) class A amplifiers are usually only appropriate for use at low levels of power output (e.g. 1 W, or less).

Power amplifiers, whether they be operated in class A, class AB, or class B, invariably employ a 'push-pull' output stage. Figure 4.34 shows the basic arrangement of a low-power class A output stage which uses complementary (n-p-n and p-n-p) output transistors. This circuit is ideal for driving headphones or a low impedance line but requires the use of a split (nominally ±9 V) supply rail. As with all output stages of this type, the output power delivered to the load is a function of the supply voltage and load impedance. The maximum theoretical output power (assuming perfect transistors and neglecting the effect of the two emitter resistors, R_3 and R_4) is given by the relationship:

$$P = \frac{V_s^2}{8R}$$

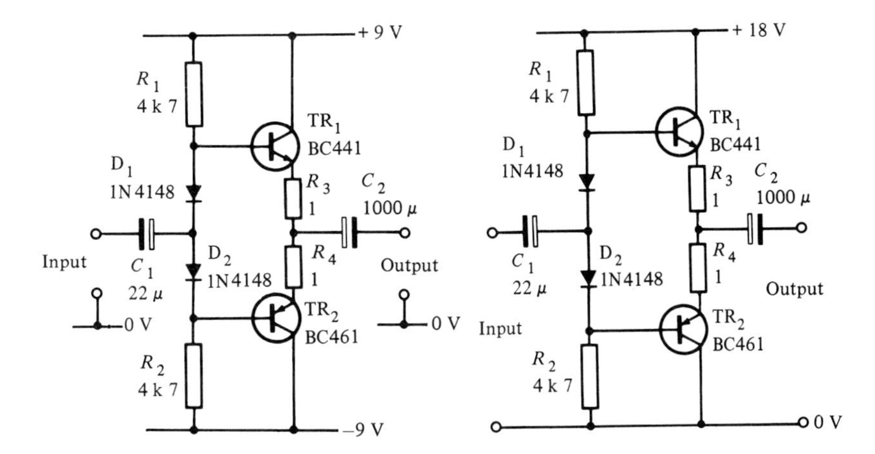

Figure 4.34 *Low-power class A output stage*

Figure 4.35 *Single supply rail version of Figure 4.34*

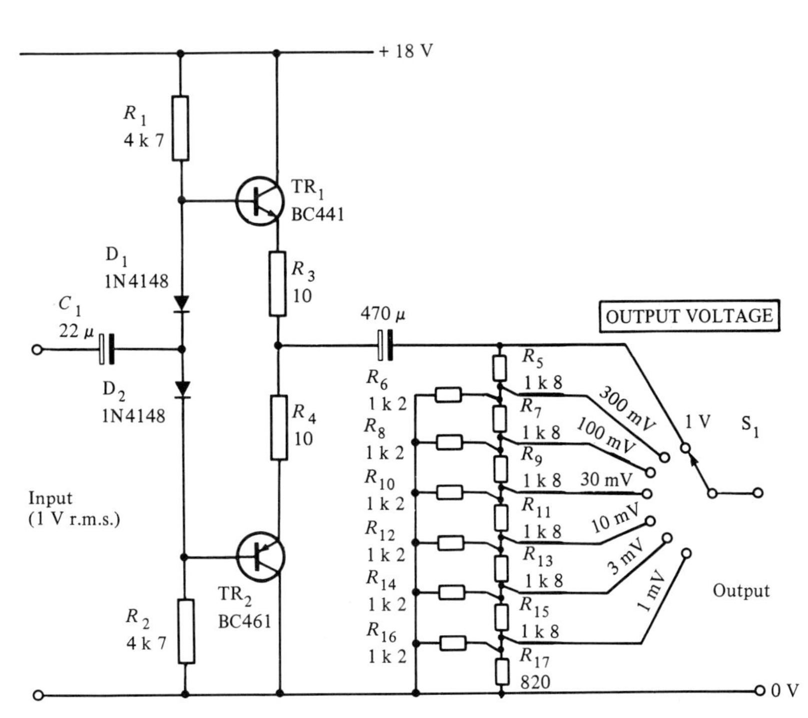

Figure 4.36 *Class A output stage with attenuated output*

Where V_s is the total supply voltage (18 V in the case of the circuit shown in Figure 4.34) and R is the impedance of the load.

The circuit of Figure 4.34 provides a maximum output power of approximately 500 mW (r.m.s. sinewave at 1 kHz) into 8 Ω. The circuit is not, however, recommended for load impedances of less than 8 Ω. Frequency response extends from approximately 40 Hz to over 400 kHz.

The obvious disadvantage of the arrangement shown in Figure 4.34, at least as far as battery operated circuits is concerned, is that the circuit requires both positive and negative supply rails. This problem can be

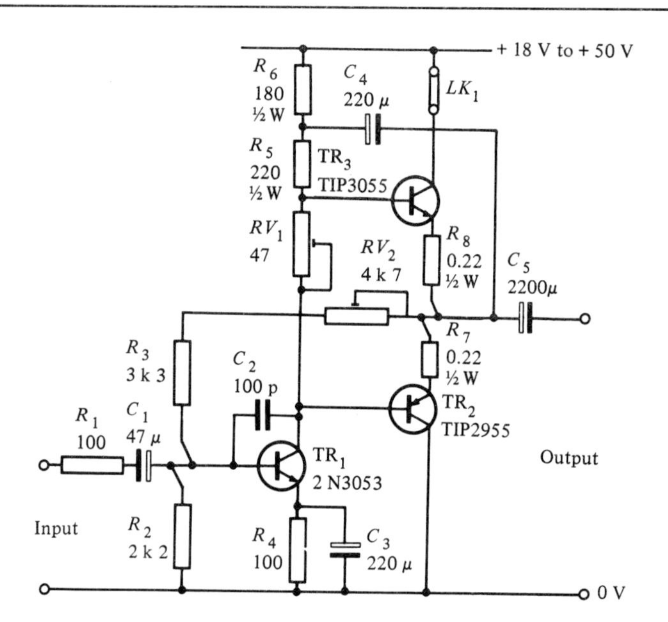

Figure 4.37 *Low-cost 25 W power amplifier*

readily overcome using the arrangement shown in Figure 4.35. A typical application for this circuit would be the output stage of an audio frequency signal generator. Figure 4.36 shows the circuit of such an arrangement incorporating a switched output attenuator.

Since all three of the previous circuits provide unity voltage gain, the input voltage swing (peak-peak) must be equal to the total supply voltage in order to produce maximum output. A sufficiently large voltage swing can be readily obtained using an operational amplifier (a typical circuit is included in Chapter 5) alternatively a single-stage common-emitter transistor amplifier (operating in class A) can be used as a driver, as shown in Figure 4.37.

The circuit of Figure 4.37 is ideal for low-cost applications and will operate with supply rail voltages of between 18 V and 50 V. The output transistors will require a heatsink of 2 °C/W or less and, since the metal tabs are connected to the collector of both TR_2 and TR_3, both should be fitted with insulating washers and bushes. Where the d.c. supply rail exceeds 40 V, TR_1 should also be fitted with a heatsink. This should be a push-fit TO5 type rated at 48 °C/W, or better.

The voltage gain of the driver stage is increased by raising the effective collector load impedance seen by TR_1 using bootstrapping provided by C_3. The two pre-set resistors should both be adjusted under quiescent (no signal) conditions. RV_1 is adjusted so that the voltage at the junction of R_5 and R_6 is exactly half the supply voltage (this ensures symmetrical operation of the output transistors). LK_1 should then be broken and an ammeter inserted to measure the output stage collector current. RV_2 should be adjusted for a quiescent collector current of 30 mA (class AB operation). The link should then be replaced.

It should be noted that the circuit of Figure 4.37 has a low input

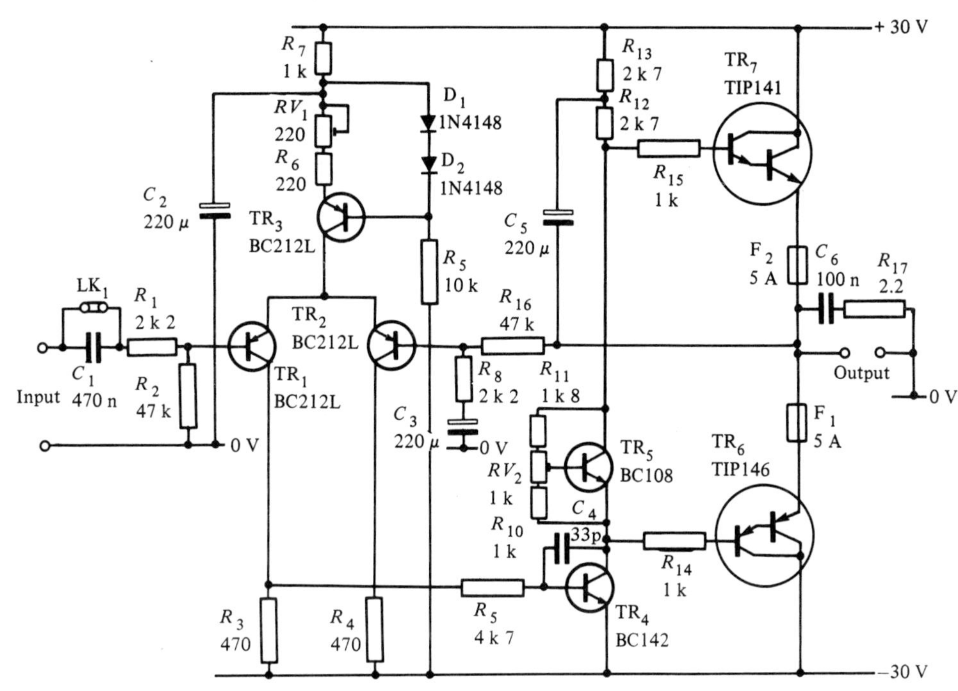

Figure 4.38 *High-performance 40 W power amplifier*

impedance and will require approximately 1 V r.m.s. to drive it to full output. The circuit will produce maximum output powers of 10 W, 16 W and 25 W into a load impedance of 8 from nominal supply rails of 30 V, 40 V and 50 V d.c. (off-load). The distortion produced by the circuit is typically less than 1% at full-output.

While the performance of the circuit of Figure 4.37 is adequate for uncritical applications, it needs further refinement to be suitable for 'high-fidelity' applications. Figure 4.38 shows the circuit of a high performance amplifier capable of delivering an output of 40 W r.m.s. into 8 Ω when operated from positive and negative supply rails of nominally 30 V. The output transistors, TR_6 and TR_7, are complementary power Darlington devices. These devices are ideal for use as output stages since their relatively high input impedance ensures that the driver operates with an acceptable voltage gain.

Since this is a fairly complex circuit, it is worth spending some time explaining the function of each stage. TR_1 and TR_2 constitute a differential amplifier (also known as a 'long-tailed pair'). The output voltage of this stage (developed across R_3, the collector load of TR_1) is proportional to the difference between the input voltage (applied to the base of TR_1) and feedback voltage (applied to the base of TR_2). The feedback voltage is governed by the potential divider R_{16}/R_8 and hence these two resistors determine the overall voltage gain of the amplifier (equal to the ratio of R_{16} to R_8).

The base voltage of TR_3 is held constant by the two forward biased diodes, D_1 and D_2. TR_3 thus acts as a constant current source to the emitters of TR_1 and TR_2. This ensures symmetry since the total collector current of TR_1 and TR_2 remains constant, i.e. if the collector current of TR_1 increases by a certain amount (due to an increase in signal current) the collector current of TR_2 must fall by the same amount.

RV_1 sets the emitter current of TR_1 and TR_2 and, since TR_1 is d.c. coupled to the base of TR_4, it can be used to adjust the symmetry of the output stage in the same manner as that provided by RV_1 in the circuit of Figure 4.37.

Bootstrapped feedback is applied via C_5 while C_4 provides negative feedback to reduce the voltage gain at high frequencies in order to promote stability. TR_5 operates as a simple shunt bias voltage regulator and provides temperature compensation for the output stage by reducing the bias voltage as temperature increases. It is thus desirable for TR_5 to be mounted in close proximity to the output transistors to provide efficient thermal coupling. Ideally, it should be bonded to the heatsink midway between TR_6 and TR_7. RV_2 sets the quiescent collector current for the output transistors (measured by removing one, or other, of the emitter fuses and inserting an ammeter).

If desired, the entire amplifier may be operated d.c. coupled from input to output by inserting LK_1. In this case, the frequency response will extend from d.c. to approximately 50 kHz. Where LK_1 is removed, the lower cut-off frequency is approximately 15 Hz. Total harmonic distortion measured at 30 W output into 8 is typically less than 0.05% and noise −90 dB referred to full output. With the stated values for R_3 and R_4, the voltage gain is 20 and hence an input of approximately 1 V r.m.s. will be required for full output.

Integrated circuit power amplifiers

The LM380 is an audio amplifier IC having a fixed gain of 50 enclosed in either an 8-pin (LM380N-8) or 14-pin (LM380N-14) DIL package. The device incorporates output current limiting as well as thermal shutdown circuitry.

The LM380 operates with supply voltages of up to 22 V. The device is, however, unsuitable for operation below 8 V and hence, for applications in which a battery supply is to be used, this should be at least 9 V. Furthermore, in order to avoid premature output current limiting, it is recommended that the minimum load impedance is 8 Ω.

Stability can be assured by placing a decoupling capacitor in close proximity to the positive supply rail input. (This simple precaution is considered good practice where most IC power amplifiers are concerned.)

Figures 4.39 and 4.40 show how an LM380 can be used in complete low-power audio amplifier stages which require an absolute minimum of additional components. Both arrangements provide an output of approximately 500 mW into 8 Ω and are perfectly adequate for equipment such as intercoms, radio receivers, etc.

Since the LM380 is contained within a plastic package, the device relies

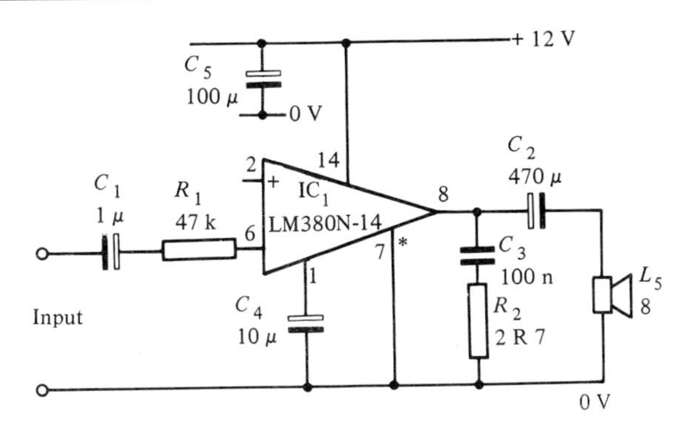

Figure 4.39 *Simple LM380 audio frequency amplifier stage*

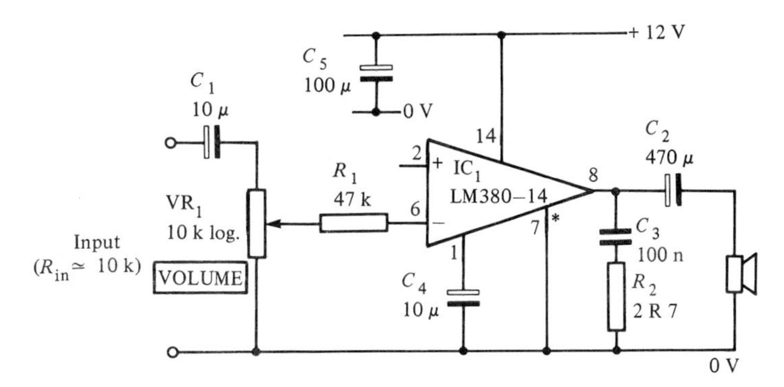

Figure 4.40 *Variation of the circuit of Figure 4.39 incorporating a volume control*

on thermal conductivity through its pins to the PCB foil in order to dissipate heat. For high power operation (e.g. 1 W to 2 W from supply rails of between 12 V and 22 V) it is, therefore, essential that an adequate area of PCB foil is provided. Ideally, this should be 12 cm^2, or more connected directly to pins 3, 4, 5, 10, 11 and 12 are also electrically connected to 0 V. Where PCB space is restricted or in applications where sustained high-power operation is envisaged, DIL bond-on or clip-on heatsinks (or 24 °C/W) should be fitted directly to the plastic package.

An alternative to the use of an LM380, at the cost of a few additional components, is the use of a TBA820M. This device is housed in an 8-pin DIL plastic-package and provides a maximum output power of 2 W into 8 Ω when operated from a 12 V d.c. supply. Apart from smaller physical size, the TBA820M offers the following advantages over the simpler LM380 device:

(a) Voltage gain is programmable using a single external resistor.
(b) Operation at lower supply voltages (3 V minimum as against 8 V minimum for the LM380).

The circuit of a TBA820M based output stage is shown in Figure 4.41. This circuit provides an output of 1 W r.m.s. into 8 Ω and has a frequency response extending from 20 Hz to 20 kHz at the −3 dB points. Where it is permissible to return the loudspeaker to the positive supply rail rather than the 0 V rail, it is possible to effect a few economies and the circuit of

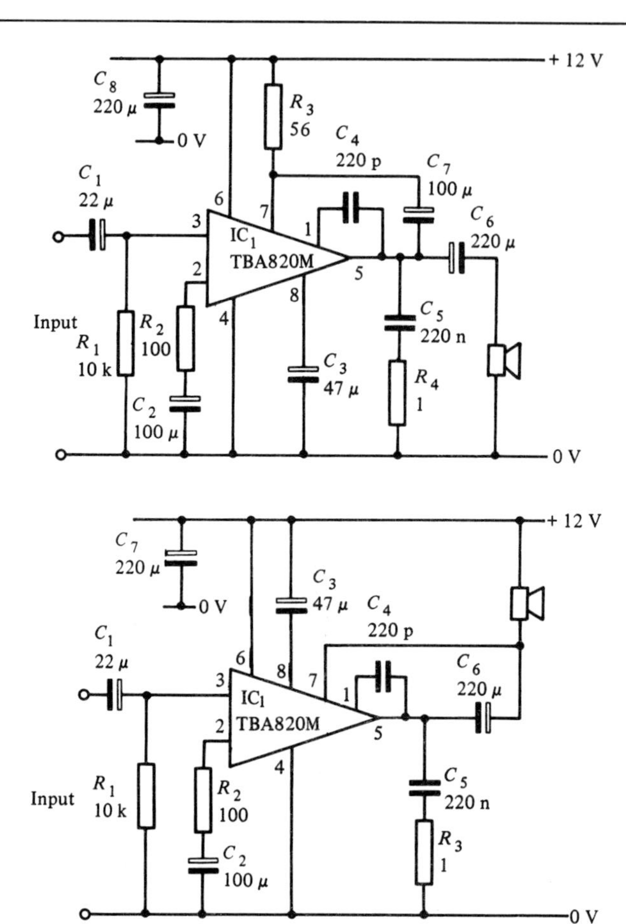

Figure 4.41 *TBA820M audio frequency amplifier*

Figure 4.42 *Economy version of the circuit of Figure 4.41*

Figure 4.42 may be used. This circuit provides almost identical performance to its conventional counterpart.

The TDA2004 is a dual audio power amplifier contained in an 11-lead tab mounting plastic package. The two amplifiers may be connected separately (e.g. for stereo applications) or wired in a bridge configuration to achieve a four-fold increase in output power. The device incorporates thermal shut down as well as short circuit and safe operating area protection.

The TDA2004 will operate over a range of supply voltages (from 8 V to 18 V maximum) and can provide a typical output power of 6.5 W into 4 Ω at 0.2% THD when operated from a 14.5 V supply rail. Typical stereo and mono (bridge configured) amplifiers using the TDA2004 are shown in Figures 4.43 and 4.44 respectively.

The voltage gain of the circuit of Figure 4.45 is determined by the ratio R_3/R_1 and R_4/R_2. With the values quoted, each stage operates with a voltage gain of approximately 200. The frequency response extends from 1 Hz to 15 kHz at 3 dB points and each channel is capable of delivering 4.5 W into an 8 Ω loudspeaker when operating from a 12 V supply rail.

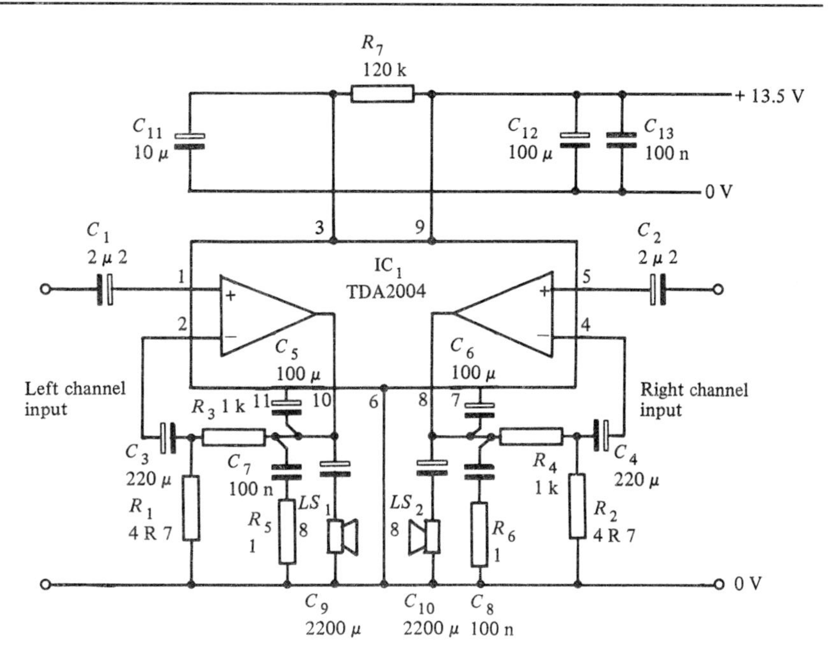

Figure 4.43 *TDA2004 stereo audio frequency amplifier*

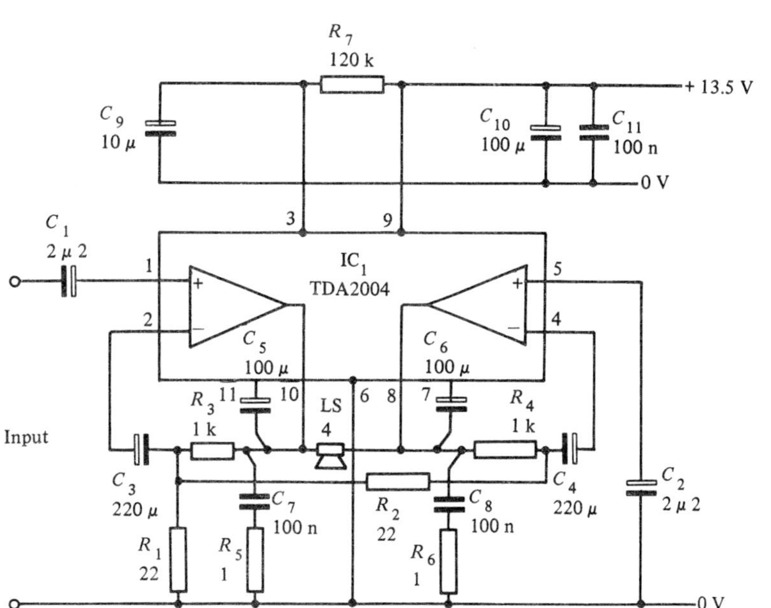

Figure 4.44 *TDA2004 mono (bridge configured) audio frequency amplifier*

The bridge configuration is capable of delivering an output of up to 20 W into a 3 Ω load when operating from a 15 V supply. In either case a heatsink of 4 °C/W, or less is essential (note that the heatsink tab is internally connected to 0 V).

The TDA2030 is an integrated circuit power amplifier capable of producing output powers of up to 20 W into a 4 Ω load. The device

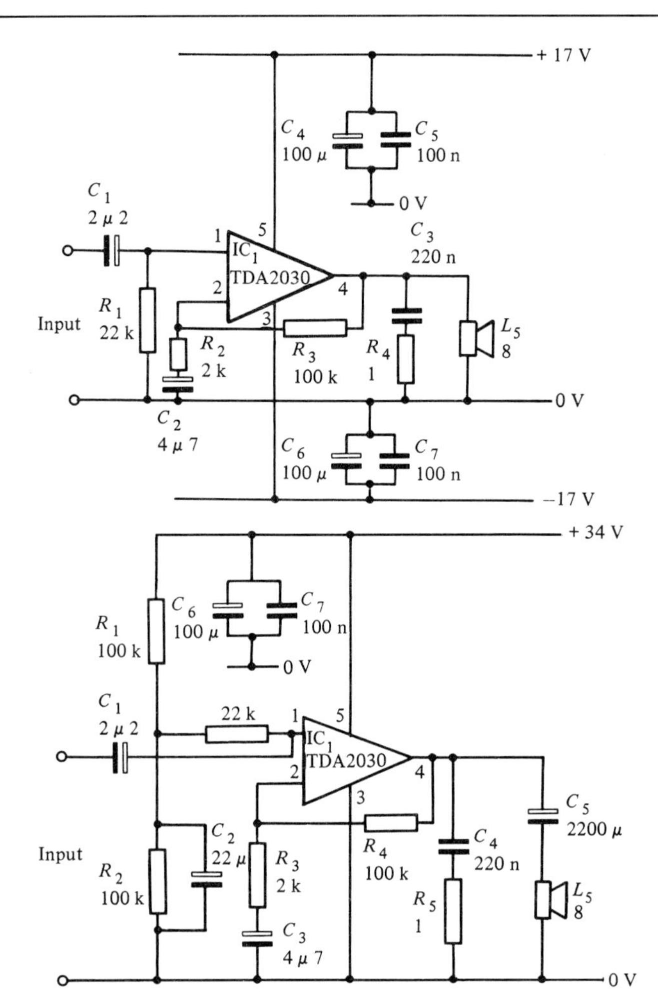

Figure 4.45 *TDA2030 audio frequency amplifier*

Figure 4.46 *Single-supply rail version of the circuit of Figure 4.45*

incorporates short circuit protection, thermal shutdown and safe operating area protection. The TDA2030 is housed in a 5-pin TO220 package, the metal tab of which is bolted to a suitably rated heatsink (note that, like the TDA2004, the tab is internally connected to the negative supply rail).

The voltage gain of the TDA2030 is fixed by two external resistors. The supply voltage may range from ± 6 V and ± 18 V for dual supply rail operation and 12 V to 36 V for single rail operation.

Figures 4.43 and 4.44 respectively show the TDA2030 operating with dual and single supply rails. The performance of these circuits is almost identical and, using the supply rails quoted, outputs of 6.5 W into 15 Ω, 10 W into 8 Ω, and 13 W into 4 Ω may be realized at a THD of less than 0.1%. Both circuits exhibit a voltage gain of 50 and have a frequency response extending from approximately 10 Hz to over 100 kHz.

The TDA20302 requires a heatsink rated at 4 °C/W or less. Furthermore, it should be noted that since the heatsink tab is internally connected to pin-3 (the negative supply rail), it will be necessary to use an insulating kit in conjunction with the circuit of Figure 4.46.

5 Operational amplifiers

Designed originally for analogue computer and control applications, the operational amplifier has found its way into almost every field of electronics. Today's integrated circuit operational amplifiers offer many advantages over their discrete component predecessors. Circuit design is greatly simplified with the added bonus that the characteristics of the latest generation of operational amplifiers far exceed those of their predecessors.

Operational amplifier types and characteristics

Before itemizing the characteristics of some commonly available operational amplifier, it is worth introducing some of the terms commonly applied to these devices.

(a) *Open-loop voltage gain.* This is the ratio of output voltage to input voltage measured with no feedback applied. It may thus be thought of as the 'internal' voltage gain of the device. In practice, this value is exceptionally high (typically greater than 100,000) but is liable to considerable variation from one device to another.

(b) *Closed-loop voltage gain.* This is the ratio of output voltage to input voltage measured with feedback applied. The effect of providing negative feedback is to reduce the loop voltage gain to a value which is both predictable and manageable. Practical closed-loop voltage gains range from 1 to 10,000. (High values of voltage gain may, however, impose unacceptable restrictions on bandwidth, see page 105.)

(c) *Input resistance.* This is the ratio of input voltage to input current expressed in ohms. It is often expedient to assume that the input of an operational amplifier is purely resistive though this is not the case at high frequencies where shunt capacitive reactance may become significant. The input resistance of operational amplifiers is very much dependent on the semiconductor technology employed. In practice values range from about 2 MΩ for common bipolar types to over 10^{12} Ω for FET and CMOS devices.

(d) *Output resistance.* This is the ratio of open-circuit output voltage to short-circuit output current expressed in ohms. Typical values of output resistance range from less than 10 Ω to around 100 Ω depending upon the configuration and amount of feedback employed.

(e) *Input offset voltage.* An ideal operational amplifier would provide zero output voltage when 0 V is applied to its input. In practice, due to imperfect internal balance, there may be some small voltage present at the output. The voltage that must be applied differentially to the operational amplifier input in order to make the output voltage exactly zero is known as the input offset voltage.

Offset voltage may be minimized by applying relatively large amounts of negative feedback or by using the 'offset null' facility provided by a number of operational amplifier devices. Typical values of input offset voltage range from 1 mV to 15 mV. Where a.c., rather than d.c., coupling as employed, offset voltage is not normally a problem and can be happily ignored.

(f) *Full-power bandwidth.* This is equivalent to the frequency at which the maximum undistorted peak output voltage swing falls to 0.707 of its low frequency (d.c.) value (the sinusoidal input voltage remaining constant). Typical full-power bandwidths range from 10 kHz to over 1 MHz for some high-speed devices.

(g) *Slew rate.* This is the rate of change of output voltage with time, corresponding to a rectangular step input voltage. Slew rate is measured in V/s (or V/μs) and typical values range from 0.2 V/μs to over 20 V/μs. Slew rate imposes a limitation on circuits in which large amplitude pulses rather than small amplitude sinusoidal signals are likely to be encountered.

(h) *Common-mode rejection ratio.* This is the ratio of differential voltage gain to common-mode voltage gain. Common-mode rejection ratio is thus a measure of an operational amplifier's ability to ignore signals simultaneously present on both inputs (i.e. common-mode) in preference to signals applied differentially. Common-mode rejection ratio is usually specified in decibels and typical values range from 80 dB to 110 dB.

The desirable characteristics for an operational amplifier are summarized below:

(a) The open-loop voltage gain should be very high (ideally infinite).
(b) The input resistance should be very high (ideally infinite).
(c) The output resistance should be very low (ideally zero).
(d) Full-power bandwidth should be as wide as possible.
(e) Slew-rate should be as large as possible.
(f) Input offset should be as small as possible.
(g) Common-mode rejection ratio should be as large as possible.

The characteristics of modern IC operational amplifiers come very close to those of an 'ideal' operational amplifier, as Table 5.1 will testify.

Operational amplifiers are available packed singly, in pairs (dual types), or in fours (quad types). The 081, for example, is a single general purpose BIFET operational amplifier which is also available in dual (082) and quad (084) forms. Where four operational amplifiers are required in a circuit, it obviously makes good sense to use a quad device rather than

Table 5.1

Type	*741*	*355*	*081*	*3140*	*7611*
Technology	Bipolar	JFET	BIFET	MOSFET	CMOS
Open loop voltage gain (dB)	106	106	106	100	102
Input resistance	2 MΩ	10^{12} Ω	10^{12} Ω	10^{12} Ω	10^{12} Ω
Full-power bandwidth (kHz)	10	60	150	110	50*
Slew rate (V/μs)	0.5	5	13	9	0.16*
Input offset voltage (mV)	1	3	5	5	15
Common mode rejection ratio (dB)	90	100	76	90	91*

*Depends on quiescent current.

four single devices. Not only will this minimize the total number of components but it will also reduce the space required on the PCB, help simplify the PCB layout, improve reliability, and reduce overall cost. Furthermore, it should go without saying that the characteristics of individual operation amplifiers packaged on the same chip will be very closely matched.

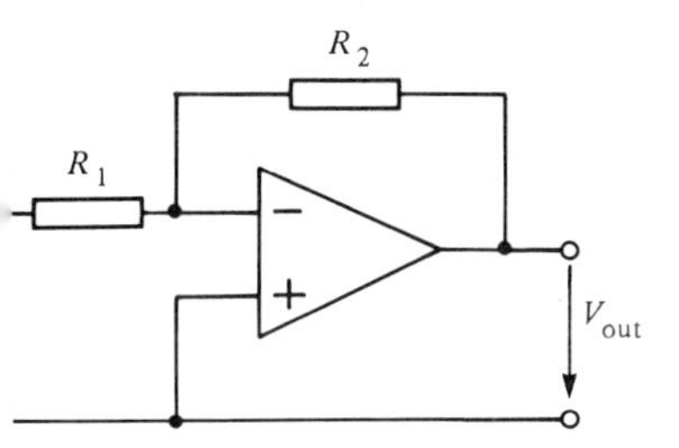

Figure 5.1 *Basic inverting amplifier configuration*

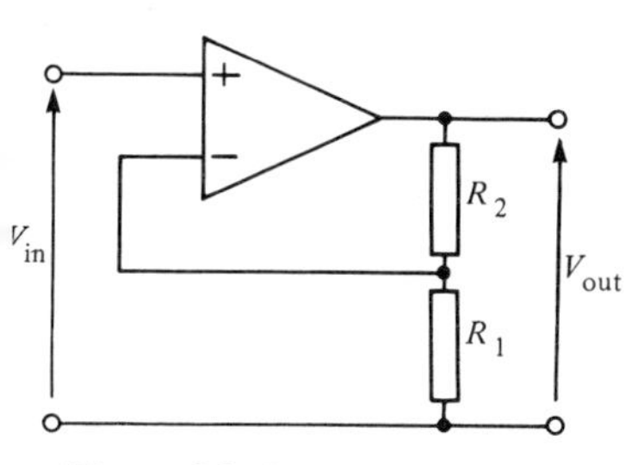

Figure 5.2 *Basic non-inverting amplifier configuration*

Operational voltage amplifiers

The three basic configurations for operational voltage amplifiers are shown in Figures 5.1, 5.2 and 5.3. Supply rails have been omitted from these diagrams for clarity but are assumed to be symmetrical about 0 V. All three of these basic arrangements are d.c. coupled and the following characteristics apply.

Table 5.2

	Input resistance	*Voltage gain*	*Phase shift*
Inverting amplifier (Figure 5.1)	R_1	$\frac{R_2}{R_1}$	180°
Non-inverting amplifier (Figure 5.2)	$R_{in} \times \frac{A_{ol}}{1+\frac{R_2}{R_1}}$*	$1+\frac{R_2}{R_1}$	0°
Differential amplifier (Figure 5.3)	$2R_1$	$\frac{R_2}{R_1}$	180°

*Where R_{in} is the input resistance of the operational amplifier, and A_{ol} is the open loop voltage gain of the operational amplifier.

A practical d.c. coupled inverting amplifier with a voltage gain of 10 and an input resistance of 10 kΩ is shown in Figure 5.4. To preserve

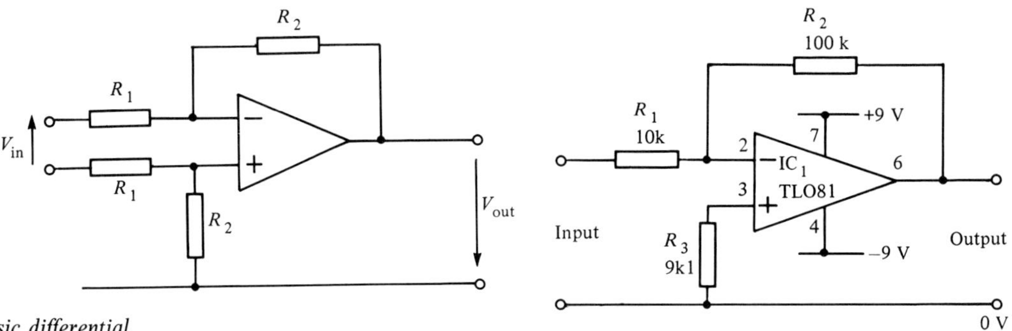

Figure 5.3 *Basic differential amplifier configuration*

Figure 5.4 *Practical direct coupled inverting amplifier with a voltage gain of 10 and an input resistance of 10 kΩ*

symmetry and minimize offset voltage, a third resistor is included in series with the non-inverting input. The value of this resistor should be equivalent to the parallel combination of R_2 and R_1. Hence:

$$R_3 = \frac{R_1 \times R_2}{R_1 + R_2}$$

In the case of high gain amplifiers, R_2 will be very much larger than R_1 and the optimum value for R_3 will be very nearly equal to that of R_1. Hence, where voltage gains exceed about 10, it is usually expedient to make R_3 equal to R_1.

The gain of the inverting amplifier shown in Figure 5.4 can be readily modified over the range 1 to 1000 by suitable choice of component values. It is important to note that, since the product of gain and bandwidth is a constant for any particular device, an increase in gain can only be achieved at the expense of bandwidth. Typical voltage gains and corresponding bandwidths for the circuit of Figure 5.4 are shown in Table 5.3 below:

Table 5.3

Voltage gain	*Bandwidth*
1	d.c. to 2 MHz
10	d.c. to 200 kHz
100	d.c. to 20 kHz
1000	d.c. to 2 kHz

The relationship between voltage gain and bandwidth for a typical operational amplifier is illustrated graphically in Figure 5.5. The gain × bandwidth product can be considered to be numerically equal to the frequency at which the voltage gain falls to unity.

Where a variable, rather than fixed, gain is required, the circuit of Figure 5.6 may be adopted. This circuit has a voltage gain which is

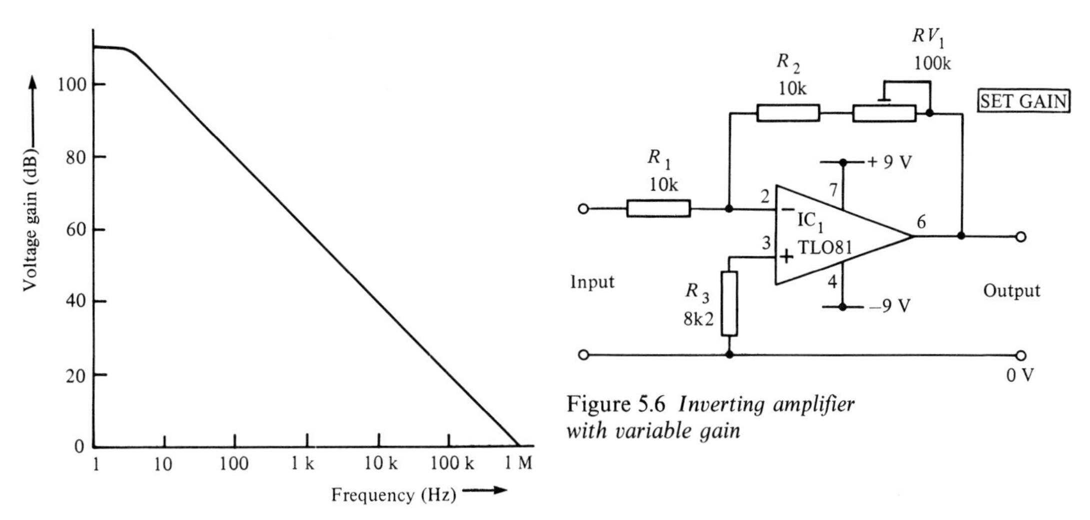

Figure 5.5 *Relationship between gain and bandwidth for a 741 operational amplifier*

Figure 5.6 *Inverting amplifier with variable gain*

adjustable from 1 to just over 10. At maximum gain the frequency response ranges from d.c. to approximately 200 kHz.

Where a very high value of input resistance is required, the non-inverting configuration will usually be preferred. Figure 5.7 shows a circuit which offers a similar performance to that of Figure 5.6 but exhibits virtually infinite input impedance (equivalent to the input resistance of the operational amplifier multiplied by the ratio of open loop to closed loop voltage gain). The input and output voltages are, of course, in phase.

Despite the obvious advantage of having a very much higher input impedance than its inverting courterpart, the non-inverting amplifier has a notable disadvantage. Since the input resistance is exceptionally high, the stray shunt reactance present at the operational amplifier's input now becomes significant. This in turn leads to a reduction in slew rate.

In order to minimize the offset present in the non-inverting configuration, an additional resistor may be inserted in series with the input, as shown in Figure 5.8. As before, the optimum value of this resistor is equal to the parallel combination of the resistors connected to the inverting input (R_1 and R_2).

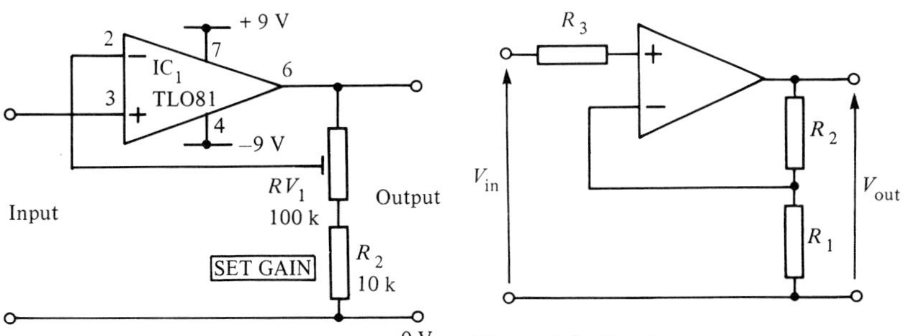

Figure 5.7 *Non-inverting amplifier with variable gain*

Figure 5.8 *Non-inverting amplifier incorporating offset voltage minimization*

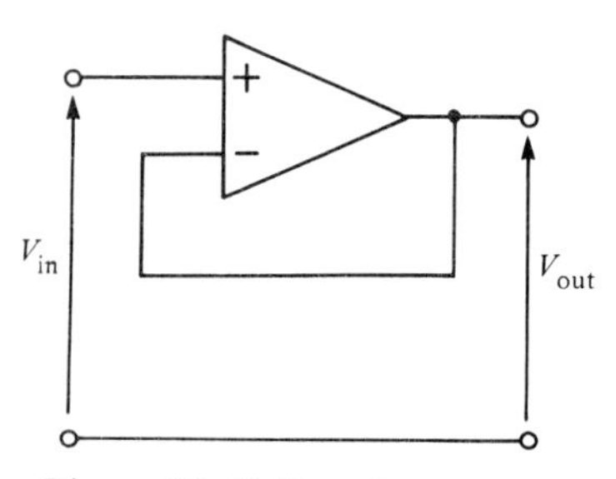

Figure 5.9 *Unity gain operational amplifier buffer*

Even when offset minimization is unimportant, a fixed resistor can be usefully incorporated in series with the input of a non-inverting amplifier. This resistor (of typically between 47 kΩ and 220 kΩ) will provide some protection against excessive voltage applied to the input. It will not, however, materially affect the voltage gain of the stage.

A unity gain non-inverting voltage amplifier is shown in Figure 5.9. This voltage follower configuration employs 100% voltage negative feedback and combines a very high input resistance with a very low output resistance. The obvious application for such a circuit is where some form of buffer is required between stages to minimize the effects of loading. The operation of the circuit is analogous to that of an emitter follower (see Chapter 2).

All of the amplifier circuits described previously have used direct coupling and thus have frequency response characteristics which extend to d.c. This, of course, is undesirable for many applications, particularly where a wanted a.c. signal may be superimposed on an unwanted d.c. voltage level. In such cases a capacitor of appropriate value may be inserted in series with the input, as shown in Figure 5.10. The value of this capacitor

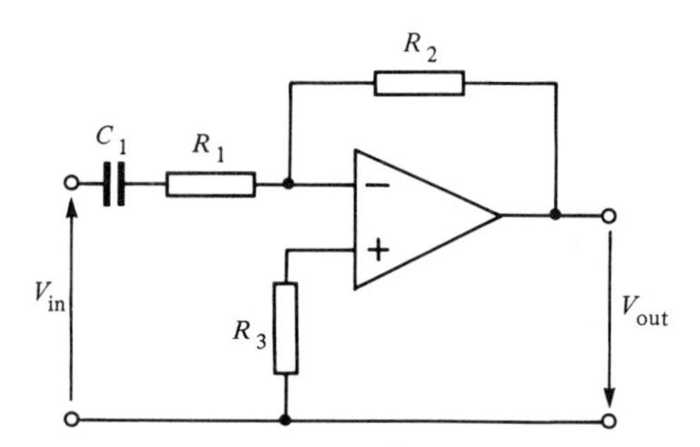

Figure 5.10 *Amplifier with restricted low frequency response*

should be chosen so that its reactance is very much smaller than the input resistance at the lower applied input frequency. Where the non-inverting configuration is employed, this capacitor need only have a relatively small value since the input resistance will be very high and a 100 nF non-electrolytic component will be quite adequate for most applications. In the case of inverting amplifiers (which exhibit very much smaller input resistances) the capacitor should be proportionately larger. However, it is worth avoiding the use of electrolytic coupling capacitors if at all possible. The effect of the capacitor on an amplifier's frequency response is shown in Figure 5.11.

The frequency response of an inverting operational voltage amplifier may be very easily tailored to suit individual requirements, as shown in Figure 5.12. The lower cut-off frequency is determined by the value of the input capacitance, C_1, and input resistance, R_1. The lower cut-off frequency is given by:

$$f_2 = \frac{1}{2\pi C_1 R_1} = \frac{0.159}{C_1 R_1}\,\text{Hz}$$

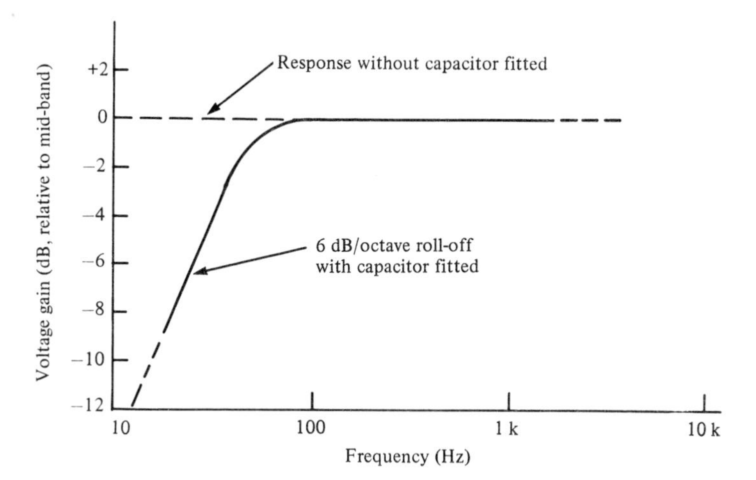

Figure 5.11 *Typical effect of a coupling capacitor on amplifier low frequency response*

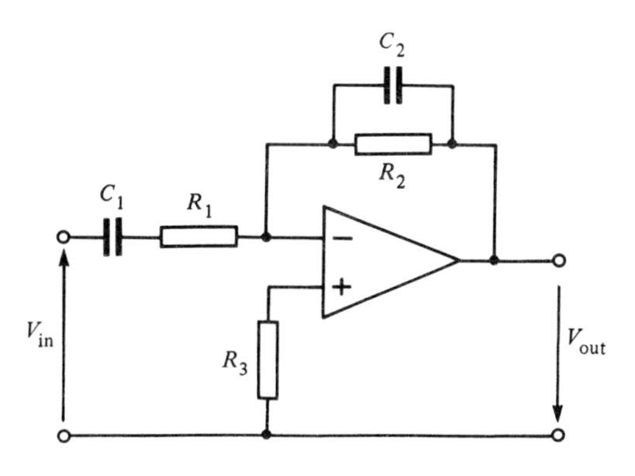

Figure 5.12 *Inverting amplifier with tailored frequency response*

Where C_1 is in farads and R_1 is in ohms.

Provided the upper frequency response it not limited by the gain × bandwidth product, the upper cut-off frequency will be determined by the feedback capacitance, C_2, and feedback resistance, R_2 such that:

$$f_2 = \frac{1}{2\pi C_2 R_2} = \frac{0.159}{C_2 R_2} \text{Hz}$$

Where C_2 is in farads and R_2 is in ohms.

Figure 5.13 shows an inverting operational voltage amplifier with a tailored frequency response extending from 100 Hz to 5 kHz and providing a mid-band voltage gain of 10 and nominal input impedance of 10 kΩ. The frequency response of this circuit is shown in Figure 5.14.

All of the circuits considered thus far have employed symmetrically split supply rails. This arrangement may not always be convenient, especially in circuits which derive their supplies from a battery. In such cases, and provided the supply rail is of sufficient magnitude, it is usually possible to use a potential divider to set a mid-point voltage which will define the quiescent potentials at the operational amplifier's inputs. Figure 5.15 shows how the previous circuit can be adapted for single supply rail operation.

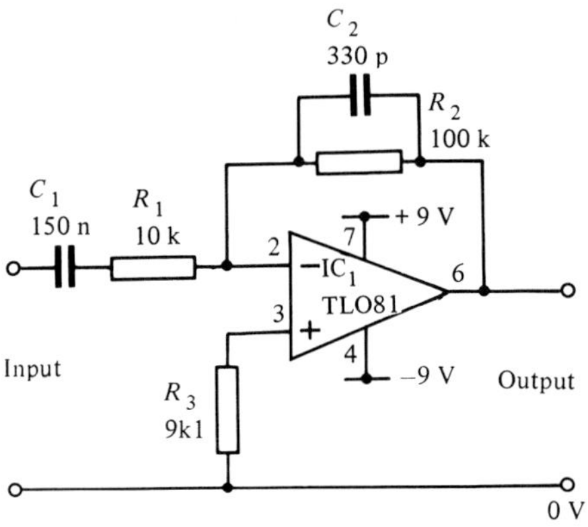

Figure 5.13 *Amplifier having a mid-band voltage gain of 10 (inverting) and frequency response tailored for the range 100 Hz to 5 kHz (approximately)*

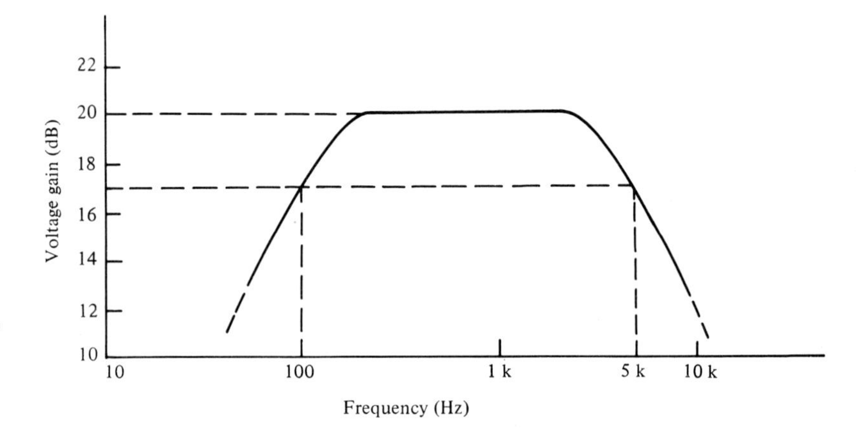

Figure 5.14 *Frequency response of the circuit shown in Figure 5.13*

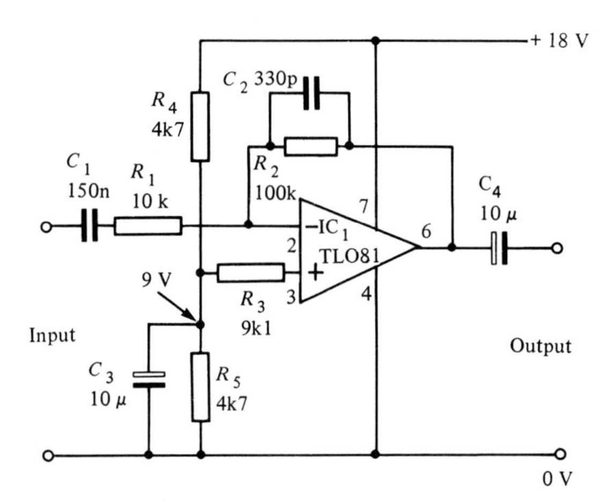

Figure 5.15 *Single supply rail version of the circuit shown in Figure 5.13*

The available supply voltage should be greater than twice the minimum single supply rail voltage but less than twice the maximum single supply rail voltage. A 741, for example, will operate correctly in the circuit of Figure 5.15 over a range of supply voltages extending from 6 V to 30 V.

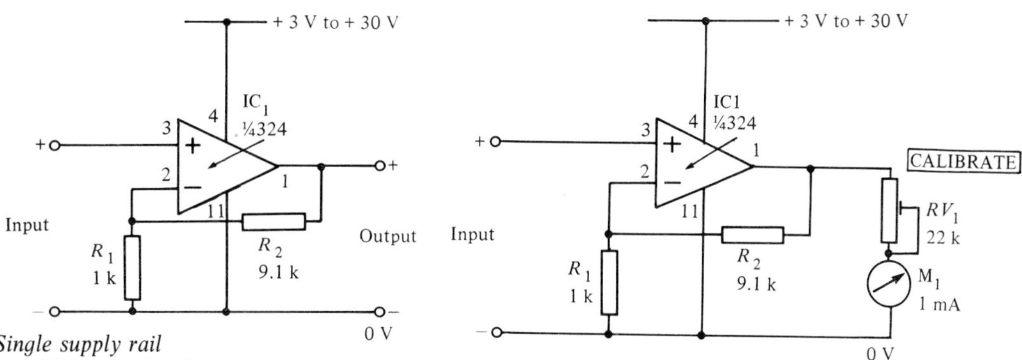

Figure 5.16 *Single supply rail amplifier using a 324 operational amplifier*

Figure 5.17 *Meter driver based on a 324 operational amplifier*

Whereas most operational amplifiers are intended only for operation using dual supply rails, a number of operational amplifiers will operate successfully from a single supply rail without the need for centre tapping. Examples of such devices include the 324 (quad) and 358 (dual) operational amplifiers.

Figure 5.16 shows how a 324 (or 358 with suitable changes to the pin numbers) may be used as a d.c. amplifier operating from a single-supply rail. It should be noted that the circuit is only suitable for unipolar operation (the circuit will only function correctly with an input having a positive polarity). The supply voltage for the stage can be anything between 3 V and 30 V and the output can swing between 0 V and approximately 1.2 V less than the positive supply rail voltage. This circuit will not, of course, cope with a.c. signals (the stage acts as a half-wave rectifier when an a.c. input is applied). Furthermore, it is inadvisable to allow the input voltage to fall much below 0 V or the IC may become internally damaged.

Figure 5.17 shows a meter driver based on a 324. This stage has a voltage gain of just slightly greater than 10 and an input impedance of around 10 MΩ. With the addition of an appropriate transducer and some simple signal conditioning, the circuit can be modified to indicate a variety of physical parameters.

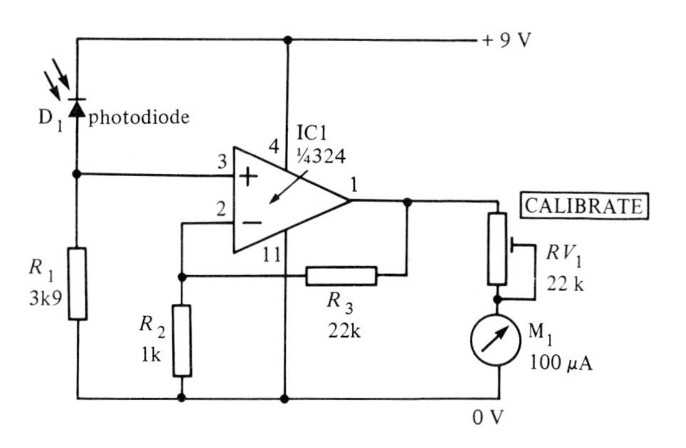

Figure 5.18 *Sensitive light level meter*

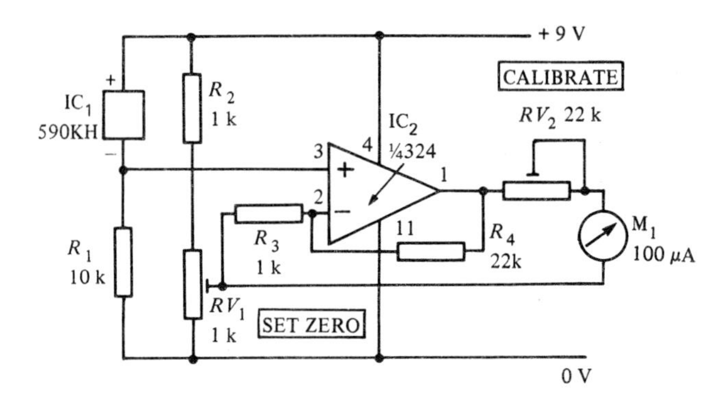

Figure 5.19 *Sensitive electronic thermometer*

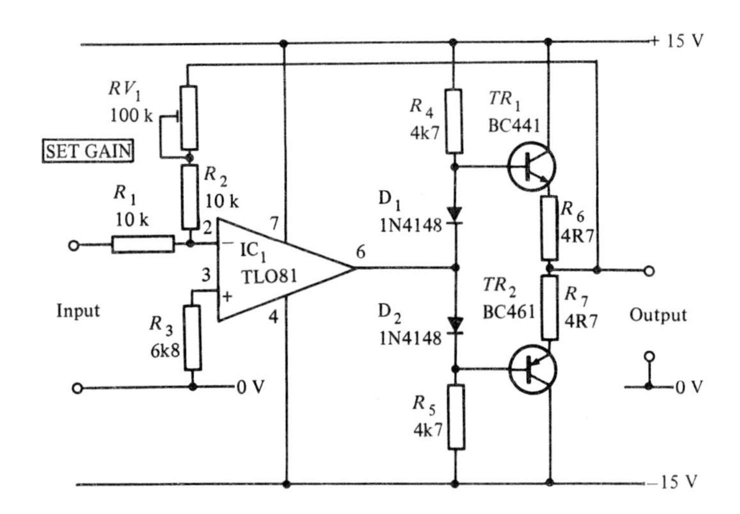

Figure 5.20 *Typical augmented operational amplifier output stage (inverting configuration)*

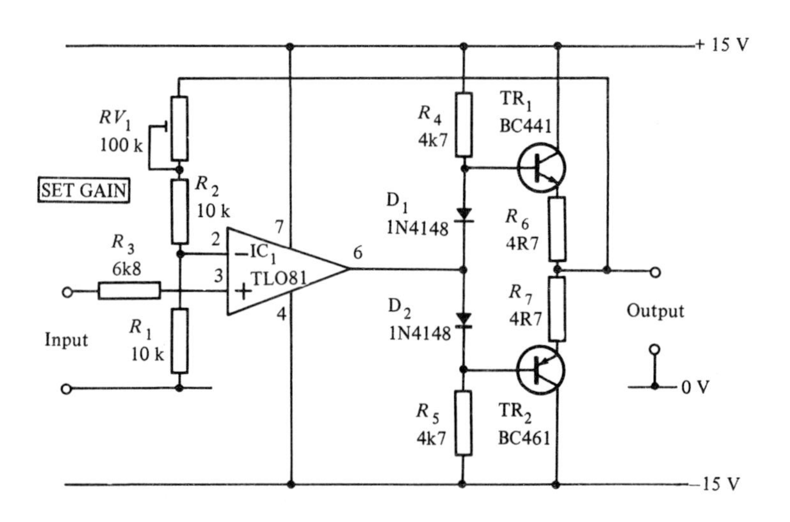

Figure 5.21 *Typical augmented operational amplifier output stage (non-inverting configuration)*

Figure 5.18 shows a sensitive light level meter which employs a photodiode transducer (note that such devices have a spectral response which is significantly at variance with that of the human eye) while Figure 5.19 shows a sensitive electronic thermometer. In the latter case, an offset ('set zero') adjustment is provided by means of RV_1. This should be used to zero the meter at 0 °C. Full-scale calibration should then be provided by RV_2. If necessary, the sensitivity of either circuit may be increased or decreased by making appropriate changes to the feedback resistor (R_3 in the case of Figure 5.18 and R_4 in the case of Figure 5.19).

Figures 5.20 and 5.21 show how a low-power complementary output stage can be used to augment the performance of basic inverting and non-inverting operational amplifier configurations and provide extra current drive at the output. Emitter followers, TR_1 and TR_2, provide current and power gain but do not, of course, affect the voltage gain of the arrangement.

Figure 5.22 shows a very high gain differential amplifier arrangement which produces output pulses having a repetition frequency which is proportional to the angular velocity of a rotating magnet. The differential configuration is employed in order to minimize the effects of common mode noise and hum (both of which could cause problems when using a conventional inverting or non-inverting configuration).

The circuit of Figure 5.22 incorporates a balancing potentiometer (RV_1) which is used to minimize the offset voltage (not all operational amplifiers incorporate such a facility). A sensitive d.c. voltmeter should be connected to the output and RV_1 then adjusted for an output of exactly 0 V in the absence of any input (rotation stopped).

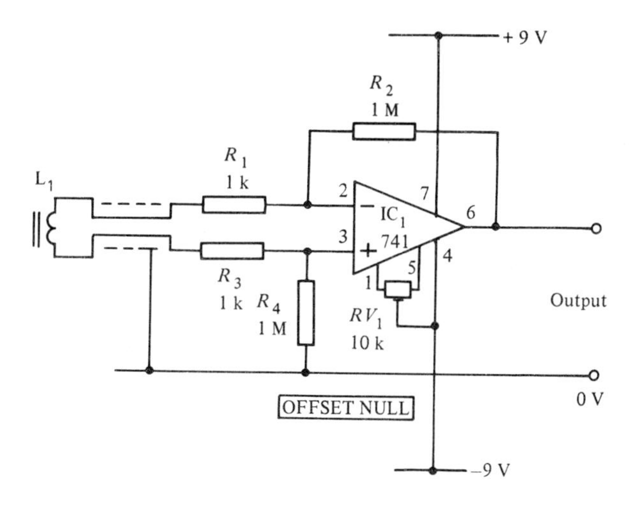

Figure 5.22 *High gain differential amplifier stage used in conjunction with a magnetic sensor*

Figures 5.23 and 5.24 shows two useful a.c. coupled amplifier stages which exhibit deliberately non-linear transfer characteristics. For low-level signals, the circuit of Figure 5.23 provides an inverting voltage gain of 100. Output voltage limiting occurs whenever the input exceeds approxi-

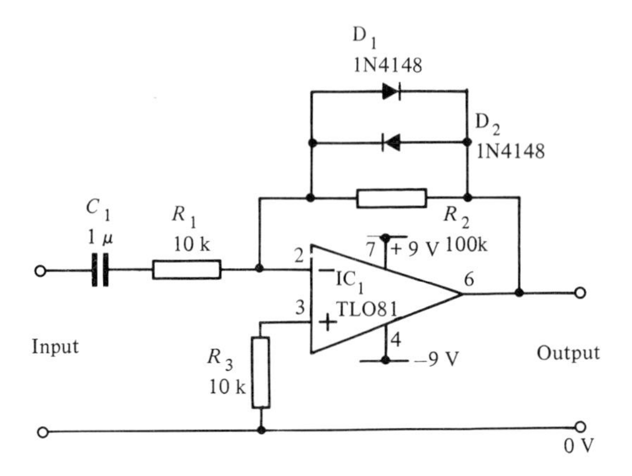

Figure 5.23 *Limiting amplifier stage*

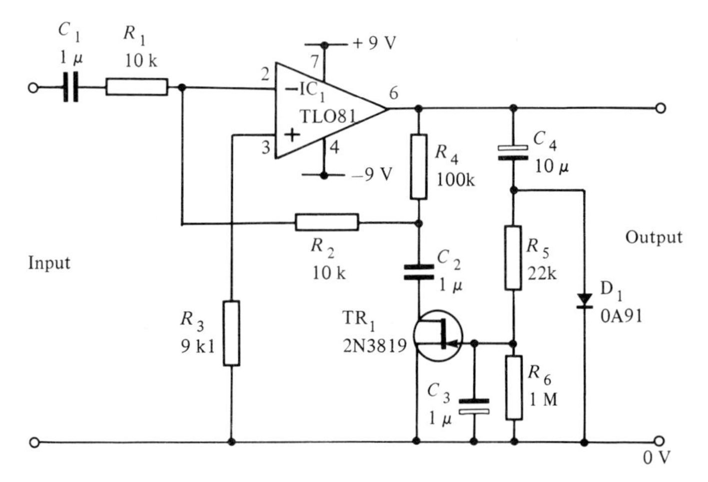

Figure 5.24 *Audio frequency amplifier incorporating compression*

mately 10 mV peak-peak. The output voltage is then limited to a maximum of 1.2 V peak-peak regardless of the amplitude of the input. This, of course, has the effect of squaring a sinusoidal input and hence a subsequent low-pass filter may be required in order to reduce the amplitude of the unwanted harmonics generated.

Figure 5.24 provides a somewhat more sophisticated compressive action. Instead of limiting the input signal by means of a drastic reduction in gain whenever one or other of the feedback diodes conducts, this circuit uses a junction gate FET as a variable resistive element which controls the feedback loop. The compressive automatic gain control (AGC) action is thus progressive over a range of increasing input amplitude. The voltage gain is again approximately 100 for low amplitude signals of less than 10 mV or so. Above this level, the voltage gain progressively falls to about 10 for signals of around 1 V peak-peak.

The 'attack' and 'decay' time constants of the AGC arrangement are determined by the time constants $R_3 \times C_4$ and $R_4 \times C_4$ respectively, the values shown representing a resonable compromise for general speech applications. Some experimentation may, however, be required in order to achieve satisfactory transient response from the arrangement. Whereas the harmonic components produced by this circuit are considerably less than its predecessor, levels of distortion and transient performance dictate that the circuit should only be used for non-critical applications, such as speech, telephony and radio communications.

Filters

Operational amplifiers are ideal for use in active filter circuits and make filter design a relatively simple task. Simple low-pass and high-pass Sallen and Key filters are shown in Figures 5.25 and 5.26. These second order filters respectively provide an attenuation of 12 dB/octave below and above a cut-off frequency of approximately 1 kHz, as shown in Figures 5.27 and 5.28. Both circuits provide unity gain and can be readily modified for use at other frequencies by appropriate modification of the capacitor and/or resistor values. As an example, if R_1 and R_2 are both left at 10 k, doubling the capacitor values will halve the cut-off frequency. Conversely, halving the capacitor values will double the cut-off frequency.

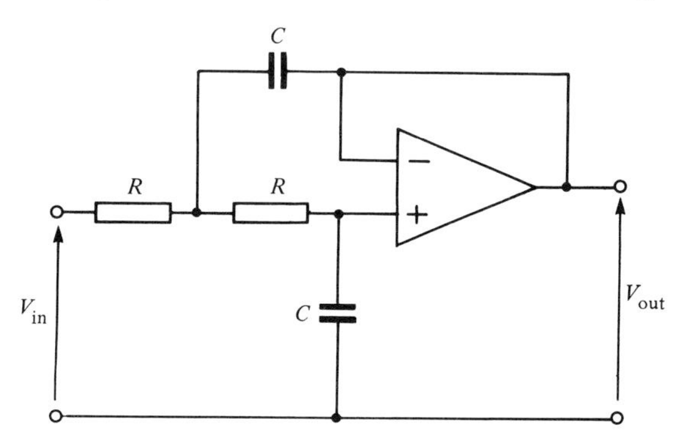

Figure 5.25 *Low-pass Sallen and Key filter*

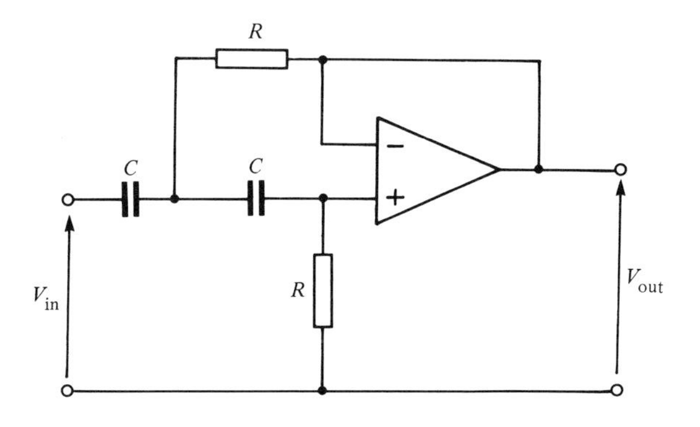

Figure 5.26 *High-pass Sallen and Key filter*

Figures 5.29 and 5.30 show how the Sallen and Key filter can be made variable. Figure 5.29 is a low-pass filter which has a cut off frequency adjustable from 8.5 kHz to 85 kHz. Figure 5.30 shows a high-pass filter which has a cut-cff frequency adjustable from 200 Hz to 2 kHz.

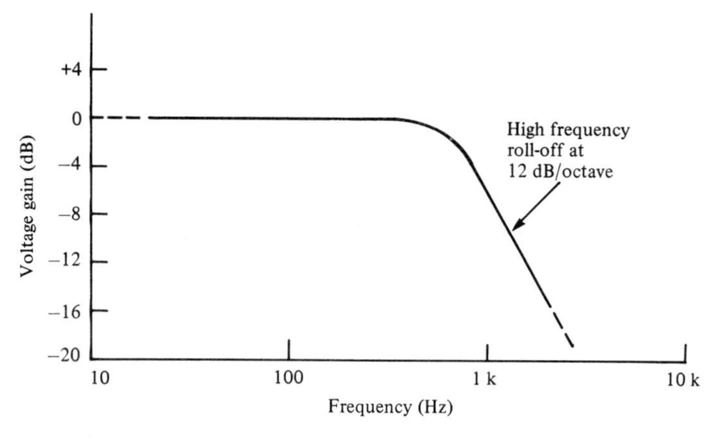

Figure 5.27 *Frequency response of the circuit shown in Figure 5.25*

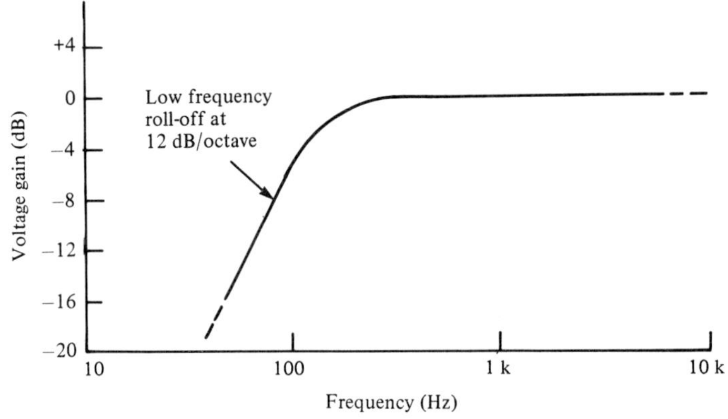

Figure 5.28 *Frequency response of the circuit shown in Figure 5.26*

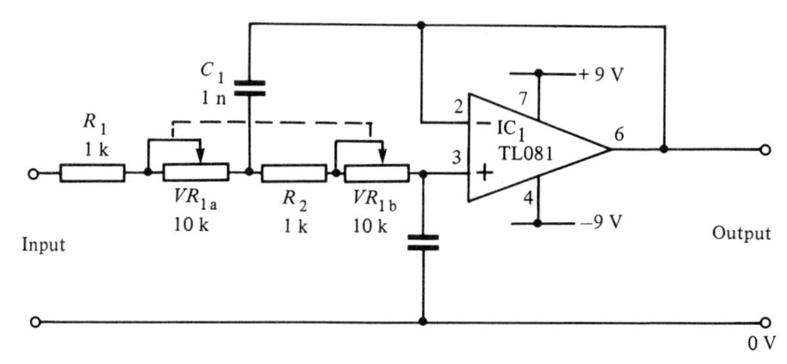

Figure 5.29 *Variable low-pass Sallen and Key filter*

If desired, a band-pass filter can be realized by cascading Sallen and Key low-pass and high-pass filters. An alternative to this technique, where a high Q-factor and narrow bandwidth is required, is the use of a Wien bridge filter of the type shown in Figure 5.31. This simple but very effective filter is capable of Q-factors of up to around 100 and is both predictable and reliable. The centre frequency of the filter is given by:

$$f = \frac{1}{2\pi CR}$$

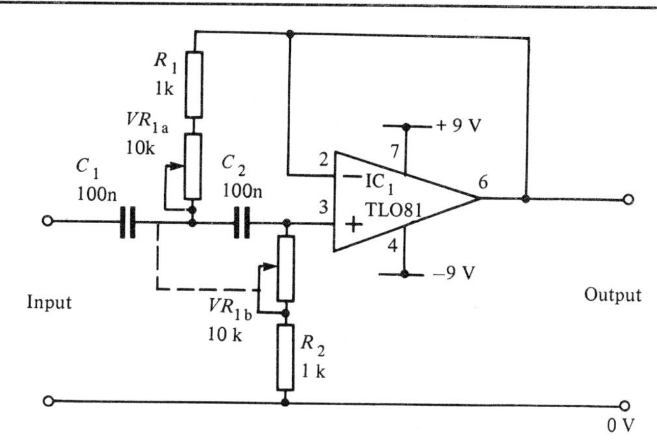

Figure 5.30 *Variable high-pass Sallen and Key filter*

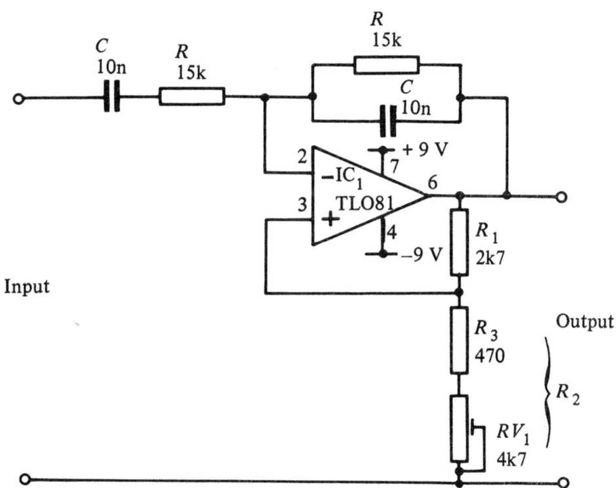

Figure 5.31 *Wien bridge filter*

Where f is in Hz, C is in farads, and R is in ohms.

The voltage gain of the circuit (at the centre frequency) is numerically equal to:

$$A = \frac{R_1 + R_2}{R_2 - 2R_1}$$

It should, however, be noted that the input and output signals are 180° out of phase at the centre frequency. Continuous oscillation will occur when R_1 is made equal to (or is greater than) $R_2/2$. Hence to ensure stability:

$$R_2 < 2R_1$$

The pre-set resistor, RV_1, is used to set the required gain, Q-factor and bandwidth of the filter. The circuit may be readily modified for use at other frequencies by suitably changing the values of R and C. Since the impedance of the signal source will appear in series with the series branch of the Wien bridge network, it is important to use a low impedance source in order to maintain predictable performance. In practice a source

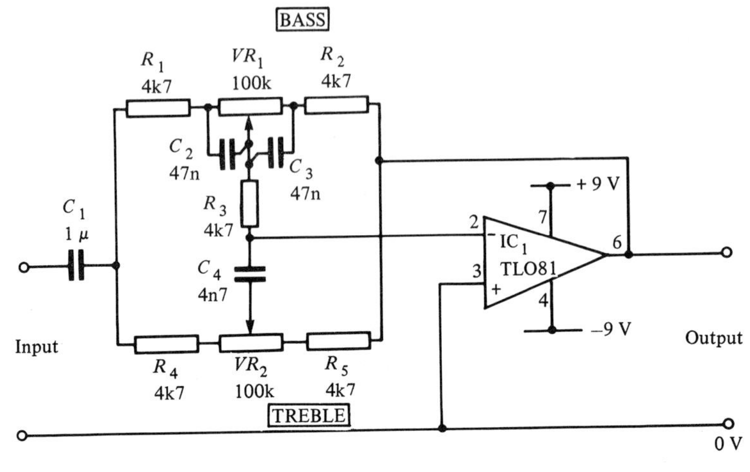

Figure 5.32 *Baxandall audio frequency tone control*

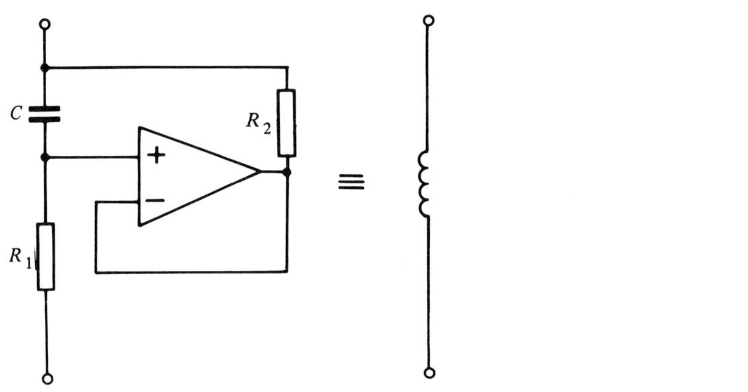

Figure 5.33 *Simulated inductor (gyrator)*

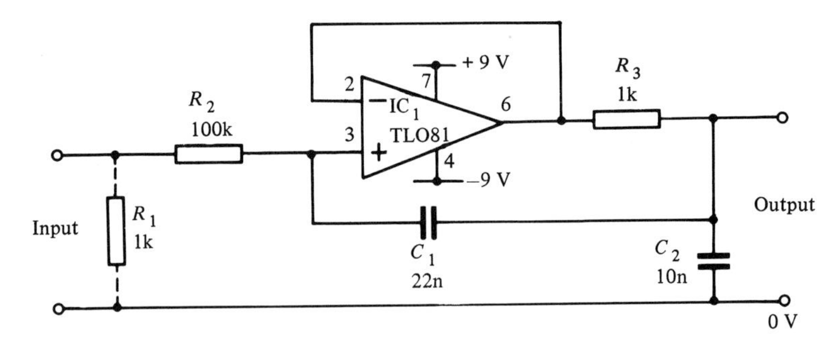

Figure 5.34 *Quasi-tuned circuit filter stage based on a gyrator*

impedance of 100 Ω, or less, should prove adequate for most purposes. This same consideration also applies to the four Sallen and Key filter circuits described previously.

Perhaps the most common application of variable filters is the common or garden audio frequency tone control. These often make use of a Baxandall arrangement like that shown in Figure 5.32. The arrangement shown uses linear 100 kΩ carbon potentiometers and has a 'flat' frequency response extending from 10 Hz to 50 kHz with unity voltage gain.

Maximum 'cut' and 'boost' achievable with the circuit is approximately ± 18 dB at 100 Hz and 12 kHz using the 'bass' and 'treble' controls respectively.

Operational amplifiers can also be used to simulate inductive elements in filters, as shown in Figure 5.33. The arrangement (known as a 'gyrator') is suitable for operation below about 40 kHz and allows the construction of filters which behave in a similar manner to conventional L-C tuned circuits.

Figure 5.34 shows a quasi-tuned circuit filter based on a gyrator. The Q-factor of this circuit is approximately 4 and the centre frequency of the filter is a little over 1 kHz. The tuning capacitor, C_2, may be varied to provide frequency adjustment. Increasing C_2 to 100 nF will reduce the resonant frequency to around 500 Hz while decreasing C_2 to 1 nF will increase the resonant frequency to approximately 5 kHz. The arrangement of Figure 5.34 requires a low impedance source which also provides a d.c. return path to 0 V. Where the filter is to be a.c. coupled at the input, a 1 k resistor may simply be placed in shunt with the input.

Precision rectifiers

Another useful application of operational amplifiers is in minimizing the forward voltage drop associated with diode rectifiers. By incorporating a rectifier in the feedback path of an inverting operational amplifier it is possible to reduce the forward voltage drop by a factor equal to the open-loop voltage gain of the amplifier.

Figure 5.35 shows a precision half-wave rectifier which provides a d.c. output which not only integrates its rectified output but also provides

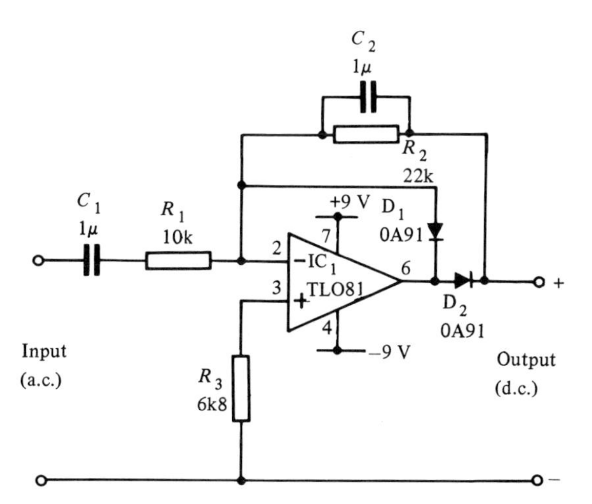

Figure 5.35 *Precision half-wave rectifier*

form-factor correction for a sinusoidal input. The d.c. output is thus made equal to the r.m.s value of the sinusoidal input voltage. This circuit is suitable for operation over a frequency range extending from around 10 Hz to 100 kHz and has an input impedance of 10 kΩ. In practical a.c. meter applications the circuit of Figure 5.35 would usefully be preceded by a high-impedance input stage (using a non-inverting configuration at the expense of a little bandwidth).

A more versatile precision rectifier arrangement is shown in Figure 5.36. This circuit uses a bridge rectifier and incorporates a 100 μA moving coil meter which reads full-scale when a 20 mV r.m.s. sinusoidal input is applied.

If the meter scale is calibrated over the range '0' to '100' readers may prefer an input sensitivity of 100 mV for full-scale deflection. In this case, R_1 should be increased to 50 kΩ (easily achieved by connecting two 100 kΩ resistors in parallel). This will also have the beneficial effect of increasing the input resistance by a factor of five.

The frequency response of the circuit shown in Figure 5.36 extends from approximately 10 Hz to 10 kHz which is adequate for most low frequency and audio applications. The input sensitivity can be easily switched to allow different voltage ranges, each providing an appropriate value for R_1 determined from the equation:

Full-scale r.m.s. voltage sensitivity $= 2 \times R_1$ mV

Where R_1 is specified in kΩ.

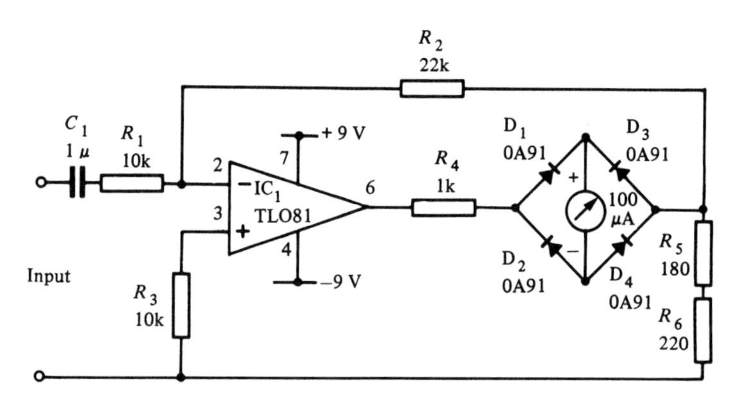

Figure 5.36 *Versatile precision rectifier arrangement used to provide a sensitive r.m.s. sine wave calibrated a.c. voltmeter*

Waveshaping circuits

A traditional application of operational amplifiers is associated with providing output voltages which are proportional to the differential function or integral function of the input voltage. In practical terms (and leaving aside the world of analogue computing) these two circuit configurations can be simply considered as a means of performing simple waveshaping operations.

Figure 5.37 shows the circuit of a typical operational differentiator while Figure 5.38 shows the circuit of a typical operational integrator. Representative input and output waveforms for these two circuits are shown in Figures 5.39 and 5.40 respectively. The criterion for achieving effective differentiation of a waveform by the circuit of Figure 5.37 is that:

$t \ll C \times R$

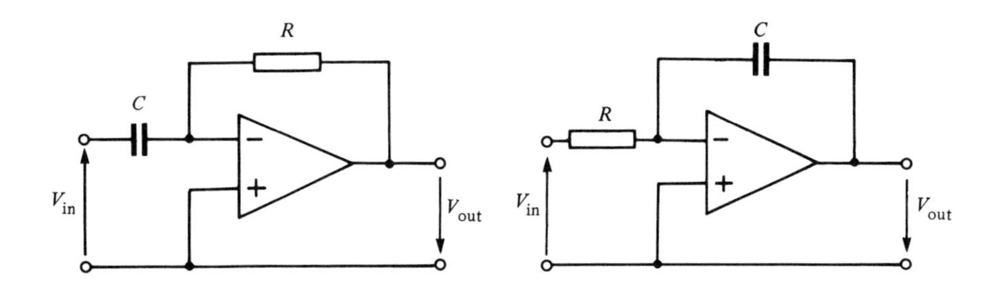

Figure 5.37 *Operational differentiator*

Figure 5.38 *Operational integrator*

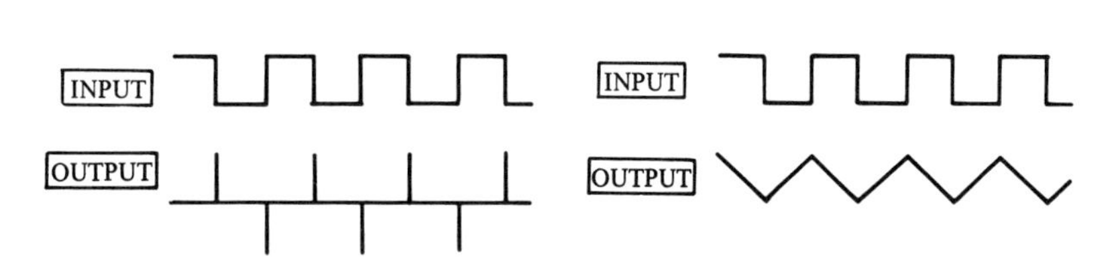

Figure 5.39 *Representative waveforms for the circuit shown in Figure 5.37*

Figure 5.40 *Representative waveforms for the circuit shown in Figure 5.38*

Where t is the periodic time of the input voltage.

Conversely, the criterion for achieving effective integration of a waveform by the circuit of Figure 5.38 is that:

$$t \gg C \times R$$

Where t is the periodic time of the input voltage.

Comparators

One final application of operational amplifiers is worthy of mention. This usually involves the device operating in an open-loop configuration (i.e. without negative feedback applied). The output voltage can thus be in one of two states: either nearly equal to that of the negative supply or nearly equal to the positive supply. The advantage of this arrangement is that, by virtue of the very high open loop voltage gain, it is possible to detect very small changes in input voltage and the circuit becomes ideal for driving LED indicators, relays, and other switching devices.

Figure 5.41 shows a typical comparator arrangement used in a light operated switch. The input to the circuit is, in fact, a bridge configuration. Two arms of the bridge are formed by a pre-set 'threshold' control, RV_1, while the remaining arms are formed by a fixed resistor and photodiode. The relay operates (and, where fitted, the LED becomes illuminated) whenever the light level falls below the threshold level set by RV_1. The sense of the circuit may be reversed (i.e. the relay will operate when the light level rises above the threshold level set by RV_1) by one of the two methods:

(a) Swapping the positions of R_1 and D_1,
(b) Swapping the input connections to IC_1.

Rather than use a conventional linear operational amplifier in Figure

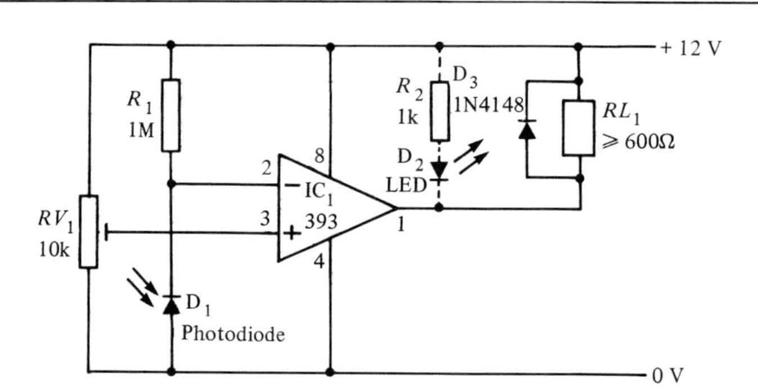

Figure 5.41 *Light operated switch using a photodiode*

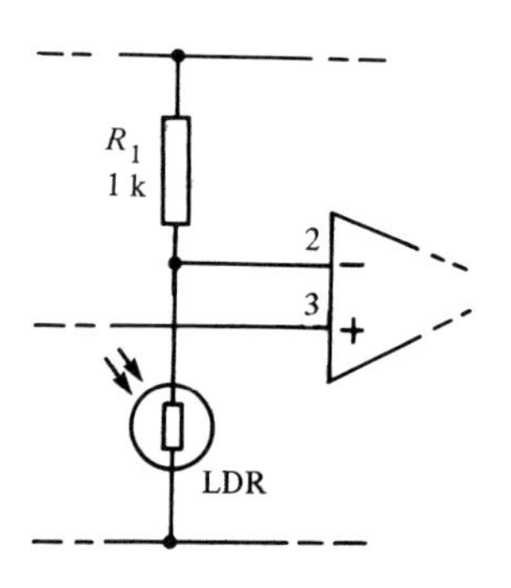

Figure 5.42 *Modification of the circuit shown in Figure 5.41 to permit the use of a light dependent resistor (LDR)*

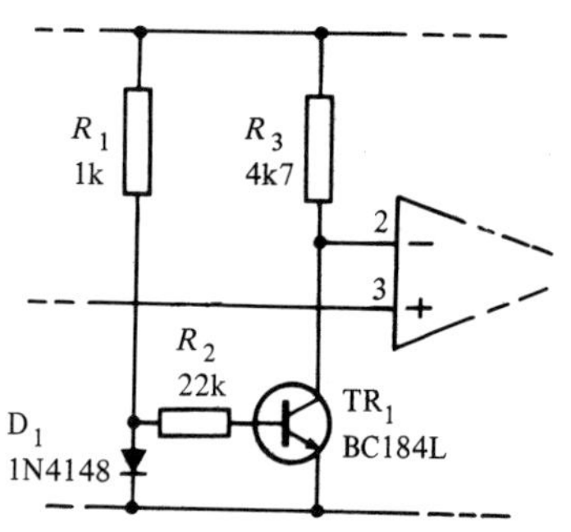

Figure 5.43 *Modification of the circuit shown in Figure 5.41 to function as a temperature operated switch*

5.41, the circuit employs a specialized comparator device. This offers high sensitivity coupled with a very fast switching characteristic (equivalent to an operational amplifier having a very high slew rate). One other important difference is that the device has an 'open collector' output stage which permits the direct connection of a relay type load provided that it demands an operating current of less than 20 mA.

Figure 5.42 shows how the comparator circuit can be modified for use with a conventional light dependent resistor (LDR). This transducer has a spectral response which more closely follows that of the human eye. The circuit performs in the same sense as that of Figure 5.42 (i.e. falling light level triggers the circuit and causes the relay to operate).

Figure 5.43 shows how the circuit can be modified to operate as a temperature operated switch. Since linearity and absolute calibration are not essential, a cheap silicon transistor, TR_1, is used as the temperature sensor (suitably mounted in proximity, and in good thermal contact with, the controlled heat source). The relay will operate whenever the temperature falls below the pre-set threshold. If the sense of operation has to be reversed, this should be achieved by swapping the input connections to IC_1.

The 20 mA maximum sink current of the 393 may present problems when a low resistance relay is to be driven. In such cases, the modification shown in Figure 5.44 should be adopted. If the comparator is to provide an audible warning rather than operate a relay or LED, a piezo-electric warning device may be connected as shown in Figure 5.45.

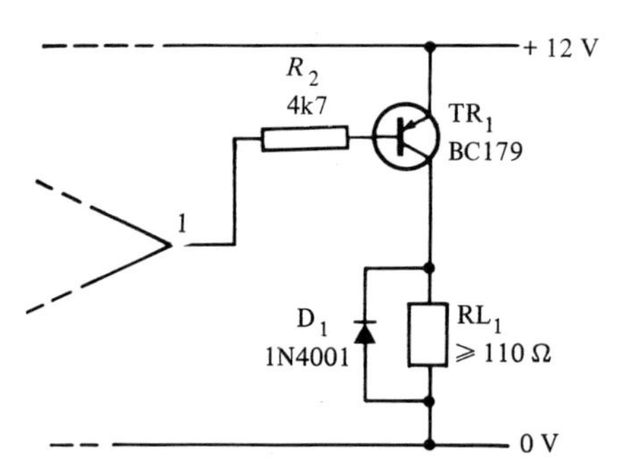

Figure 5.44 *Modification of the circuit shown in Figure 5.41 to drive a low resistance relay*

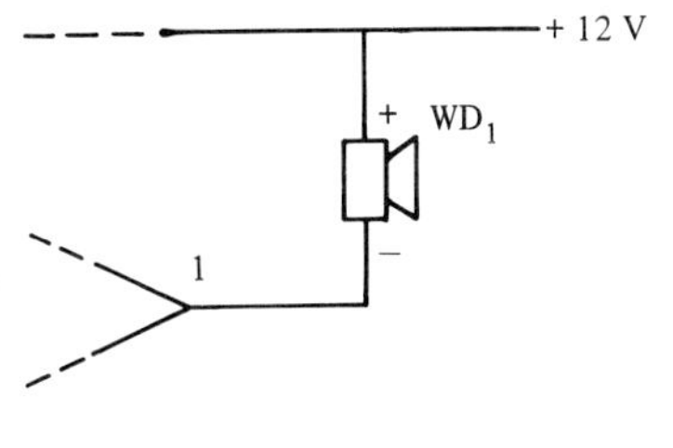

Figure 5.45 *Modification of the circuit shown in Figure 5.41 to drive a piezo-electric warning device*

One final application of the comparator is shown in Figure 5.46. This shows how both halves of the 393 can be utilized in a simple logic probe for use in TTL and CMOS logic circuits. D_1 will become illuminated (indicating a logic 1 condition at the probe tip) whenever the input voltage is greater than 2.5 V. D_2, on the other hand, will become illuminated (indicating a logic 0 condition at the probe tip) whenever the input voltage is less than 1.2 V. Neither of the LEDs is illuminated when the input voltage falls within the range 1.2 V to 2.5 V (corresponding to an indeterminate or tri-state condition).

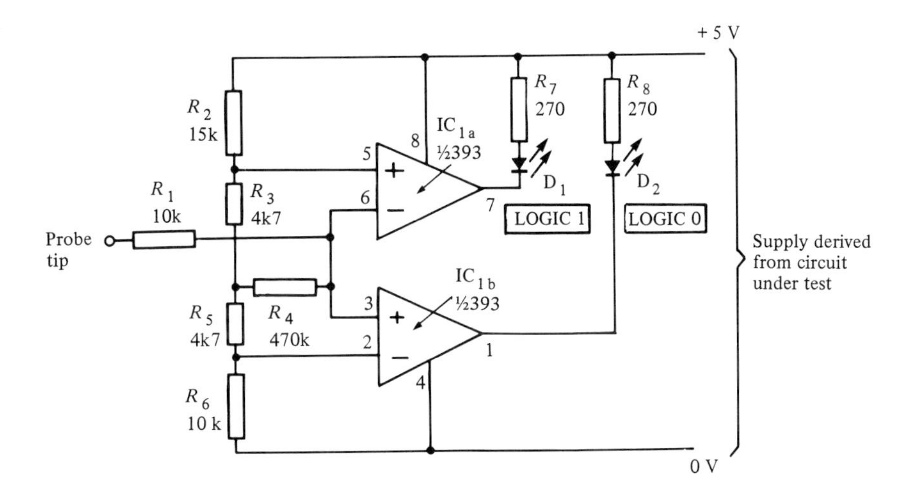

Figure 5.46 *TTL/CMOS logic probe based on a dual comparator*

So that pulses of very short duration can be detected, a practical logic probe would benefit from the addition of a pulse stretching circuit. Such a device can be incorporated by capacitively coupling from the outputs of IC_1 (pins 1 and 7) to the trigger input of a 555 timer connected as a monostable pulse generator (see Chapter 7 for more information). The 555's output (pin 3) should be taken to a separate 'pulse' LED.

6 Logic circuits

The growth of digital electronics has been both continuous and unrelentingly spectacular. Today's pocket calculators are both many times more powerful and more compact than the first generation of electronic computers of a quarter of a century ago. Furthermore, when one considers the relative cost of these devices (a differential in real terms of perhaps 1,000,000:1), one cannot fail to marvel at the advances in technology.

The major contributory factor in all of this, of course, is the availability of increasingly more sophisticated digital integrated circuit devices. The thermionic valve and relay logic of twenty-five years ago is today replaced by semiconductor logic comprising many tens of thousands of transistors fabricated on a single slice of silicon. The power and flexibility of such devices represents a real challenge to the circuit designer.

Logic families and characteristics

Happily, it is not usually necessary to have a detailed understanding of the internal circuitry of a digital integrated circuit in order to be able to make effective use of it. Rather, we have to be aware of some basic ground rules concerning the supply voltage rails, and input and output requirements. Furthermore, with digital circuitry we are primarily concerned with the logical function of logic elements rather than their precise electrical characteristics.

The integrated circuit devices on which modern digital circuitry depends belong to one or other of several 'logic families'. The term simply describes the type of semiconductor technology employed in the fabrication of the integrated circuit. This technology is instrumental in determining the characteristics of a particular device which encompasses such important criteria as supply voltage, power dissipation, switching speed, and immunity to noise.

The most popular logic families, at least as far as the more basic general purpose devices are concerned, are complementary metal oxide semiconductor (CMOS) and transistor transistor logic (TTL). TTL also has a number of subfamilies including the popular low power Schottky (LS-TTL) variants.

The most common range of conventional TTL logic devices is known as the '74' series. These devices are, not surprisingly, distinguished by the prefix number 74 in their coding. Thus devices coded with the numbers

7400, 7408, 7432, and 74121 are all members of this family which is often referred to as 'standard TTL'. Low power Schottky variants of these devices are distinguished by an LS infix. The coding would then be 74LS00, 74LS08, 74LS32, and 74LS121.

The 74 standard TTL series is rated for operation over the temperature range 0 °C to +70 °C. Military specification devices (having identical internal circuitry and logical function of their standard counterparts) are available in the 54 series.

Other common infix coding for TTL devices is given in Table 6.1:

Table 6.1

Infix letters	*Meaning*
C	CMOS version of a corresponding TTL device
F	'Fast' – a high speed version of the device
H	High-speed/high power version
S	Schottky (a name resulting from the input circuit configuration)
HC	High-speed CMOS version (with CMOS compatible inputs)
HCT	High-speed CMOS version (with TTL compatible inputs)
L	Low-power
LS	Low-power with Schottky input configuration

A comparison of the typical characteristics of some common subfamilies is given in Table 6.2:

Table 6.2

Family	*Propagation delay (ns)*	*Power (per gate) (mW)*	*Maximum switching frequency (MHz)*
Standard	10	10	35
H	6	22	50
L	33	1	3
S	3	19	125
LS	10	2	45

For most applications, LS devices should be used in preference to all other types. They offer a good compromise between speed of operation and power consumption and furthermore are commonly available at lower cost than other types (some standard TTL devices are no longer in production). If power consumption is crucially important, then CMOS devices should be used rather than L or LS TTL.

Popular CMOS devices form part of the '4000' series and are coded with an initial prefix of 4. Thus 4001, 4174, 4501 and 4574 are all CMOS devices. CMOS devices are sometimes also given a suffix letter; A to denote the 'original' (now obsolete) unbuffered series, and B to denote the improved (buffered) series. A UB suffix denotes an unbuffered B-series device.

Logic gate symbols and truth tables

The British Standard (BS) and American Standard (MIL/ANSI) symbols for some basic logic gates are shown in Figure 6.1. The MIL/ANSI standard has overwhelming support (even in the UK) and is often preferred by equipment manufacturers. It is important to be aware of *both* standards and be prepared to present logic diagrams using both sets of symbols. They should not, however, be mixed within the same logic diagram.

We shall now briefly consider the action of each of the basic logic gates shown:

Buffers

While buffers do not affect the logical state of a digital signal (i.e. a logic 1 input results in a logic 1 output whereas a logic 0 input results in a logic 0 output) they provide extra current drive at the output and can also be used to regularize the logic levels present at an interface.

Inverters

Inverters are used to complement the logical state (i.e. a logic 1 input results in a logic 0 output and vice versa). Inverters also provide extra current drive and, like buffers, can be used to regularize the logic levels present at an interface.

AND gates

AND gates will only produce a logic 1 output when all inputs are simultaneously at logic 1. Any other input combination results in a logic 0 output.

OR gates

OR gates will produce a logic 1 output whenever any one or more inputs are at logic 1. Putting this another way, an OR gate will only produce a logic 0 output whenever all of its inputs are simultaneously at logic 0.

NAND gates

NAND gates will only produce a logic 0 output when all inputs are simultaneously at logic 1. Any other input combination will produce a logic 1 output. A NAND gate, therefore, is simply an AND gate with its output inverted. The circle shown at the output denotes this inversion.

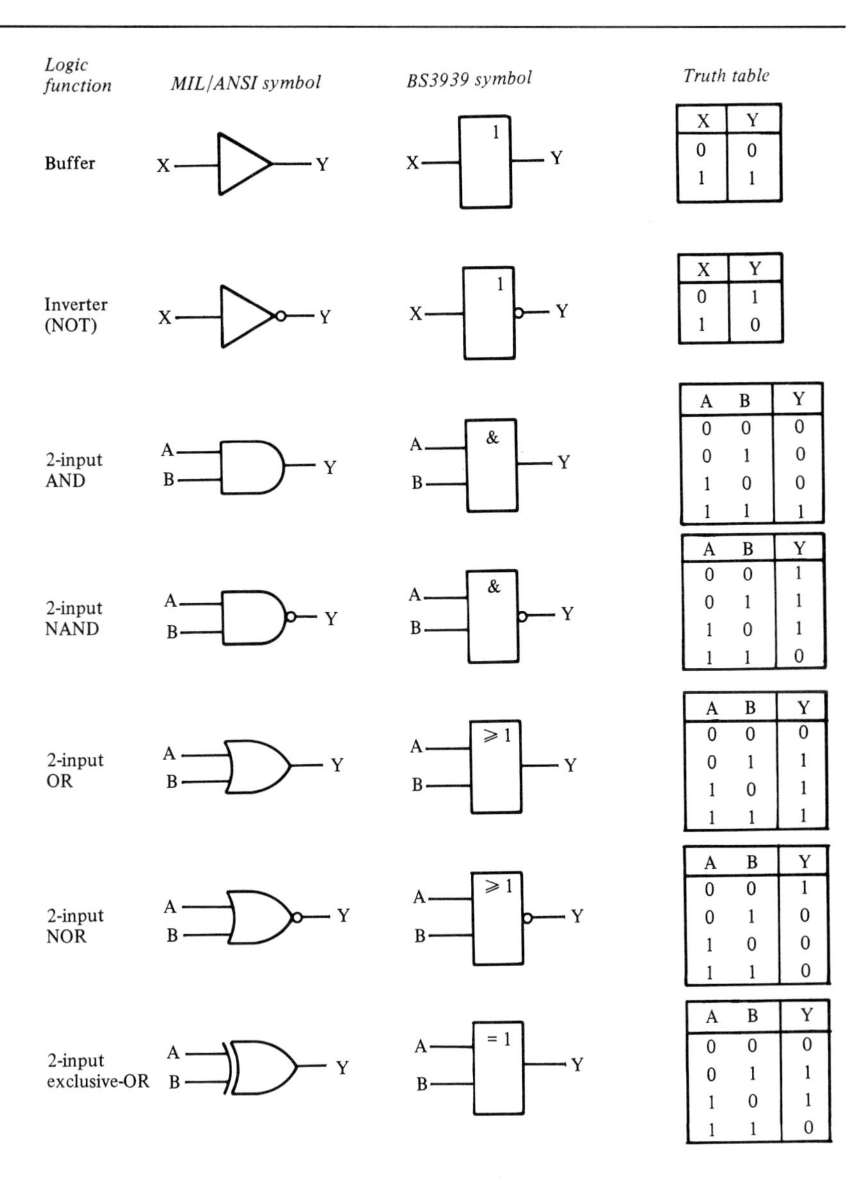

Figure 6.1 *Logic gate symbols and truth tables*

NOR gates

NOR gates will only produce a logic 1 output when all inputs are simultaneously at logic 0. Any other input combination will produce a logic 0 output. A NOR gate, therefore, is simply an OR gate with its output inverted. A circle is again used to indicate inversion.

Exclusive-OR gates

Exclusive-OR gates will produce a logic 1 output whenever either of the inputs is at logic 1 and the other is at logic 0. Exclusive-OR gates produce a logic 0 output whenever both inputs have the same logical state (i.e. when both are at logic 0 or both are at logic 1).

Readers should note that, while inverters and buffers each have only one input, exclusive-OR gates have two inputs and the other basic gates (AND, OR, NAND and NOR) are commonly available with up to eight inputs.

Truth tables provide a handy method of illustrating the function of a logic gate and show the state of the output of the gate resulting from all possible input conditions. For a logic gate with n inputs, there are 2^n possible input conditions. Hence a two-input gate will have four possible input states, a three-input gate will have eight possible input states, and so on.

Power supply requirements of logic circuits

Most TTL and CMOS logic systems are designed to operate from a single supply voltage rail of nominally + 5 V. With TTL devices, it is important for this voltage to be very closely regulated. Typical TTL IC specifications call for regulation of better than ± 5% (i.e. the supply voltage should not fall outside the range 4.75 V to 5.25 V).

In fairness, most TTL devices will operate happily outside this range; 4 V to 5.5 V being not untypical. The logical function of a gate is the same when operated from a 4 V supply as it is when operated from 5 V. However, the switching characteristics are considerably different since the 'propagation delay' (in effect, the time taken for a logical change to pass through the logic gate) increases considerably as the supply voltage falls. While this may not be important in many applications it does explain why such circuits as counters and frequency dividers cease to operate to their full specification when operated from reduced supply voltages.

The upper limit (or 'absolute maximum') supply voltage for TTL devices is + 7 V. If the supply voltage ever exceeds this value any TTL devices connected to the supply rail are liable to 'self destruct' very quickly.

CMOS logic devices are fortunately very much more tolerant of their supply voltage than their TTL counterparts. Most CMOS devices will operate from a supply rail of anything between + 3 V and + 15 V. This, coupled with a minimal requirement for supply current (a CMOS gate typically requires a supply current of only a few microamps in the quiescent state) makes them eminently suited to battery powered equipment. Indeed, there is little need for any form of supply regulation in most portable CMOS-based equipment; the equipment will often continue to operate until the battery voltage falls below 3 V.

Like their TTL counterparts, CMOS devices offer inferior switching speeds when operated on reduced supply voltages and, while it is common practice to operate CMOS devices from a + 5 V rail, switching speeds can typically be doubled by operating them from + 9 V, + 12 V or + 15 V rails.

TTL devices require considerably more supply current than their CMOS equivalents. A typical TTL logic gate requires a supply current of around

8 mA; approximately 1000 times that of its CMOS counterpart when operating at a typical switching speed of 10 kHz.

When specifying a power supply for a logic circuit, it is important to cater for the maximum load current to be supplied. Data books normally give supply figures for TTL devices in either mA or mW, quoted either per-gate or per-package. Any mW figures should first be converted to mA (by dividing by the supply voltage which will invariably be 5 V). Thus:

$$\text{Supply current per-gate (mA)} = \frac{\text{Power consumption per-gate (mW)}}{5}$$

The total current for each gate can then be calculated by multiplying the number of gates within the package (whether or not they are used) by the per-gate supply current.

This process should be repeated for each package present and the results added to find the total current required by the integrated circuit devices. The current required by any other circuit loads (e.g. relays, LED, etc.) should then be added to this figure in order to provide an estimate of the total supply current. A margin of around 20% should then be added to the result before selecting an appropriate circuit from Chapter 3.

Adequate decoupling of the supply rail is an essential consideration when designing logic circuits. The impedance of the supply must be very low (typically 0.1 Ω or less) over a wide range of frequencies (up to 35 MHz for standard TTL and CMOS and up to 150 MHz for 'fast' and Schottky TTL. Several representative power supplies are described in Chapter 3.

High frequency decoupling should be provided by ceramic disc or plate capacitors having values in the range 10 n to 100 n. To be effective, such components should be fitted in close proximity to the supply pins of an integrated circuit. Furthermore, at least one capacitor should be fitted for every two to four IC devices.

Low frequency decoupling should employ suitably rated electrolytic capacitors. Radial lead devices will normally be preferred though axial lead devices may also be used where PCB space is not critical. Tantalum bead types are also ideal for decoupling and require very little space. They are, however, relatively expensive and this consideration will often preclude their use for large scale production. Electrolytic decoupling capacitors should be fitted at regular intervals along the main supply rails and should typically have values of between 1 μF and 47 μF. At least one capacitor should be fitted for every six to ten IC devices.

In some cases it may be necessary to incorporate additional inductance in the positive supply lead in the form of one, or more, ferrite bead inductors. These should be arranged (in conjunction with two or more capacitors of appropriate value) in a π-section low-pass filter configuration as shown in Figure 6.2.

Before moving on, it should be stressed that there is little point in adding extra decoupling unless an adequate 0 V return path has been provided. For decoupling to be effective it is absolutely essential to ensure that the common 0 V rail has a very low impedance. This can only be achieved by

L
4 turns 30 SWG on ferrite bead
+5 V
+5 V
Input
C_1
100 n
C_2
100 n
Output
0 V

Figure 6.2 *π-section low-pass filter for supply decoupling*

paying careful attention to the PCB track layout and the 0 V rail in particular. Now, to summarize the main points concerning decoupling:

(a) Use h.f. decoupling capacitors (e.g. disc ceramic) of between 10 nF and 100 nF fitted at the rate of one capacitor for every two to four devices.
(b) Use l.f. decoupling capacitors (e.g. radial electrolytic) of between 1 μF and 47 μF fitted at the rate of one capacitor for every six to ten devices.
(c) For *any* circuit include a minimum of one h.f. and one l.f. decoupling capacitor.
(d) Distribute capacitors carefully around the PCB at strategic points along the supply rail; place h.f. types close to supply pins and l.f. types at regular intervals.
(e) Pay special attention to the PCB 0 V common rail; decoupling will be ineffective unless this provides an adequate return path.
(f) In 'difficult' cases consider additional π-section supply rail filters in the form series ferrite bead inductors and shunt ceramic capacitors.

Logic levels and noise margins

Logic levels are simply the range of voltages used to represent the logic states 0 and 1. Not surprisingly, the logic levels for CMOS differ markedly from those associated with TTL. In particular, CMOS logic levels are relative to the supply voltage used (recall that this can be anything from 3 V to 15 V) while the logic levels associated with conventional TTL are absolute. Table 6.3 generally applies:

Table 6.3

	TTL	*CMOS*
Logic 1	More than 2 V	More than 2/3 V_{DD}
Logic 0	Less than 0.8 V	Less than 1/3 V_{DD}
Indeterminate	Between 0.8 V and 2 V	Between 1/3 V_{DD} and 2/3 V_{DD}

Note: V_{DD} is the CMOS positive supply rail voltage.

In a perfect world there would be no uncertainty nor any ambiguity about the logic levels present in a digital circuit. Unfortunately, this is seldom the case since spurious signals (or 'noise') are invariably present to some degree. The ability to reject noise is thus an important property of logic devices. This is, of course, particularly true where the digital system is to be used in a particularly noisy environment (such as a steelworks or shipyard).

The ability of a logic device to reject noise is measured in terms of its 'noise margin' and is defined as the difference between:

(a) The minimum values of high state output and input voltage.
(b) The maximum values of low state output and input voltage.

Standard 7400 series TTL usually exhibit noise margins of 400 mV while CMOS offer somewhat superior performance with a noise margin equivalent to one third of the positive supply rail voltage, as shown in Figure 6.3.

Since CMOS devices do not generate large current transients on the power rails, very little noise is produced by the logic and less decoupling will be required. Since the transition time (i.e. time for a change of output to occur) of a CMOS device tends to be greater than the propagation delay within the gate, these devices tend to be relatively impervious to noise and glitches.

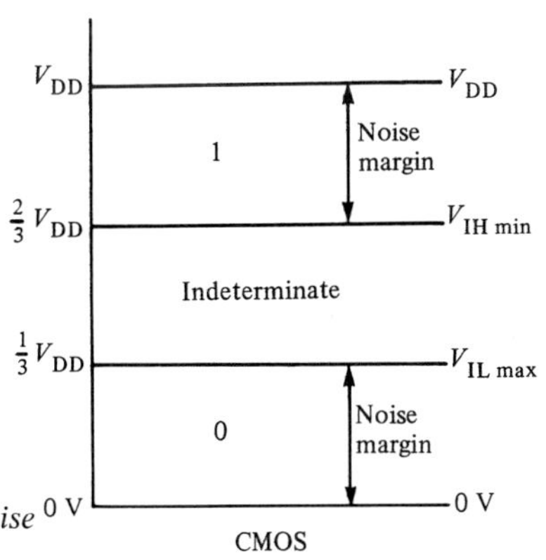

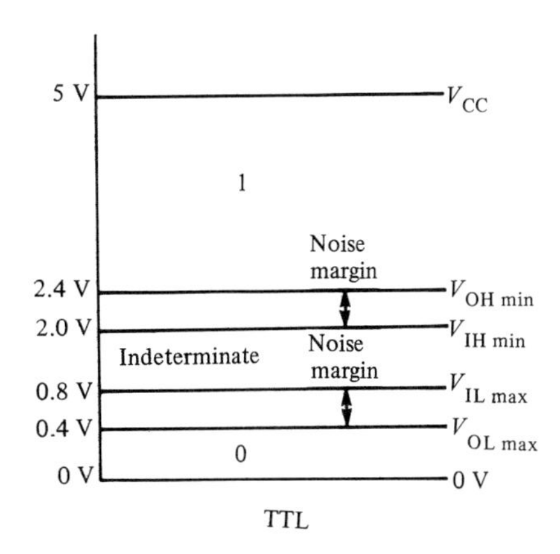

Figure 6.3 *Logic levels and noise margins for TTL and CMOS devices*

Generating logic states

Having dealt with logic levels, and before discussing some typical logic gate arrangements, it is worth mentioning the methods by which the logical states, 0 and 1, can be generated. We shall deal first with the permanent logic states required, for example, to configure the operation of a particular device within a logic arrangement.

While a permanent logic 0 state can be generated by simply connecting an input directly to the 0 V rail, it is not good practice to generate a logic 1 input by linking directly to the positive supply rail. A 'pull-up' resistor should be incorporated in the input lead to limit the gate input current in the event of an abnormal rise in supply voltage. This will also permit the use of a 'pull-down' strap (i.e. shorting of the input to 0 V) or a logic pulser during prototype testing and fault finding on production units.

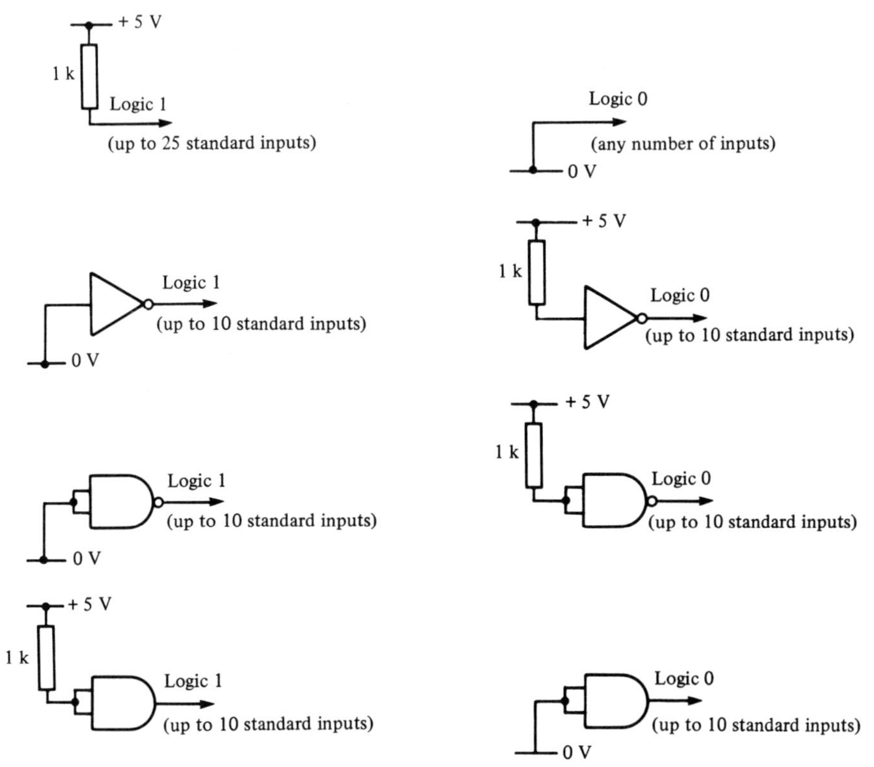

Figure 6.4 *Methods of generating logic 0 and logic 1 states*

The value used for pull-up resistors is generally uncritical and values of between 1 kΩ and 10 kΩ are typically employed. Where several inputs are to be taken to logic 1 (as will often be the case with, for example, a JK bistable arrangement) a common pull-up resistor can be employed. Figure 6.4 shows some commonly used methods of generating permanent logic 0 and logic 1 states.

Having dealt with methods of generating permanent logic 0 and logic 1 states, we shall now turn our attention to methods of producing logic 0 and logic 1 states from switches. Unfortunately the switching action of most switches is far from 'clean'. The reason for this is that the switch contacts very rarely make and break with a single action. Invariably there is a degree of 'contact bounce' when the switch is operated. This results in rapid making and breaking of the switch until it settles into its new state. Figure 6.5 shows basic logic 0/logic 1 switching arrangements while Figure 6.6 shows the waveform generated by these circuits as the contacts close. Since spurious states can cause problems (particularly if our input switching arrangement is used in conjunction with some form of counting circuitry) it is important to take positive steps to eliminate contact bounce. This is achieved using a 'de-bounce' circuit.

Immunity to transient switching states is generally enhanced by the use of active-low inputs (i.e. a logic 0 state at the input is used to assert the

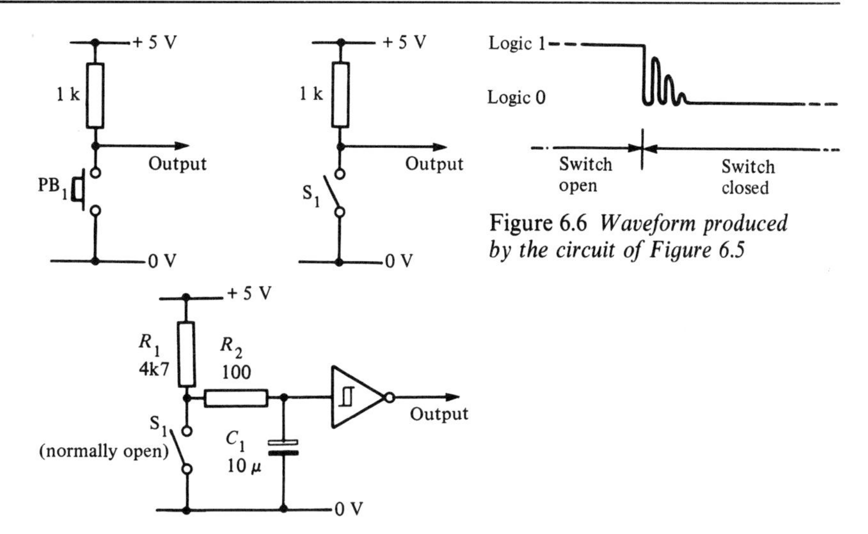

Figure 6.5 *Using a switch to generate logic states*

Figure 6.6 *Waveform produced by the circuit of Figure 6.5*

Figure 6.7 *Debounced switch input*

condition required at the output). The circuit shown in Figure 6.7 will be adequate for most toggle, slide, and push-button type switches. The value chosen for R_2 must take into account the low-state sink current required by IC_1 (normally 1.6 mA for standard TTL and 400 μA for LS-TTL). R_2 should not be allowed to exceed approximately 470 Ω in order to maintain a valid logic 0 input state. The values quoted generate an approximate 1 ms delay (during which the switch contacts will have settled into their final state). It should be noted that, on power-up, this circuit generates a logic 1 level for approximately 1 ms before the output reverts to a logic 0 in the inactive state. The circuit obeys the following state table:

Table 6.4

Switch condition	*Logic output*
Closed	1
Open	0

An alternative, but somewhat more complex, switch de-bouncing arrangement is shown in Figure 6.8. Here a single-pole double-throw (SPDT) changeover switch is employed. This arrangement has the advantage of providing complementary outputs and it obeys the following state table:

Table 6.5

Switch condition	Q	$\bar{Q}$
Position 0	0	1
Position 1	1	0

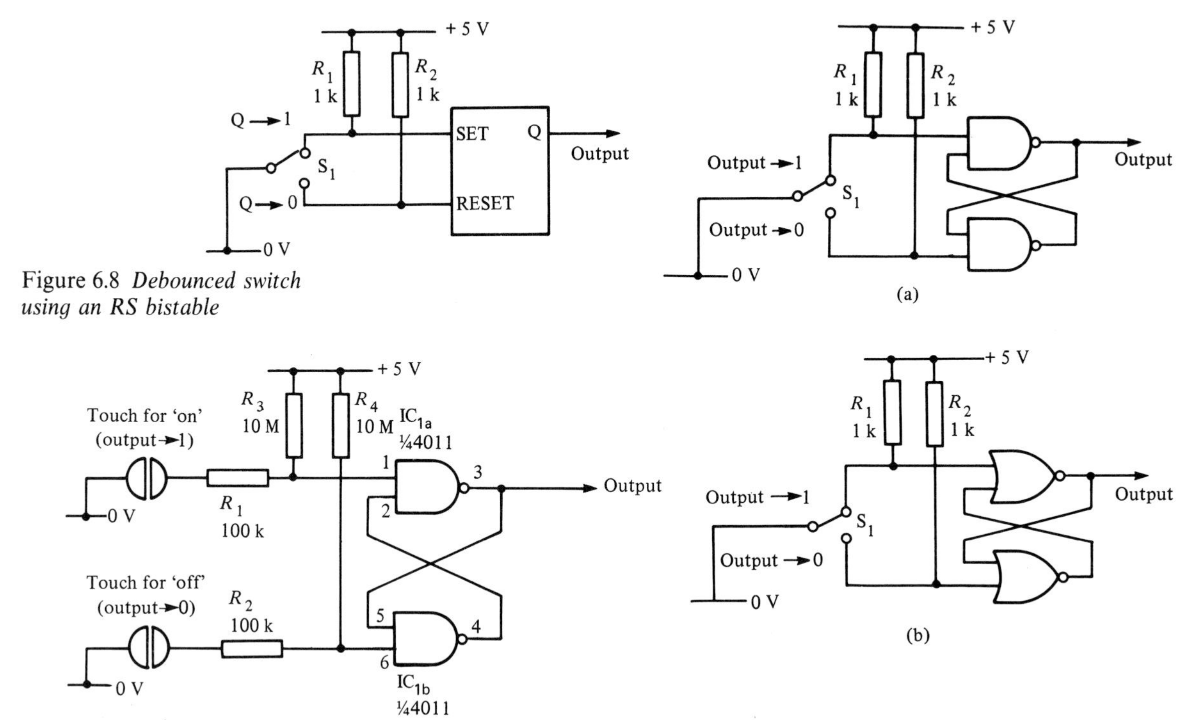

Figure 6.8 *Debounced switch using an RS bistable*

Figure 6.10 *Touch switch*

Rather than use an IC RS bistable in the configuration of Figure 6.8 it is often expedient to make use of 'spare' two-input NAND or NOR gates arranged to form bistables using the circuits shown in Figure 6.9(a) and (b), respectively. Figure 6.10 shows a rather neat extension of this theme in the form of a touch operated switch. This arrangement is based on a 4011 CMOS quad two-input NAND gate (though only two gates of the package are actually used in this particular configuration).

Finally, it is sometimes necessary to generate a latching action from a push-button switch. Figure 6.11 shows an arrangement in which a 7473 JK bistable is clocked from the output of a debounced switch. Pressing the switch causes the bistable to change state. The bistable then remains in that state until the switch is depressed a second time. If desired, the complementary outputs provided by the bistable may be used to good effect by allowing the Q output to drive an LED. This will become illuminated whenever the Q output is high. The JK bistable is discussed more fully on page 145.

Driving LED indicators

A typical red LED requires a current of around 10 mA to provide a reasonably bright display. Various methods may be employed for driving an LED from standard or LS-TTL depending upon whether the LED is

Figure 6.9 (*a*) *Debounced switch using a NAND gate bistable* (*b*) *Debounced switch using a NOR gate bistable*

to be illuminated for a logic 0 or logic 1 state and whether the gate is required to drive any subsequent gate inputs in addition to the LED. The various possibilities (together with the remaining fan-out) are shown in Figure 6.12.

When an LED is to be illuminated from the output of a CMOS gate, a different approach is necessary in order to compensate for the range of supply voltages that may be employed. If a 4049 or 4050 hex buffer is available, an arrangement of the form shown in Figure 6.13 may be employed. A logic 1 input will be required to illuminate the LED which will operate at a typical current of 15 mA. The value of R_1 should be selected according to Table 6.6:

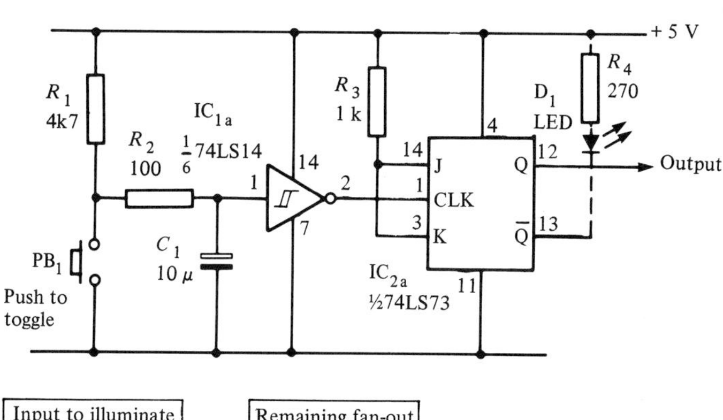

Figure 6.11 *Latching action switch*

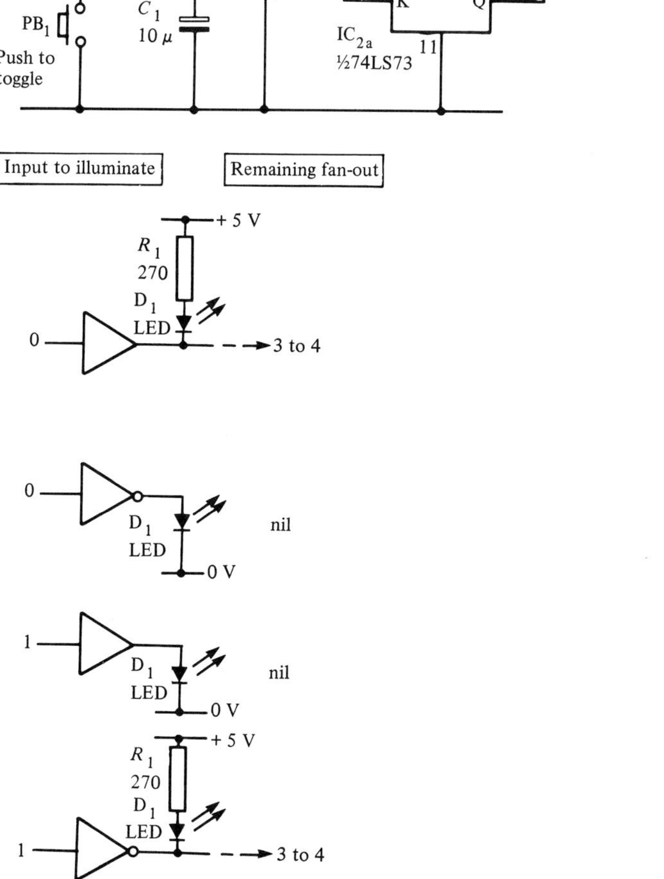

Figure 6.12 *Methods of driving an LED from a TTL gate*

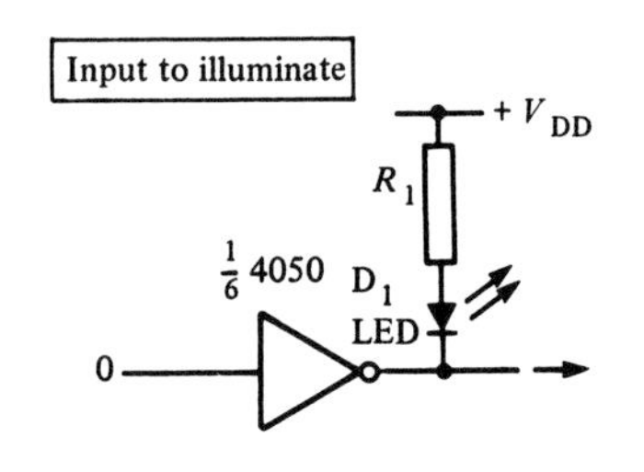

Figure 6.13 *Methods of driving an LED from a CMOS buffer*

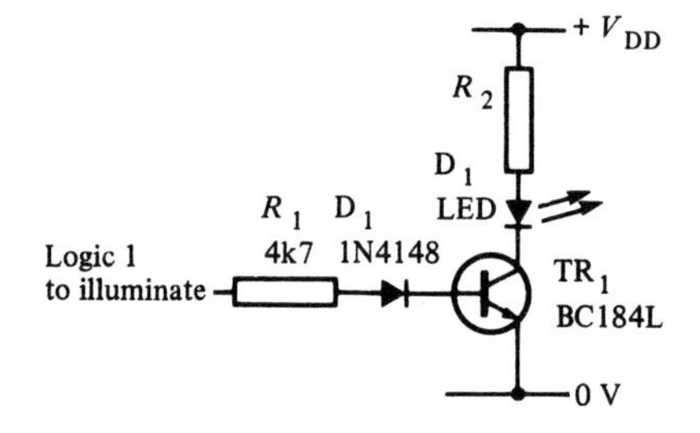

Figure 6.14 *LED driver using an auxiliary transistor*

Table 6.6

Supply voltage (V_{DD})	R_1
3 to 4	zero
4 to 5	100
5 to 8	220
8 to 12	470
10 to 15	820

Where a buffer is not available, an auxiliary transistor may be employed, as shown in Figure 6.14. The LED will again operate when the input is taken to logic 1 and the operating current will be approximately 15 mA. The value of R_2 should be chosen from Table 6.7:

Table 6.7

Supply voltage (V)	R_2
3 to 4	100
4 to 5	150
5 to 8	220
8 to 12	470
10 to 15	820

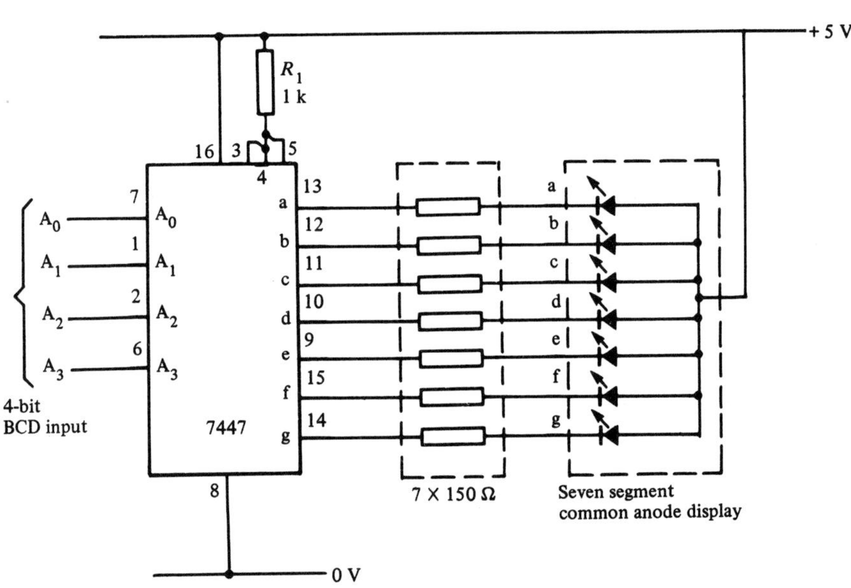

Figure 6.15 *Typical arrangement for driving a common-anode seven segment display*

BCD Input				Display
A_3	A_2	A_1	A_0	
0	0	0	0	0
0	0	0	1	1
0	0	1	0	2
0	0	1	1	3
0	1	0	0	4
0	1	0	1	5
0	1	1	0	6
0	1	1	1	7
1	0	0	0	8
1	0	0	1	9
1	0	1	0	
1	0	1	1	
1	1	0	0	
1	1	0	1	
1	1	1	0	
1	1	1	1	

Figure 6.16 *Truth table for the display produced by the arrangement in Figure 6.15*

A typical arrangement for driving a common-anode seven segment LED display is shown in Figure 6.15. Thus circuit is based on the 7447 seven segment decoder/driver. The device accepts a binary coded decimal (BCD) input and its internal logic follows the truth table shown in Figure 6.16.

Logic gate interconnection

The output of a logic gate can be connected to the input of one or more logic gates up to a maximum determined by the output capability of the gate concerned and the loading imposed by subsequent inputs to which it is connected.

The output drive capability of a gate is usually expressed in terms of its 'fan-out'. This is simply equivalent to the number of standard loads which can be driven by the gate without the logic levels becoming illegal. Where a TTL gate is to be connected to other gates of the same family, this value is almost invariably ten. Hence a TTL device can be confidently expected to drive up to ten standard inputs of the same family.

The fan-in of a logic gate is a measure of the loading effect presented by its inputs and is expressed in terms of the equivalent number of standard TTL loads that it represents. A standard TTL load (unit load) is defined as that which will produce:

(a) A high-state input current (source) of 40 μA.
(b) A low-state input current (sink) of -1.6 mA.

This relationship is illustrated in Figure 6.17.

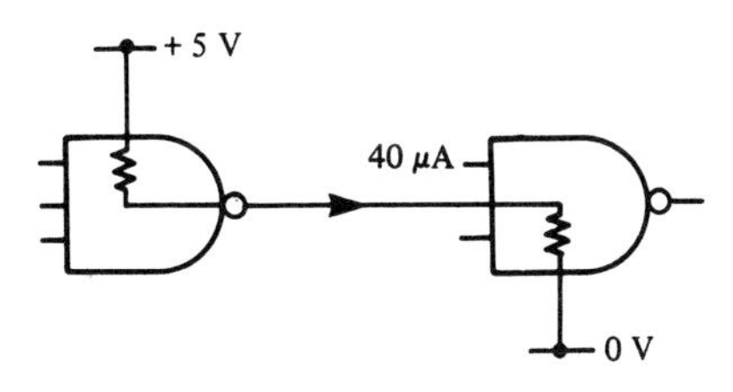

(a) High state input

Figure 6.17 *Standard TTL load*

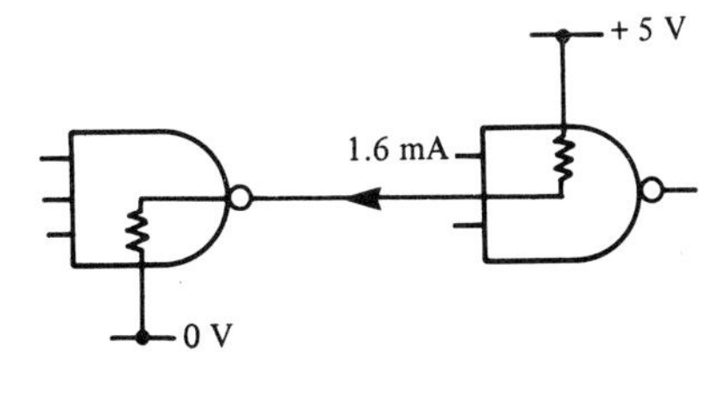

(b) Low state input

Typical values of input and output current for the various TTL subfamilies are given in Table 6.8.

Table 6.8

	Input current		*Output current*	
Family	*High state* (*μA*)	*Low state* (*mA*)	*High state* (*μA*)	*Low state* (*mA*)
7400	40	−1.6	400	−16
74H	50	−2	500	−20
74L	10	−0.18	100	−3.6
74S	50	−2	1000	−20
74LS	20	−0.4	400	−8

Each of the following TTL subfamilies has its own unique input and output characteristics optimized for a particular speed/power performance. Mixing several subfamilies together in the same logic arrangement can result in impaired performance and should be avoided if at all possible. Where mixing is unavoidable, and quite apart from any speed/power considerations, Table 6.9 should be borne in mind:

Table 6.9

	Maximum number of inputs that may be connected				
Family	*7400*	*74H*	*74L*	*74S*	*74LS*
7400	10	8	40	8	20
74H	12	10	40	10	25
74L	2	1	10	1	5
74S	12	10	40	10	50
74LS	5	4	20	4	20

It should be noted that, regardless of any loading considerations, the output of two, or more, logic gates should NEVER be directly connected together. There are just *two* exceptions to this rule:

(a) When open-collector logic gates are employed. Such devices may be connected together in a wired-NOR configuration, as shown in Figure 6.18.

(b) When tri-state logic gates are employed. Such devices have a third (high impedance) output state and their outputs may be directly linked together provided that only one output is active at any time. This generally requires that the output ENABLE signals should be non-overlapping. Most ENABLE signals are 'active-low' and this is denoted by the presence of a bar placed over the ENABLE notation or by means of a circle placed at the relevant input. In such cases, the output of a particular gate is valid when its $\overline{\text{ENABLE}}$ input is taken to logic 0. During this time, the outputs of all other gates sharing

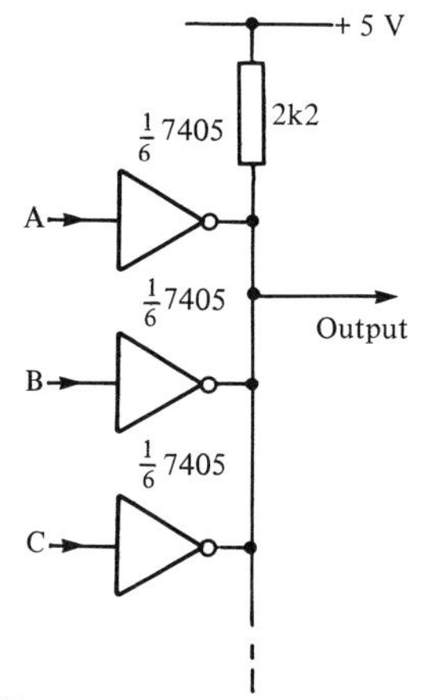

Figure 6.18 *Wired-NOR configuration based on open-collector logic gates*

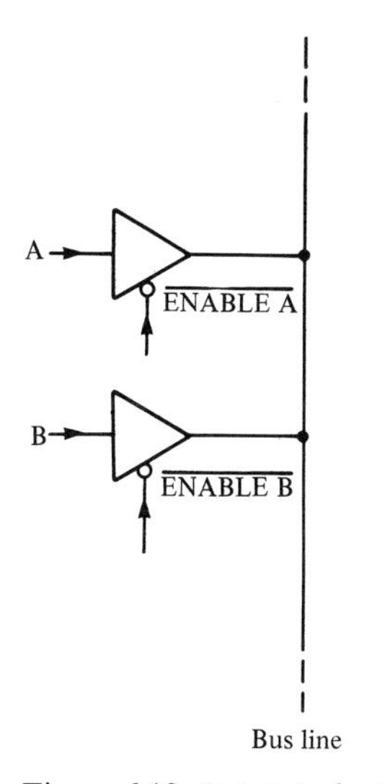

Figure 6.19 *Tri-state logic configuration*

the same output line (bus) should be driven into the high-impedance state by taking their respective $\overline{\text{ENABLE}}$ inputs to logic 1. Figure 6.19 illustrates this principle.

Where not all of the inputs of a gate are connected or where not all gates within a package are being used, one is sometimes left wondering what to do with the unused connections. A TTL input left unconnected almost invariably 'floats' to the high (logic 1) state. A CMOS input, on the other hand, may adopt either state.

Depending upon whether a logic 0 or logic 1 state is required, all unused inputs should, therefore, be tied to 0 V or (via a pull-up resistor) to the positive supply. Where a complete gate within a package is unused, it may either be left to float or, to minimize power consumption, it may be forced into the high output state. This is achieved by tying its inputs to logic 0 or logic 1 depending upon whether the gate is an inverting or non-inverting type. This arrangement may also be used as a convenient means of arriving at a 'genuine' TTL logic 1 state which can then be supplied to other gates requiring a permanent logic 1 input state. In any event, it is important to avoid the trap of connecting all of a gate's inputs and outputs together as this can result in rapid destruction of the chip.

To increase fan-out, inputs and outputs of similar gates may be paralleled. It is, however, advisable to limit the gates being paralleled to those within a single package in order to minimize transient states produced by variations in switching times between gates.

Sometimes it becomes necessary to interface the output of a TTL logic gate to one or more CMOS inputs, and vice versa. Provided a common 0 V rail is used together with similar positive supply rail voltages (it is not necessary for these to be common so long as they are both maintained fairly accurately at +5 V) this should not present much of a problem. Figure 6.20 shows various methods of achieving a satisfactory interface between TTL and CMOS devices.

Finally, it is usually desirable to make use of a limited number of logic gate types within a logic arrangement. It is, in fact, possible to realize all four of the basic logical functions (NAND, NOR, AND and OR) using arrangements based solely on NAND and NOR gates. In practice, and to minimize the number of packages present, it may be beneficial to bring into use 'spare' gates. Figure 6.21 shows a number of logic equivalents which may be useful in this process. There is, however, an important caveat which applies in high speed synchronous logic applications (such as memory address decoders). In these circumstances propagation delays within logic gate arrangements may become significant and it may be necessary to take steps to equalize the propagation delays within the network.

Monostables and bistables

The logical state of the output of a basic logic gate remains at logic 0 or logic 1 according to the logical states of their inputs. Provided the input

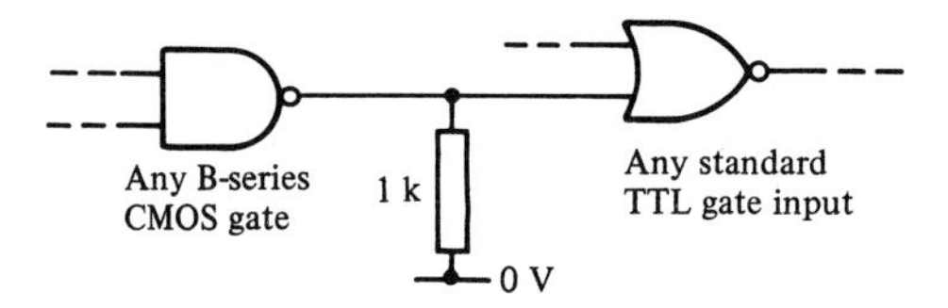

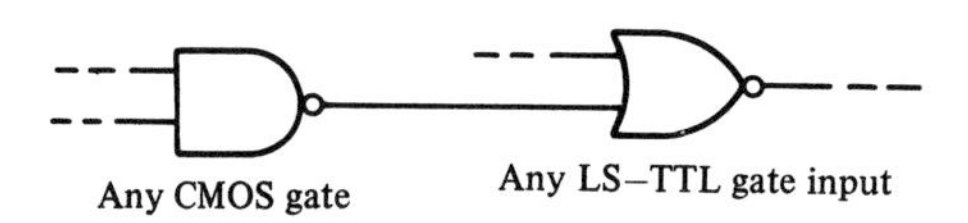

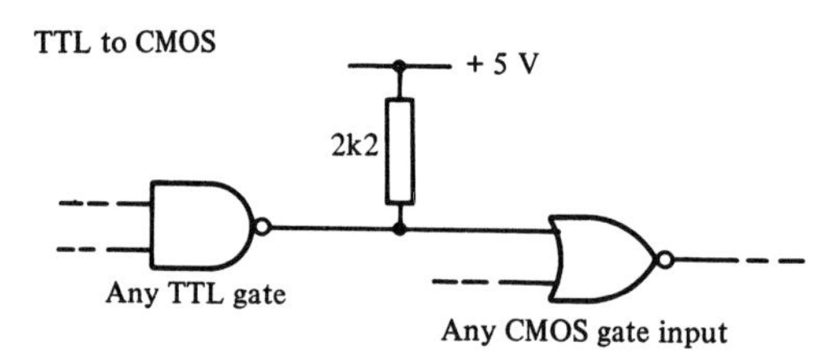

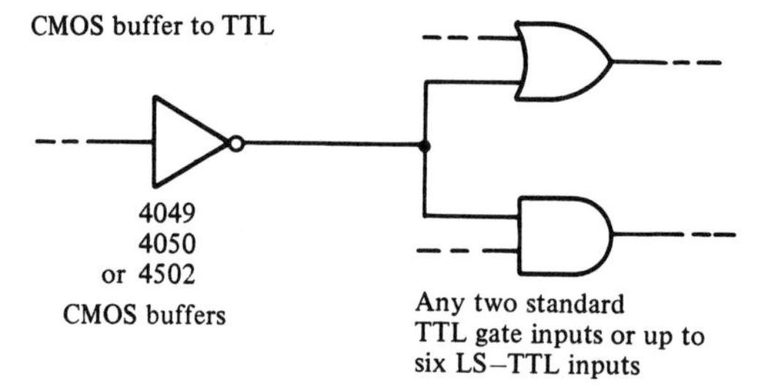

Figure 6.20 *Methods of interfacing TTL and CMOS*

states remain constant, the output state will also remain constant. There are, however, a number of applications in which a momentary pulse (i.e. a 0–1–0 or 1–0–1 transition) is required. A device which fulfils this function is said to have only one stable state and is consequently known as a monostable.

The action of a monostable is easily understood; its output is initially at logic 0 until a level or 'edge' arrives at its trigger input. This level change can be from 0 to 1 (positive edge trigger) or 1 to 0 (negative edge trigger) depending upon the particular monostable device or configuration. Immediately the trigger is received, the output of the monostable changes state to logic 1. Then, after a time interval determined by external 'timing

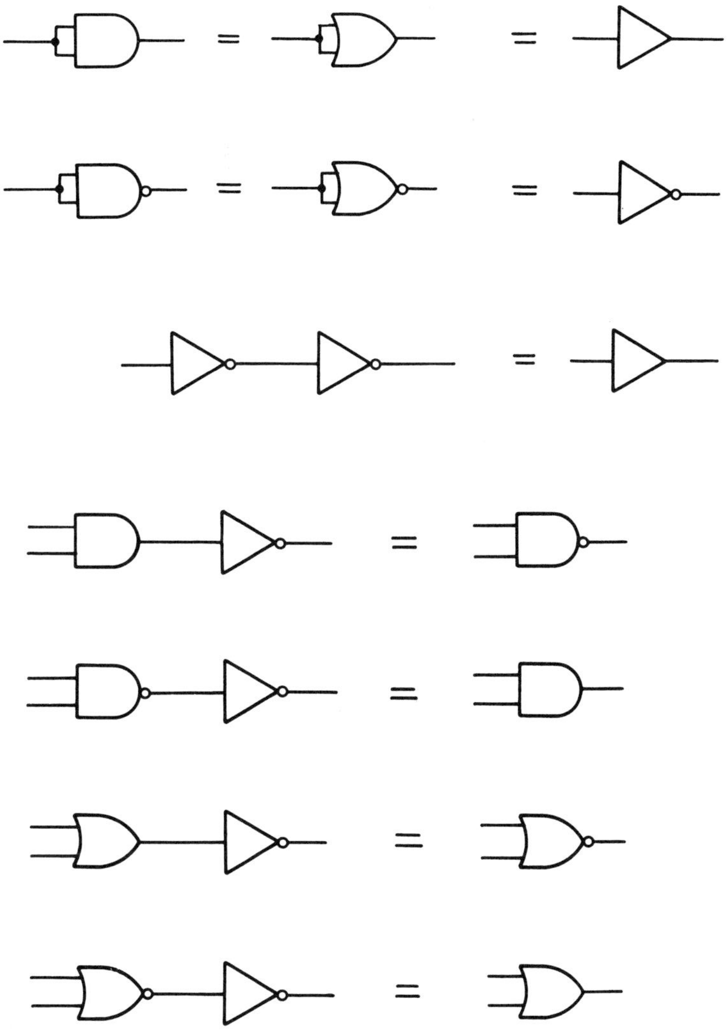

Figure 6.21 *Useful logic equivalents*

continued

components', the output reverts to logic 0. The monostable then awaits the arrival of the next trigger.

Figure 6.22 shows the arrangement of a simple negative-going (1–0–1) pulse generator which is triggered by a positive edge, as shown in the timing diagram of Figure 6.23. Since the voltage level at the input is zero before the trigger arrives, the capacitor, *C*, is initially uncharged. The inverter thus receives a logic 0 input and its output will be held high (logic 1). When the trigger edge arrives, the input voltage rises rapidly from 0 V to approximately 5 V. This voltage change is conveyed, via the capacitor, to the input of the inverting gate. The inverter recognizes a logic 1 input when its input passes through the logic 1 threshold

Figure 6.21 – *continued*

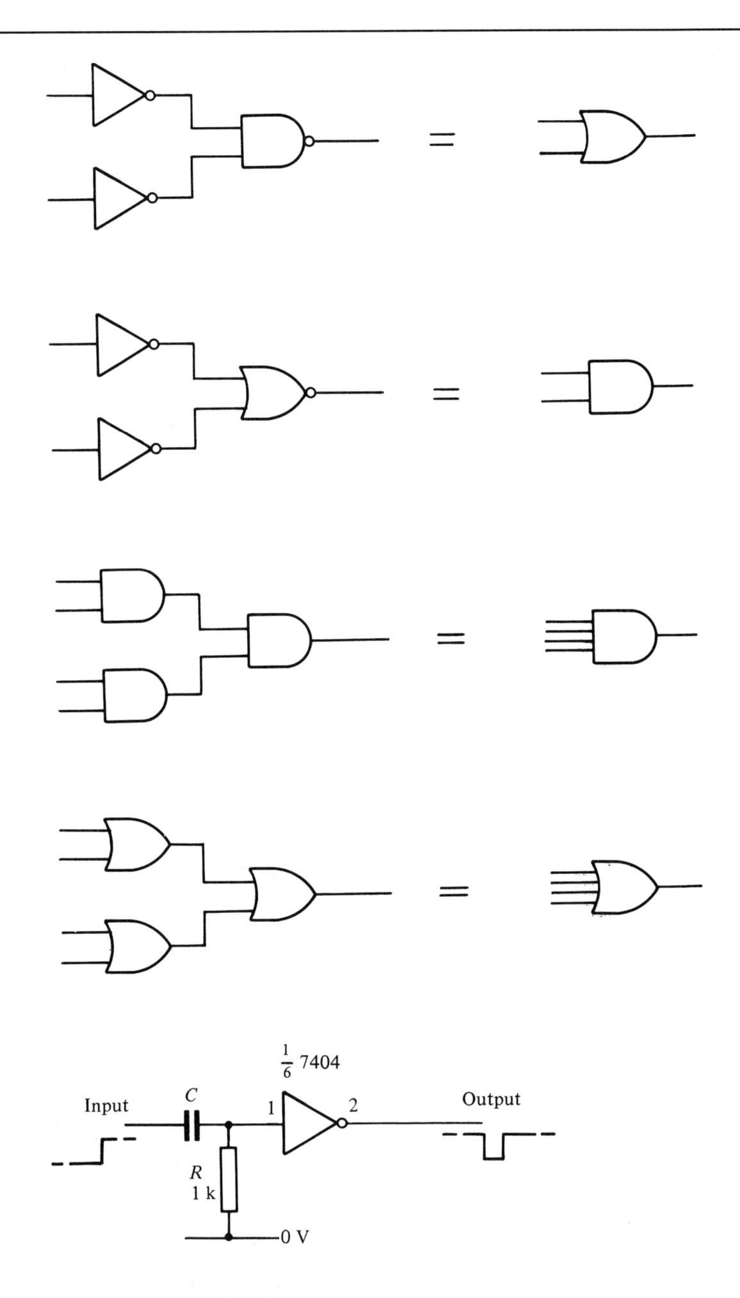

Figure 6.22 *Simple negative-going pulse generator*

(approximately 1.5 V) and its output then rapidly changes state from logic 1 to logic 0.

The capacitor then charges through the resistor, *R*, and the voltage at the input of the inverter falls exponentially back towards 0 V. When the inverter's input voltage falls below the logic 0 threshold (again at about 1.5 V) the gate once more recognizes a logic 0 input and its output state reverts to logic 1.

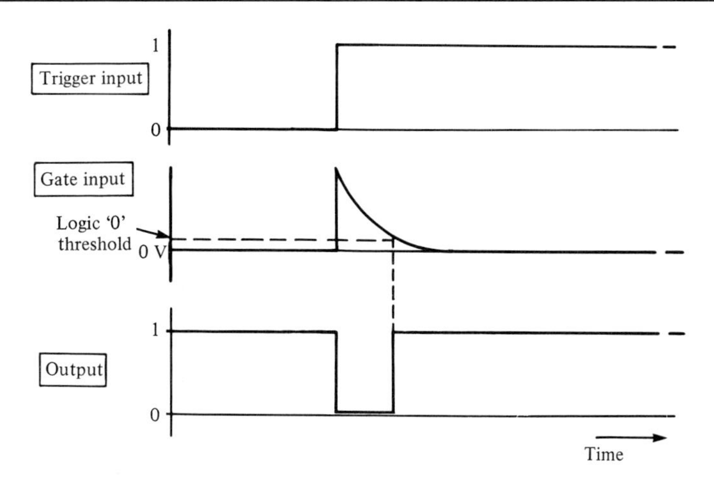

Figure 6.23 *Timing diagram for Figure 6.22*

The time taken for the capacitor to charge depends on the time constant of the circuit ($C \times R$). The duration of the output pulse can thus be fixed by a suitable choice of component values. It should, however, be noted that for conventional TTL gates an optimum value of R will be around $470\,\Omega$. Furthermore, it should not be increased much above, nor decreased much below, this value. Hence, to obtain output pulses of different duration, the capacitor rather than the resistor should be varied.

Long duration pulses may require the use of large value capacitors and these will invariably be electrolytic types. It is, therefore, essential that capacitors are low leakage types and, furthermore, close tolerance varieties are necessary if an accurately defined output pulse is required. If we need a positive-going (0–1–0) pulse rather than a negative going pulse (1–0–1) we need only add a second inverter to the output of the circuit, as shown in Figure 6.24.

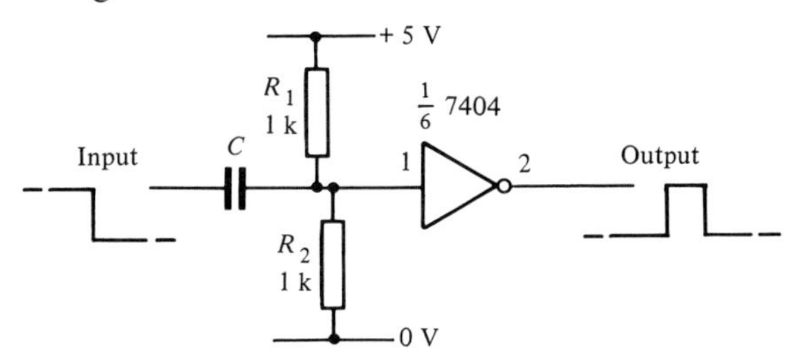

Figure 6.24 *Simple positive-going pulse generator*

Figures 6.25 and 6.26 respectively show how positive and negative output pulses may be derived from a negative edge trigger. These circuits are somewhat similar to those which operate from a positive edge trigger. In this case, however, the input of the inverter is biased into the logic 1 state by means of an additional pull-up resistor connected to the positive supply rail. Such an arrangement typically places a quiescent voltage of approximately 2.5 V at the input of the logic gate.

Whereas it is possible to make simple forms of monostable from inverters, the use of purpose-designed integrated circuit monostables is much to be preferred. The 74121 is one such device which can be triggered

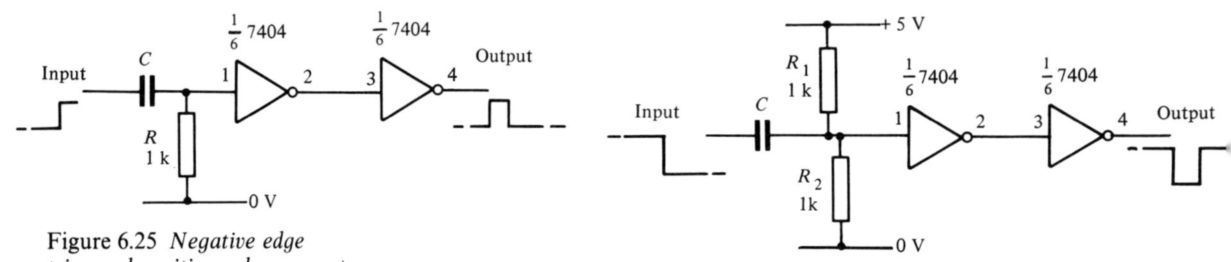

Figure 6.25 *Negative edge triggered positive pulse generator*

Figure 6.26 *Negative edge triggered negative pulse generator*

by either positive or negative edges depending upon the configuration employed. The chip has complementary outputs (labelled Q and $\bar{Q}$) and requires only two timing components (one resistor and one capacitor), as shown in Figure 6.27.

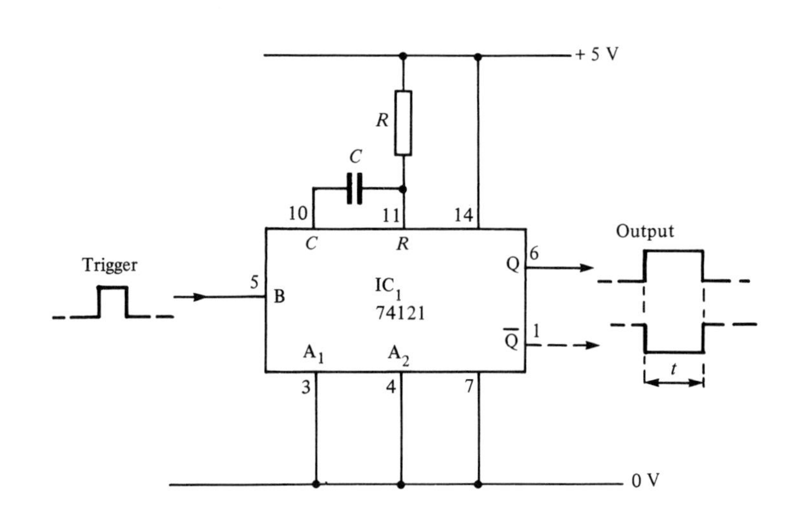

Figure 6.27 *74121 monostable configuration*

Control inputs A1, A2, and B are used to determine the 74121's trigger mode and may be connected in any one of the following three ways:

(a) A1 and A2 connected to logic 0. The monostable will then trigger on a positive edge applied to B.
(b) A1 and B connected to logic 1. The monostable will then trigger on a negative edge applied to A2.
(c) A2 and B connected to logic 1. The monostable will then trigger on a negative edge applied to A1.

It should be noted that, unlike some other monostable types, the 74121 is not re-triggerable during its monostable timing period. This simply means that, once a timing period has been started no further trigger pulse will be recognized. Furthermore, in normal use, a recovery time equal in length to the monostable pulse should be allowed before attempting to re-trigger the device.

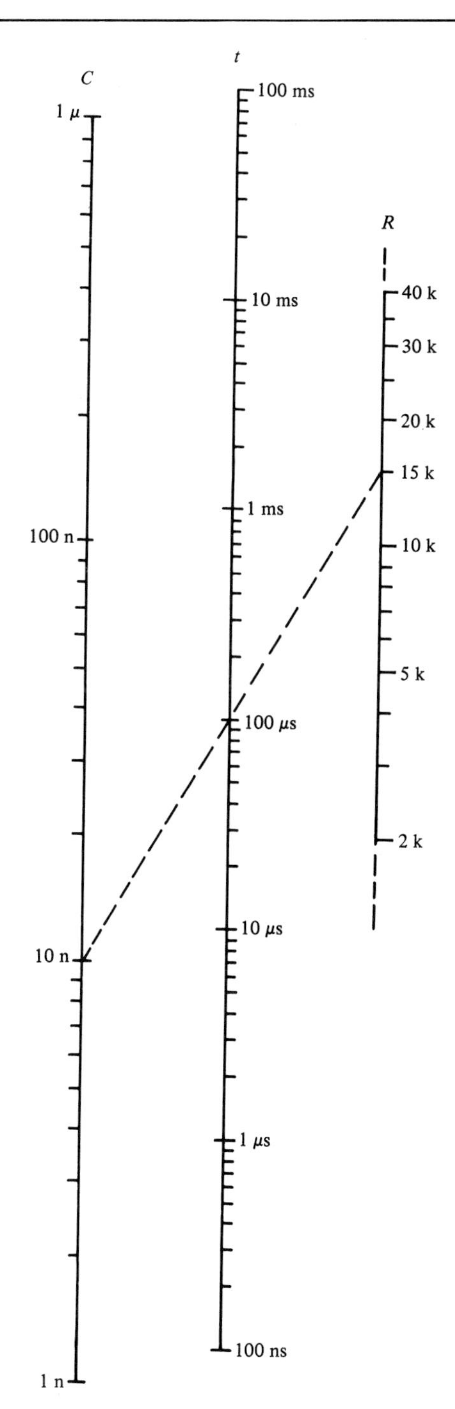

Figure 6.28 *Nomograph for the 74121 monostable pulse generator*

A typical application for a 74121 monostable device is in stretching a pulse of very short duration. Once triggered, the 74121 will continue with its fixed duration timing period long after an input pulse has reverted to its original state. The only requirement is that, to ensure reliable triggering, the input pulse should have a width of at least 50 ns. For a 74121, the values of external timing resistor should normally lie in the range 1.5 kΩ to 47 kΩ. The minimum recommended value of external capacitor is 10 pF whereas the maximum value of capacitor is only limited by the leakage current of the capacitor employed. In practice this means that, if necessary, values of several hundred μF can be used. This all leads to a monostable circuit which can provide a very much wider range of monostable periods than the simple circuits based on inverters described earlier. Typical values of 74121 monostable period for various capacitor values can be determined from the nomograph shown in Figure 6.28.

In any other than the most elementary of logic circuits, one sooner or later realizes the need for a device which can remember a logical state (ether 0 or 1) for an indefinite period of time (or at least as long as the supply remains connected). Such devices constitute a very simple form of memory and, since their output can exist in either one of two stable states (0 and 1), they are known as bistables.

Simple bistable devices can be built using nothing more than cross-coupled NAND or NOR gates, as shown in Figure 6.29(a) and (b) respectively. These arrangements have two inputs (labelled SET and RESET) and two complementary outputs (labelled Q and $\bar{Q}$). A logic 1 applied to the SET input will cause the Q output to become (or remain at) logic 1 while a logic 1 applied to the RESET input will cause the Q output to become (or remain at) logic 0. In either case, the bistable will remain in its SET or RESET state until an input is applied in such a sense as to change the state.

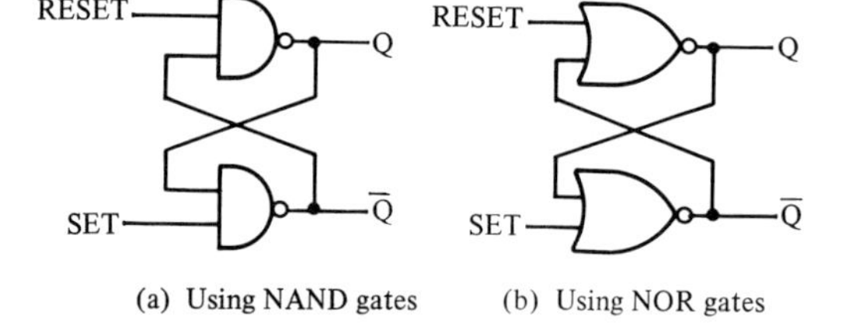

Figure 6.29 *Cross-coupled NAND and NOR gate bistables*
(a) Using NAND gates
(b) Using NOR gates

Unfortunately, simple NAND and NOR gate bistable arrangements suffer from a problem; it is not possible to predict the output state which results from the simultaneous application of a logic 1 to both the SET and RESET inputs and steps must be taken to ensure that this disallowed state does not arise.

In practice, NAND and NOR gate bistables are rarely encountered since a variety of integrated circuit bistables are available which are both more flexible and predictable in their operation. The commonly used symbols for three common bistables types (RS, D-type, and JK) are shown in Figure 6.30.

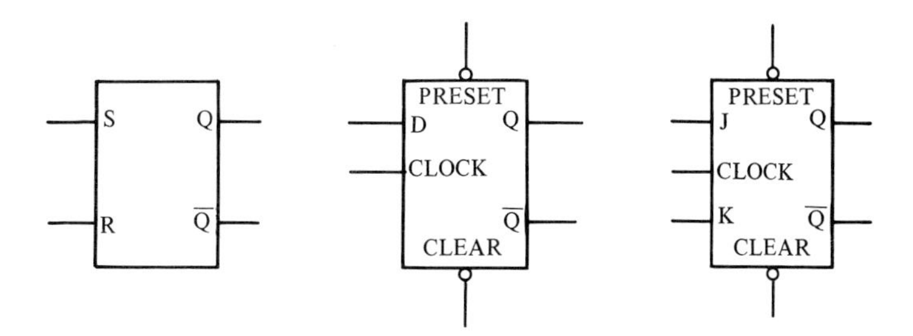

Figure 6.30 *Symbols used for RS, D-type, and JK bistables*

The D-type bistable has two principal inputs; D (standing variously for data or delay) and CLOCK. The data input (logic 0 or logic 1) is clocked into the bistable such that the output state only changes when the clock changes state. Operation is thus said to be synchronous. Additional subsidiary inputs (which are invariably active-low) are provided which can be used to directly set or reset the bistable. These are usually called PRESET (PR) and CLEAR (CLR).

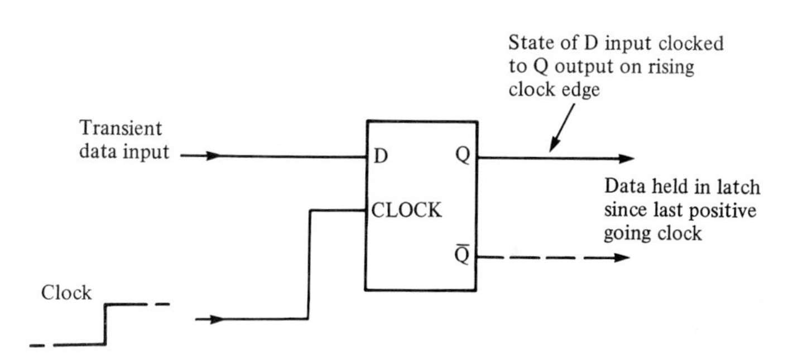

Figure 6.31 *One-bit data latch using a D-type bistable*

Figure 6.31 shows a typical application of a D-type bistable as a simple one-bit 'data latch'. The operation of the circuit is best explained by considering the timing diagram shown in Figure 6.32. Here, the state of the D input is transferred to the Q output on each rising clock edge. The Q output remains unaffected by a falling clock edge. It should be noted that, whereas most common D-type bistables (e.g. 7474, 74174, 74175) are all clocked on the rising edge of the clock waveform, this rule does not generally apply to JK bistables which invariably complete their clocking on a falling clock edge.

JK bistables, such as the 7473, have two clocked inputs (J and K), two direct inputs ($\overline{\mathrm{PRESET}}$ and $\overline{\mathrm{CLEAR}}$), a clock input, and two outputs (Q and $\overline{\mathrm{Q}}$). As with the RS bistable, the two outputs are complementary (i.e. when one is 0 the other is 1, and vice versa). Similarly, the $\overline{\mathrm{PRESET}}$ and $\overline{\mathrm{CLEAR}}$ inputs are invariably both active low (i.e. a 0 on the $\overline{\mathrm{PRESET}}$ input will set the Q output to 1 whereas a 0 on the $\overline{\mathrm{CLEAR}}$ input will set the Q output to 0). The truth tables for a JK bistables are shown in Table 6.10.

Table 6.10

J	*K*	Q_{n+1}	*Comment*
0	0	Q_n	No change
0	1	0	Output cleared
1	0	1	Output set
1	1	$\bar{Q}_n$	Output changes state

Present	*Clear*	*Q*
0	0	Indeterminate
0	1	1
1	0	0
1	1	Enables clocked operation

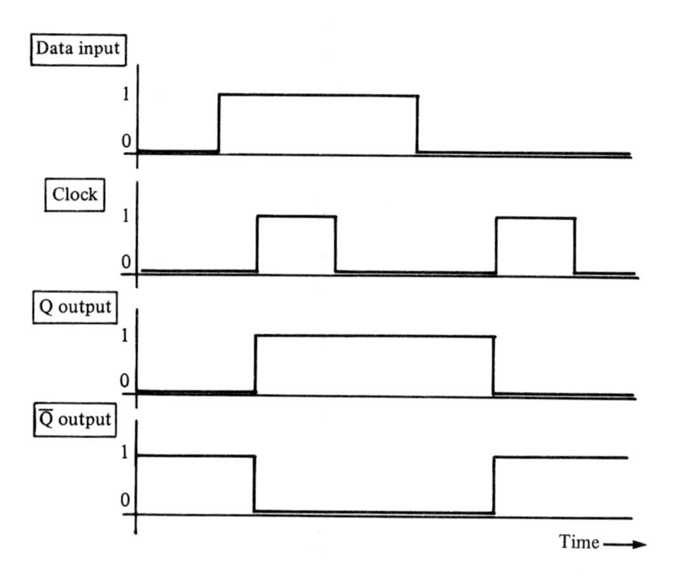

Figure 6.32 *Timing diagram for the one-bit data latch shown in Figure 6.31*

It is important to note that the $\overline{\text{CLEAR}}$ input of a 7473 bistable should never be left to float into the logic 1 state. The $\overline{\text{CLEAR}}$ input should always be pulled-up to logic 1 (using a shared pull-up resistor if necessary). Failure to observe this precaution may result in spurious output states (glitches).

Counters and shift registers

Figure 6.33 shows a typical four-stage binary counter/divider using 7473 JK bistables. Each JK bistable divides by two and thus the frequency of the final output is one sixteenth of the input as shown in the timing diagram of Figure 6.34.

Where a decade count (÷ 10) is required, it is usually more convenient to employ a dedicated decade counter (e.g. 7490) rather than modify a basic counter so that it becomes cleared after the tenth state is reached. Figure 6.35 shows a 7490 configured for decade counting (the 7490 is

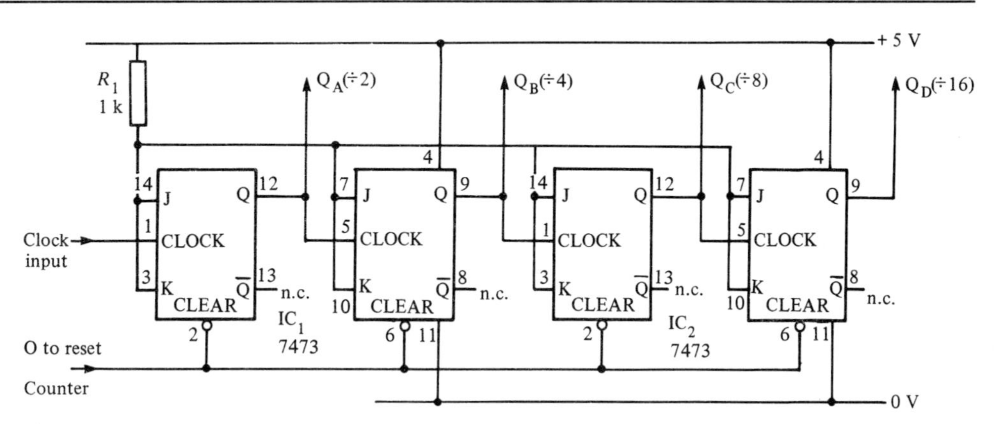

Figure 6.33 *Typical four-stage binary counter using JK bistables*

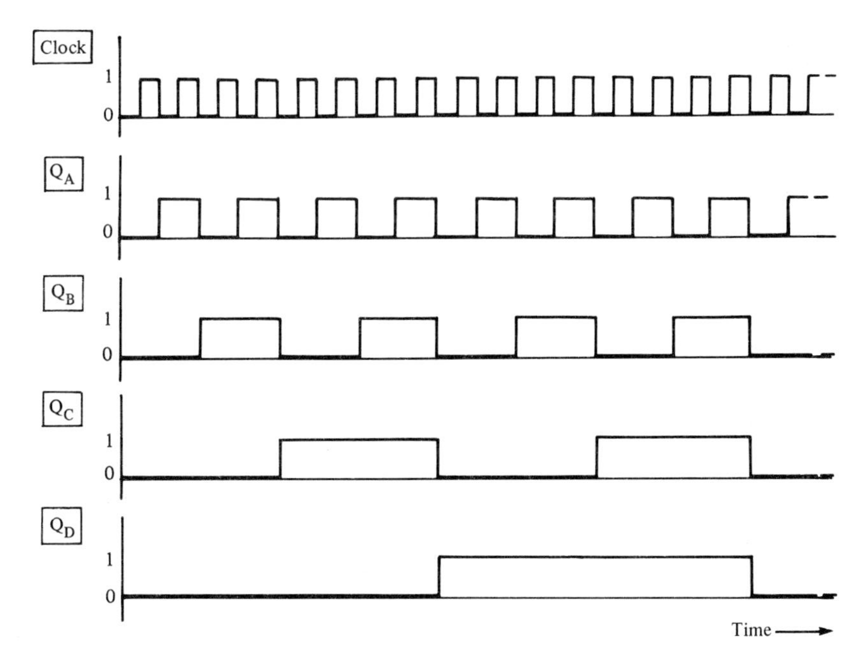

Figure 6.34 *Timing diagram for the binary counter shown in Figure 6.33*

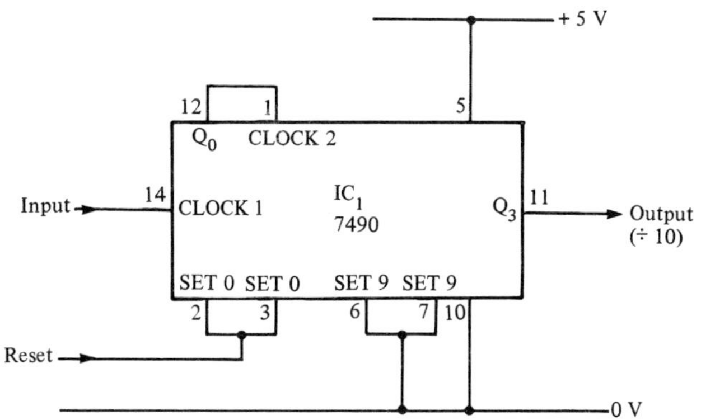

Figure 6.35 *7490 decade counter*

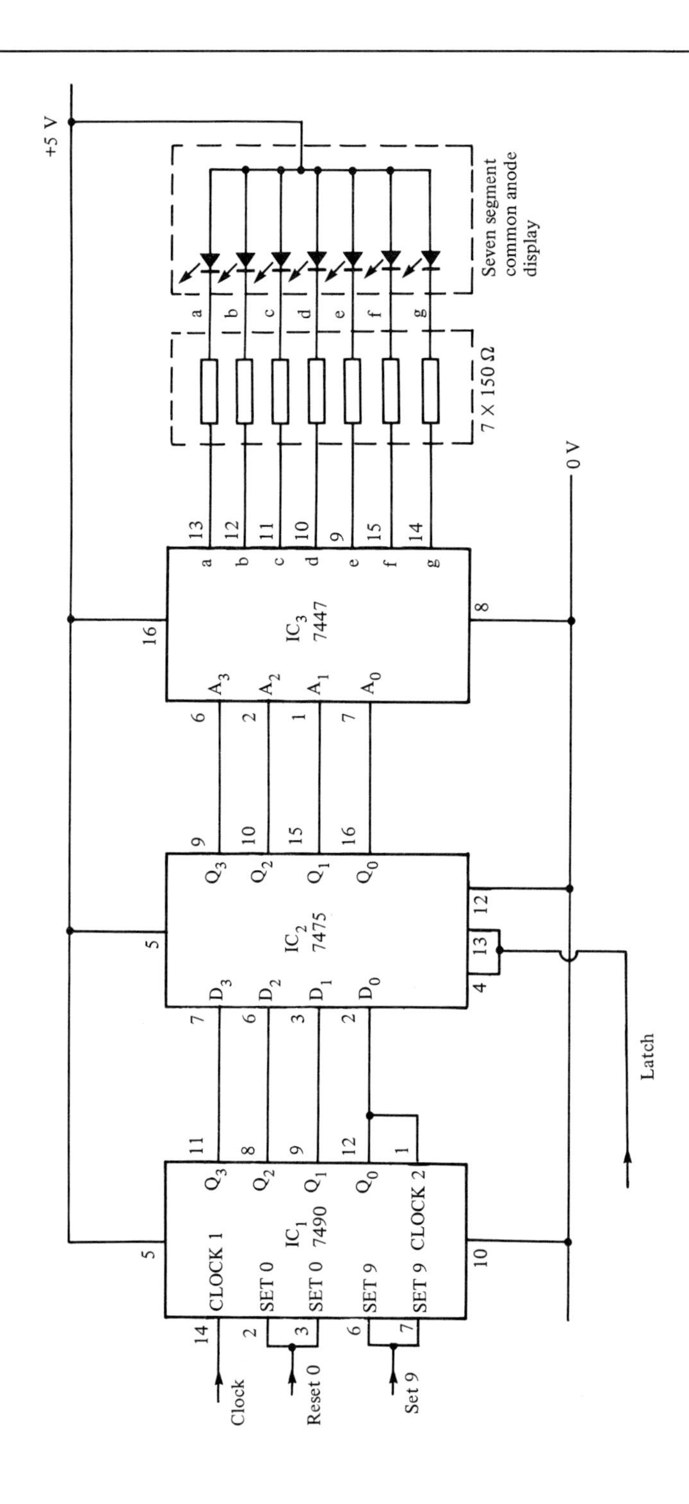

Figure 6.36 *Decade counter/latch/display*

arranged internally in two sections: ÷2 and ÷5). A complete decade counter/display stage is shown in Figure 6.36. This arrangement includes a quad D-type latch together with a seven segment decoder/driver. While this arrangement is readily cascadable, beyond two stages it is usually more cost effective to make use of an LSI decade counter chip (such as the 7216, 7217, 7224, or ZN1040E).

Where it is desirable to be able the change the direction of counting (i.e. to be able to count up or down) an arrangement of the type shown in Figure 6.37 may be employed. This circuit is based on the 74193 programmable counter. This device incorporates four bistable stages together with some sophisticated internal gating. Synchronous operation is provided by having all of the bistable stages clocked simultaneously so that output changes are mutually coincident. This mode of operation eliminates the spurious states (glitches) which are normally associated with asynchronous ripple counters. Additional external gating has been added so that the count can proceed in either direction as determined by a single UP/$\overline{\text{DOWN}}$ direction control input. The counter is also provided with a CLEAR input which restores the count to zero.

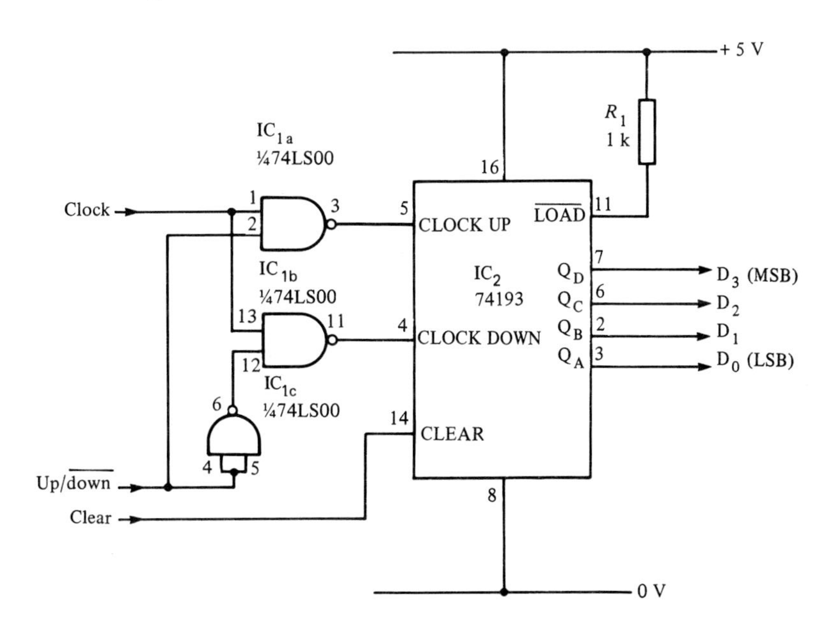

Figure 6.37 *Programmable up/down counter using a 74193*

Figure 6.38 shows a four stage shift register using 7473 JK bistables. Data is shifted from stage to stage on each falling edge of the clock. It would thus take four complete clock cycles for a logic 1 present at the input of the first stage to be transferred to the Q output of the final stage. Figure 6.39 shows a typical timing diagram for the four-stage shift register. This assumes that the register is initially in a cleared state and that the logical state of the input remains unchanged throughout the four clock cycle shifting period.

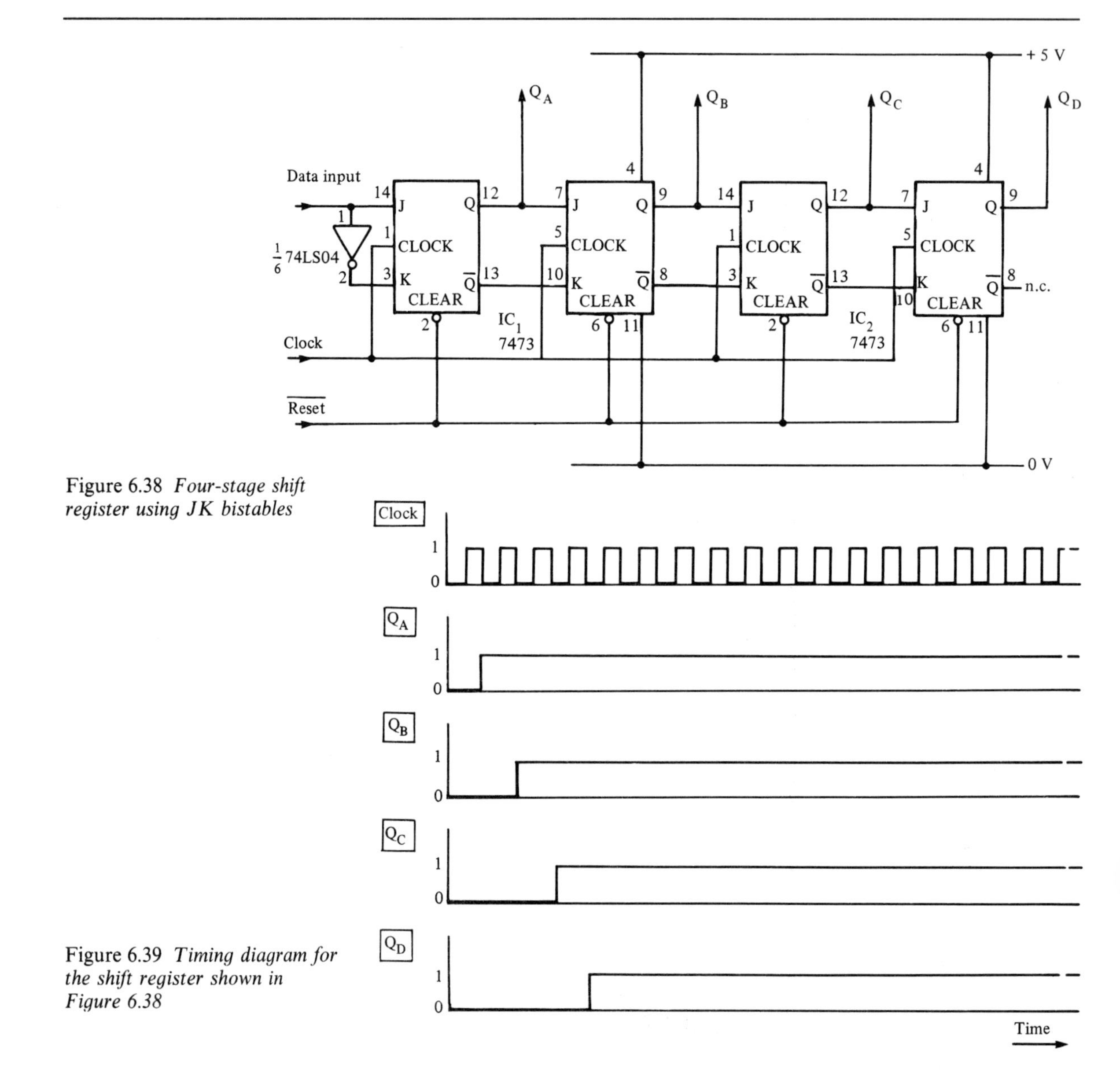

Figure 6.38 *Four-stage shift register using JK bistables*

Figure 6.39 *Timing diagram for the shift register shown in Figure 6.38*

Logic gate oscillators

Logic circuits often require the services of a locally generated clock signal. Depending upon the characteristics required, various circuit configurations are possible. The following checklist indicates the features of a clock signal which may or may not be significant in any particular application.

(a) Must the output maintain a specific frequency? If so, what frequency tolerance is required?

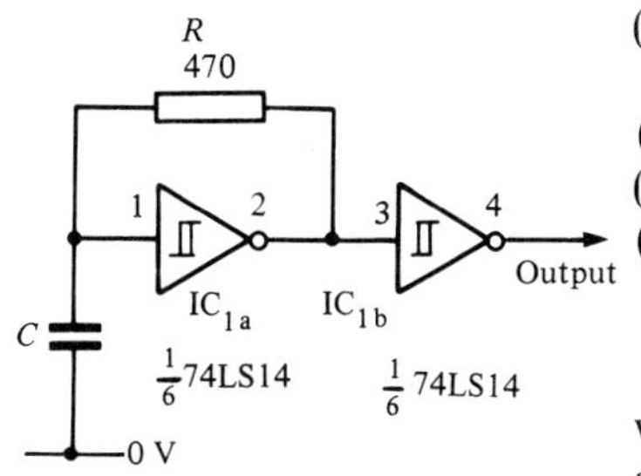

Figure 6.40 *Simple astable oscillator based on a Schmitt inverter*

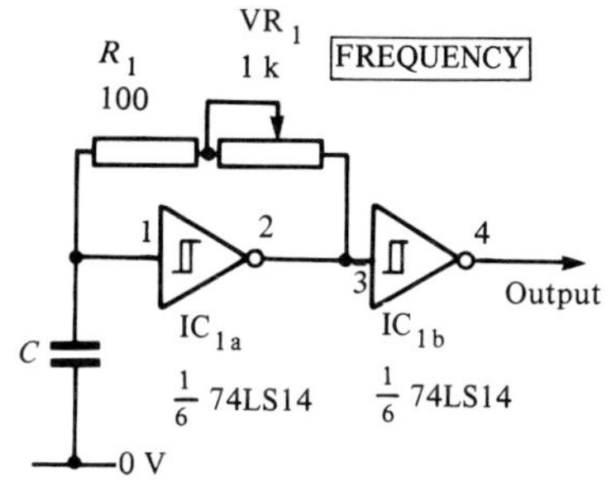

Figure 6.41 *Variable frequency Schmitt oscillator*

(b) Is the frequency stability important? If so, what limits are considered necessary?
(c) Should the output frequency be variable or adjustable?
(d) Is the output duty cycle important?
(e) What loading is imposed on the clock signal (in terms of standard input loads)?

The answers to the above questions will be instrumental in determining which type of clock circuit should be employed. For example, where the answer to (a) is 'yes', it will be necessary to use a quartz crystal as the frequency determining element.

Figure 6.40 shows a simple but very effective astable oscillator based on a Schmitt input inverter (e.g. 74LS14). The output is buffered in order to minimize the effects of loading on the output frequency and to improve the waveform of the output.

The output exhibits a mark to space ratio of approximately 2:1 and the circuit is suitable for use over a very wide range of frequencies (extending from a few Hz to around 10 MHz). The output frequency is, however, somewhat dependent upon supply voltage (increasing at a rate of about +5%/V). Some representative values for *C* and output frequency are given in Table 6.11:

Table 6.11

C	*Output frequency*
100 μ	19 Hz
10 μ	190 Hz
1 μ	1.9 kHz
100 n	19 kHz
10 n	185 kHz
1 n	1.8 MHz
100 p	9 MHz

The circuit of Figure 6.39 may be made variable over an approximate 10:1 adjustment range as shown in the arrangement of Figure 6.41. If a perfect square wave output is required, a 7473 JK bistable stage may be used in conjunction with the astable oscillator, as shown in Figure 6.42. The bistable input frequency should, of course, be twice the desired output frequency. If desired, the CLEAR input of the bistable may be used to enable or disable the clock and/or the $\bar{Q}$ output used to generate an antiphase clock signal.

Figure 6.43 shows the circuit of a low consumption CMOS clock oscillator which provides a good 50% duty cycle output. The circuit is based on a hex inverting buffer. If the duty cycle of the clock is unimportant, R_1 may be omitted. When operating from a nominal +5 V supply rail, the 4049 will reliably drive a load impedance of as low as 1 k (equivalent to two standard TTL inputs or up to six LS-TTL inputs). The circuit

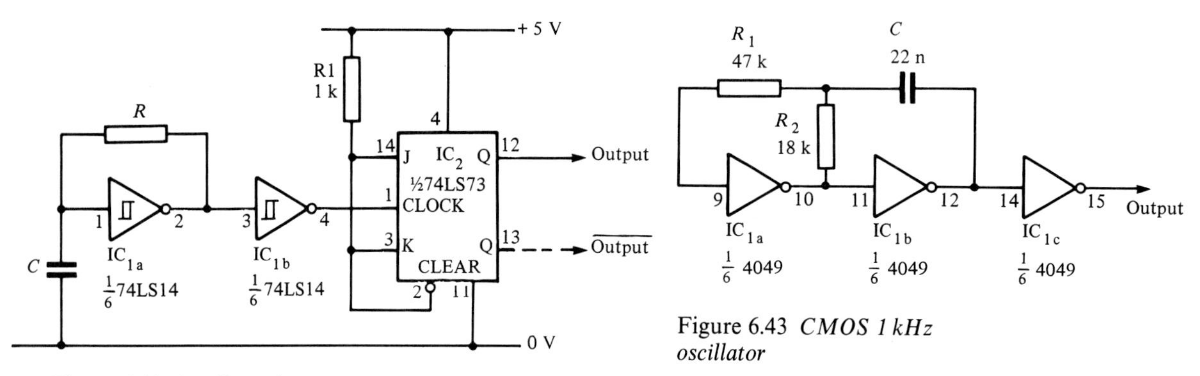

Figure 6.42 *Oscillator having precise 50% output duty cycle*

Figure 6.43 *CMOS 1 kHz oscillator*

operates reliably over a supply voltage range extending from 3 V to 15 V and the output frequency is not greatly affected by supply voltage changes (typically increasing at the rate of about +1%/V above +5 V).

When operating from a nominal 5 V supply, the values shown in Figure 6.43 provide an output frequency of 1 kHz. This may be modified by appropriate choice of values for C and R_2. Table 6.12 shows some representative values ($R_1 = 47$ k):

Table 6.12

C	R_2	*Output frequency*
10 μ	18 k	2.5 Hz
1 μ	18 k	25 Hz
100 n	18 k	250 Hz
10 n	18 k	2.5 kHz
1 n	18 k	25 kHz

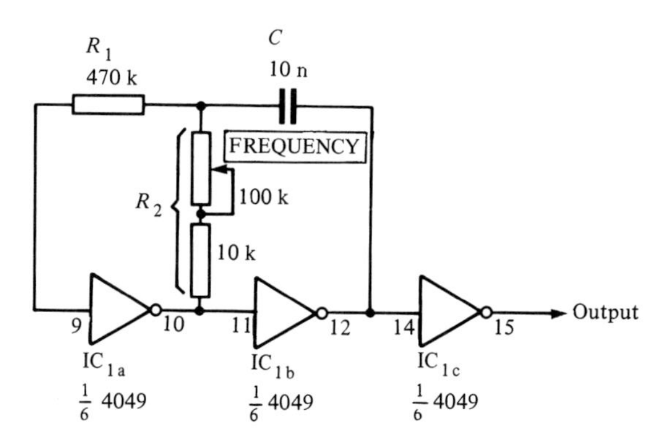

Figure 6.44 *Variable frequency CMOS oscillator*

Above 500 kHz, the output waveform is impaired (showing significant rise and fall times) and the upper limit for the circuit is approximately 5 MHz. An approximate 10:1 adjustment in the output frequency can be

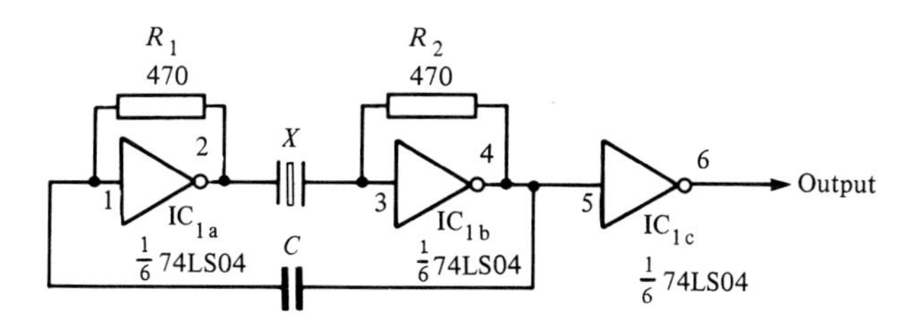

Figure 6.45 *TTL crystal oscillator*

obtained by modifying the circuit to conform with the arrangement shown in Figure 6.44. The values quoted in this circuit provide adjustment over the range 375 Hz to 3.8 kHz.

Where precise frequency control is essential as, for example, in a microprocessor system, a quartz crystal must be used as the frequency determining element. A crystal oscillator arrangement based on TTL hex inverters is shown in Figure 6.45. This circuit is suitable for operation from several hundred kHz to above 10 MHz. It should not, however, use Schmitt gates as these tend to oscillate on stray capacitance rather than locking at the crystal frequency. Above 10 MHz, 74 F or 74 H series devices should be used in preference to standard or LS types. The value of C should be determined from Table 6.13:

Table 6.13

Crystal frequency range	*C*
Below 500 kHz	100 n
500 kHz to 2 MHz	10 n
2 MHz to 8 MHz	1 n
Above 8 MHz	330 p

Finally, Figure 6.46 shows a CMOS crystal oscillator based on a 4049 hex inverter. This arrangement will operate reliably over the frequency range from 1 MHz to about 14 MHz with supply voltages from 3 V to 15 V. If precise frequency adjustment is not essential, the adjustable capacitor may be replaced by a fixed capacitor of between 68 pF and 22 pF depending upon the frequency of operation (larger values corresponding to lower frequencies).

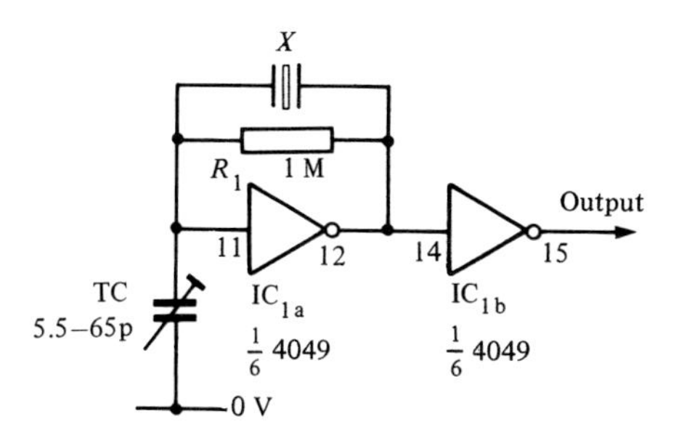

Figure 6.46 *CMOS crystal oscillator*

7 Timers

Digital circuits often require a source of accurately defined pulses. The requirement is generally for a single pulse of given duration (i.e. a 'one-shot') or for a continuous train of pulses of given frequency and duty cycle. Rather than attempt to produce an arrangement of standard logic gates to meet these requirements, it is usually simpler and more cost-effective to make use of one of the range of versatile integrated circuits known collectively as 'timers'. These devices can usually be configured for either monostable or astable operation and require only a few external components in order to determine their operational parameters.

In the case of monostable operation, the one-shot pulses produced are similar to those generated by the monostable pulse generators described in Chapter 6 but with the advantage of improved accuracy and stability when long monostable periods are required. Where astable operation is concerned (the word 'astable' simply refers to the fact that the output continuously alternates between the 'low' and 'high' states) the circuit can be considered to be a form of free-running oscillator.

Pulse characteristics

The following terms are commonly used to describe the output pulses produced by astable and monostable timer circuits:

Pulse repetition frequency (p.r.f.)

The pulse repetition frequency (p.r.f.) of a pulse waveform is simply the number of pulses which occur in a given interval of time (invariably one second). A waveform with a p.r.f. of 1 kHz will thus comprise 1000 pulses every second.

Pulse period

The period of a pulse waveform is the time taken for one complete cycle of the pulse. It is thus equal to the reciprocal of the pulse repetition frequency, i.e.:

$$\text{Pulse period, } t = \frac{1}{\text{p.r.f.}}$$

The pulse period of the example quoted earlier will thus be 1/1000 s or 1 ms.

Duty cycle

The duty cycle of a pulse waveform is the ratio of 'on' (or 'high') time to 'on' (or 'high') plus 'off' (or 'low') times. Duty cycle is often expressed as a percentage, i.e.:

$$\text{Duty cycle} = \frac{t_{on}}{t_{on} + t_{off}} \times 100\%$$

A waveform which is 'high' for 1 ms and 'low' for 1 ms thus has a duty cycle of 50% (i.e. the pulse is present for half of the period).

Mark to space ratio

The mark to space ratio of a pulse waveform is the ratio of 'on' (or 'high') time to 'off' (or 'low') time. Thus:

$$\text{Mark to space ratio} = \frac{t_{on}}{t_{off}}$$

Pulse width

The pulse width of a rectangular waveform is the time interval (measured at the 50% amplitude points) for which the pulse is 'on' or 'high'.

Rise time

The rise time of a pulse is the time interval between the 10% and 90% amplitude points of the pulse. The rise time of an 'ideal' pulse would, of course, be zero.

Fall time

The fall time of a pulse is the time interval between the 90% and 10% amplitude points of the pulse. The fall time of an 'ideal' pulse would also be zero.

Figure 7.1 illustrates a typical pulse waveform on which the various parameters discussed have been marked.

The 555 timer

The 555 timer is without doubt one of the most versatile chips ever produced. Not only is it a neat mixture of analogue and circuitry but its applications are virtually limitless in the world of digital pulse generation. In order to understand how such timer circuits operate, it is worth spending a few moments studying the internal circuitry of the 555.

The simplified internal arrangement of the 555 timer is shown in Figure 7.2. Essentially, the device comprises two operational amplifiers (used as

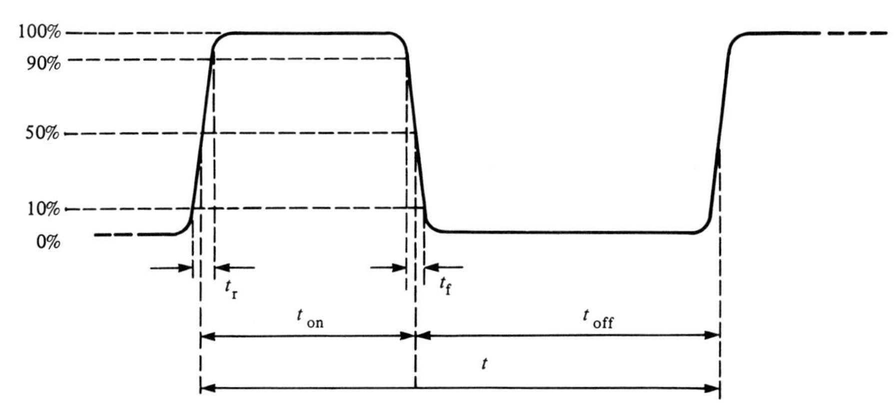

Figure 7.1 *Pulse parameters*

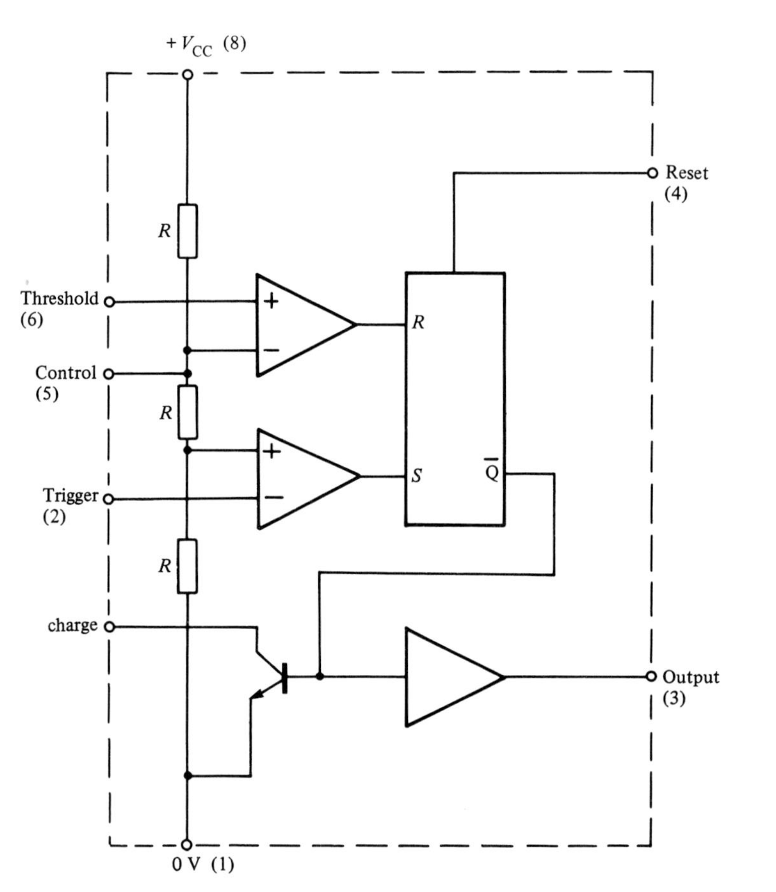

Figure 7.2 *Simplified internal equivalent of a 555 timer*

comparators) together with an RS bistable element. In addition, an inverting output buffer is incorporated so that a considerable current can be sourced or sunk to/from a load. A single transistor switch, TR_1, is also provided as a means of rapidly discharging the external timing capacitor.

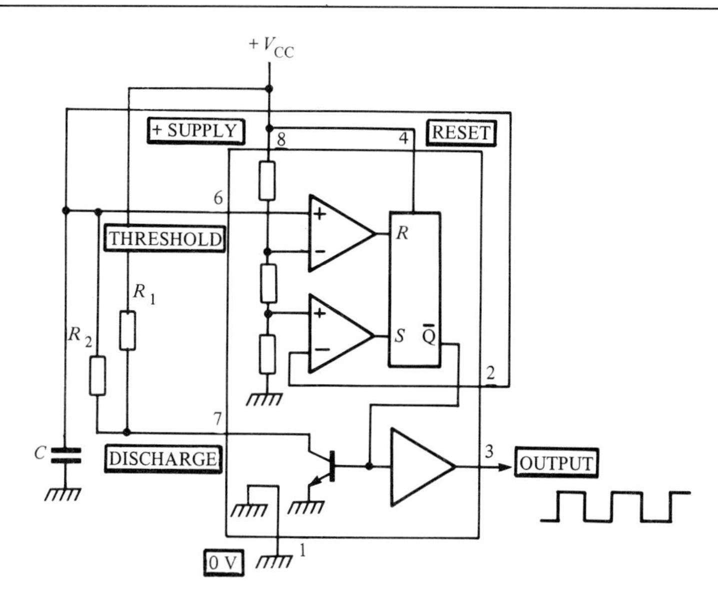

Figure 7.3 *555 astable configuration*

Figure 7.3 shows how the standard 555 can be used as an astable pulse generator. In order to understand how this circuit operates, assume that the 'output' (at pin-3) is initially 'high' and that TR_1 is in the non-conducting state. The capacitor, C, will begin to charge with current supplied by means of the series resistors, R_1 and R_2.

When the voltage at the 'threshold' input (pin-6) exceeds two thirds of the supply voltage, the output of the upper comparator will change state and the bistable will be reset, making the $\bar{Q}$ output go 'high' and turning TR_1 'on' in the process. Due to the inverting action of the buffer, the final 'output' (pin-3) will then go 'low'.

The capacitor, C, will now discharge, with current flowing through R_2 into the collector of TR_1. At a certain point, the voltage appearing at the 'trigger' input (pin-2) will have fallen back to one third of the supply voltage at which point the lower comparator will change state and return the bistable to its original set condition. The $\bar{Q}$ output of the bistable then goes low, TR_1 switches 'off', and the final 'output' (pin-3) goes high. Thereafter the entire cycle is repeated indefinitely.

The output waveform produced by the circuit of Figure 7.3 will be similar in form to that previously depicted in Figure 7.1. The essential characteristics of this waveform are:

Time for which output is 'high': $t_{on} = 0.693(R_1 + R_2)\,C$

Time for which output is 'low': $t_{off} = 0.693\,R_2C$

Period of output: $t = t_{on} + t_{off} = 0.693(R_1 + 2R_2)C$

p.r.f. of output: $$\text{p.r.f.} = \frac{1.44}{(R_1 + 2R_2)C}$$

Mark to space ratio of output: $$\frac{t_{on}}{t_{off}} = \frac{R_1 + R_2}{R_2}$$

Duty cycle of output: $$\frac{t_{on}}{t_{on}+t_{off}} = \frac{R_1+R_2}{R_1+2R_2} \times 100\%$$

Where t is in seconds, C is in farads, and R_1 and R_2 are in ohms.

It should be noted that the mark to space ratio produced by a 555 timer can never be less than unity (i.e. 1:1). However, by making R_2 very much larger than R_1 the timer can be made to produce a reasonably symmetrical square wave.

Figure 7.4 shows a standard 555 timer operating as a monostable pulse generator. The monostable timing period is initiated by a falling edge (i.e. a 'high' to 'low' transition) applied to the 'trigger input'. When such an edge is received and the 'trigger' input voltage falls below one third of the supply voltage, the output of the lower comparator goes 'high' and the bistable is placed in the 'set' state. The $\bar{Q}$ output of the bistable then goes low, TR_1 is placed in the 'off' (non-conducting) state and the final 'output' (pin-3) goes high.

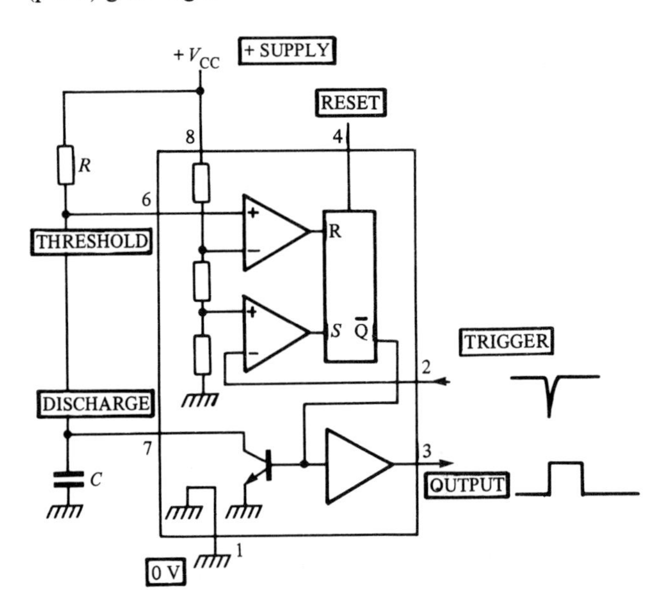

Figure 7.4 *555 monostable configuration*

The capacitor, C, then charges through the series resistor, R, until the voltage at the 'threshold' reaches two thirds of the supply voltages. At this point the output of the upper comparator changes state and the bistable is reset. The $\bar{Q}$ output then goes high, TR_1 is driven into conduction and the final 'output' (pin-3) goes low. The device then remains in the inactive state until another trigger pulse is received.

The following data refers to monostable operation:

Period for which the output is 'high': $t_{on} = 1.1\,R\,C$

Recommended 'trigger' pulse width: $t_{tr} < \frac{t_{on}}{4}$

Where t is in seconds, C is in farads, and R is in ohms.

The 555 timer family

The standard 555 timer is housed in an 8-pin DIL package and operates from supply rail voltages of between 4.5 V and 15 V. This, of course, encompasses the normal range for TTL devices and thus the device is ideally suited for use in conjunction with TTL circuitry.

The following variants of the standard 555 timer are commonly available:

Low power (CMOS) 555 (e.g. ICM7555IPA)

This device is a CMOS version of the 555 timer which is both pin and function compatible with its standard counterpart. By virtue of its CMOS technology, the device operates over a somewhat wider range of supply voltages (2 V to 18 V) and consumes minimal operating current (120 μA typical for an 18 V supply). Note that, by virtue of the low-power CMOS technology employed, the device does not have the same output current drive possessed by its standard counterpart. It can, however, supply up to two standard TTL loads.

Dual 555 timer (e.g. NE556A)

This is a dual version of the standard 555 device housed in a 14-pin DIL package. The two devices may be used entirely independently and share the same electrical characteristics as the standard 555.

Low-power (CMOS) dual 555 (e.g. ICM7556IPA)

This is a dual version of the lower-power CMOS 555 device contained in a 14-pin DIL package. The two devices may again be used entirely independently and also possess the same electrical characteristics as the low-power CMOS 555.

Monostable 555 pulse generators

Figure 7.5 shows a simple monostable timer which provides a monostable period of approximately 60 seconds. When the 'start' button (S_1) is depressed, the output Pin-3 of a standard 555) goes 'high' and remains 'high' for the duration of the timing period. The timing period can be aborted at any time by means of the 'stop' button, S_2.

The circuit can be readily adapted to drive a relay (provided the operating current is less than about 150 mA) as shown in Figure 7.6. The shunt diode, D_1, provides protection against the back e.m.f. generated by the inductance of the relay coil. D_1 can, of course, be omitted whenever the load is purely resistive.

The period of the output can be changed by simply the altering the values of the timing resistor, R, and/or timing capacitor, C. Doubling the value of R will, for example, double the timing period. Note, however,

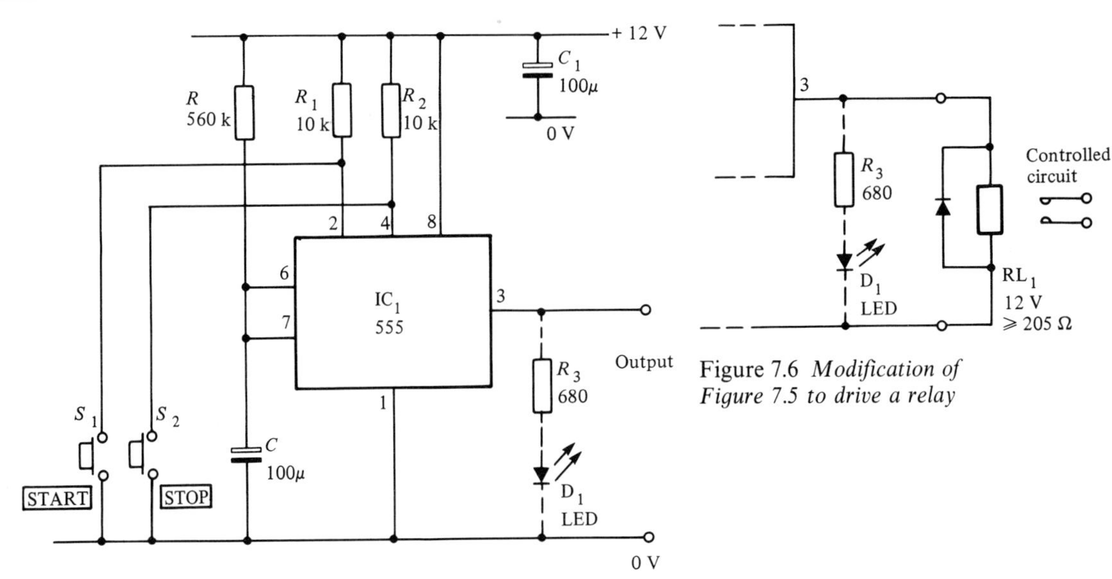

Figure 7.5 *One minute timer*

Figure 7.6 *Modification of Figure 7.5 to drive a relay*

that the performance of the timer may be unpredictable when the values of these components are outside the recommended range:

$R = 1\,\text{k}\Omega$ to $3.3\,\text{M}\Omega$
$C = 470\,\text{pF}$ to $470\,\mu\text{F}$

In situations where the output pulse duration is not as expected (particularly in circuits which employ electrolytic timing capacitors which may exhibit excessive leakage current) it may sometimes be necessary to investigate the d.c. voltages appearing at the 'threshold' and 'discharge' inputs of the timer. This, however, will *only* be a meaningful exercise when a d.c. voltmeter having *very* high internal resistance is employed. Standard bench multimeters (of 20 kohm/V, or thereabouts) are generally unsuitable for making this measurement as they may significantly alter the charge and discharge time constants.

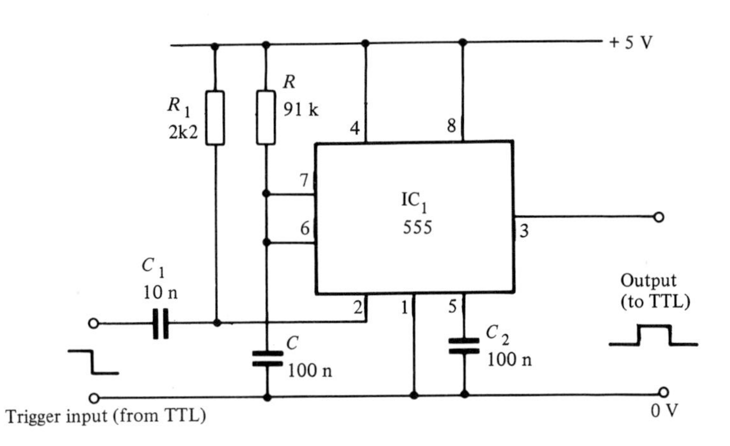

Figure 7.7 *TTL compatible 10 ms monostable pulse generator*

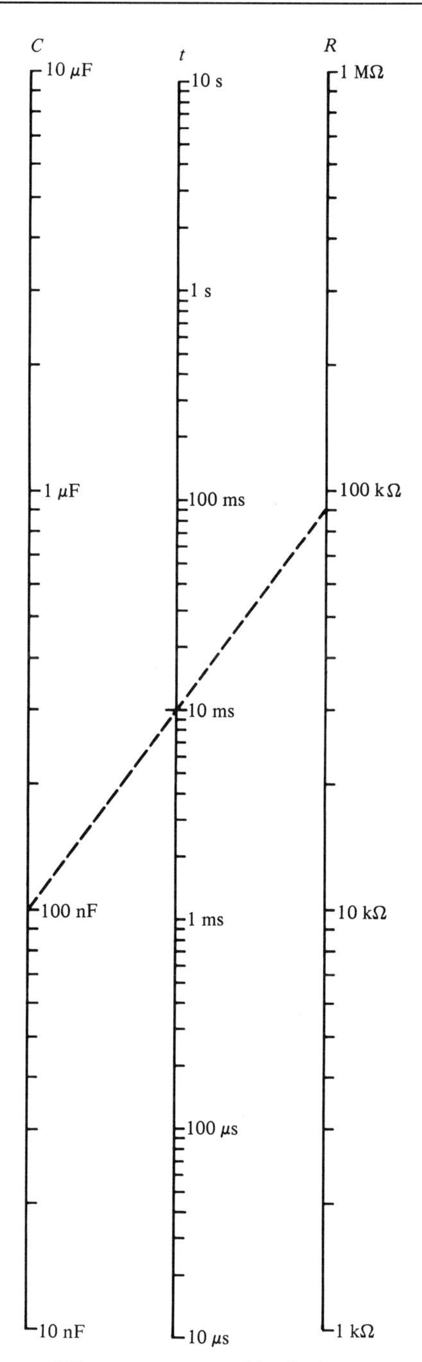

Figure 7.8 *Nomograph for 555 monostable operation. (The example shows how a 10 ms pulse width can be achieved using* $C = 100$ n *and* $R = 91$ k *approximately see Figure 7.7)*

Where a monostable timing period is to be initiated electronically rather than by means of a push-button, a negative going trigger pulse must be applied to the trigger input (pin-2 of a standard 555 timer). Figure 7.7 shows a monostable arrangement which provides a TTL compatible 10 ms

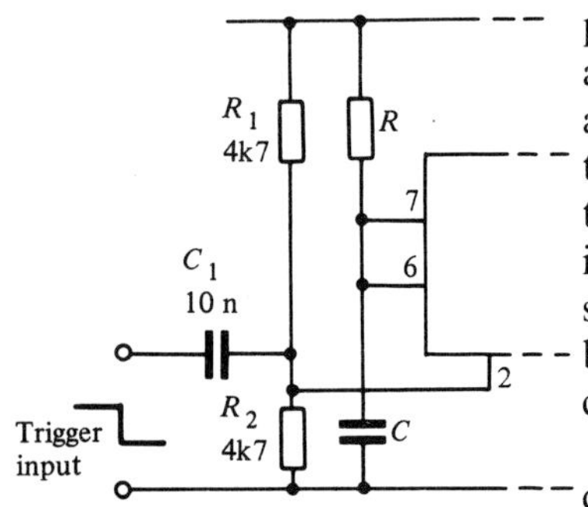

Figure 7.9 *Improving the trigger sensitivity of a 555 timer by reducing quiescent d.c. voltage present at the trigger input*

pulse. It should be noted that, in common with most timer circuits, adequate decoupling of the supply rails is essential in order to reduce the amplitude of supply borne transients. Such decoupling should normally take the form of an appropriate value capacitor (as a general rule of thumb this should be at least ten times the value of the timing capacitor) arranged in close physical proximity to the positive supply input (pin 8 of the standard 555). The negative connection of the decoupling capacitor should be returned to the 0 V rail by a low impedance path formed by a substantial conductor or appropriate width of PCB copper foil.

The monostable period of the circuit shown in Figure 7.7 can be determined using the formulae given on page 158 or with the aid of the nomograph depicted in Figure 7.8. The range of monostable periods that can be generated with this circuit extends from less than 10 μs to well over

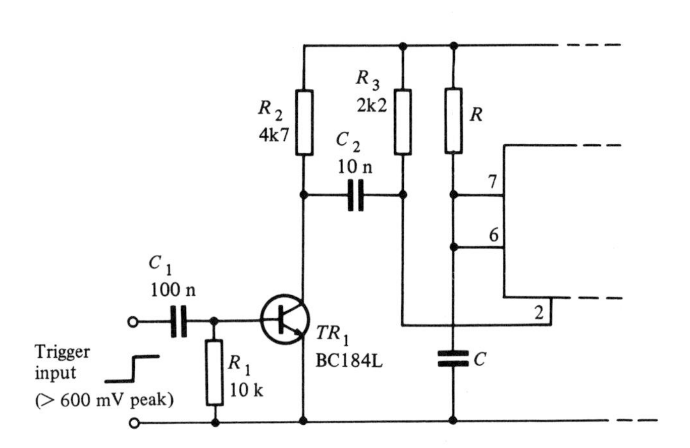

Figure 7.10 *Improving the trigger sensitivity using an additional transistor. (Note that this arrangement triggers on a positive, rather than negative, going edge)*

100 s. It should, however, be noted that, since the trigger is a.c. coupled to the trigger input of the timer, the falling edge pulse *must* have sufficient amplitude for the voltage at pin-2 to fall below one third of the supply voltage. Figures 7.9 and 7.10 show two methods of improving the trigger sensitivity of basic monostable timers.

Astable 555 pulse generators

Figure 7.11 shows a 555 timer operating in astable mode and producing a 5 V TTL compatible output with a p.r.f. of 100 Hz and duty cycle of approximately 67%. This circuit can be easily modified for other p.r.f.'s using the formulae quoted on page 159 or with the aid of the nomograph shown in Figure 7.12. The range of p.r.fs that can be generated with this circuit extends from less than 0.01 Hz to well over 100 kHz.

As mentioned earlier, a limitation of the basic 555 astable is that it is not possible to obtain a square wave output having an exact 50% duty cycle. However, in order to produce a reasonable approximation to a square wave, R_1 can be made very much smaller than R_2.

Figure 7.13 shows a neat method of obtaining a near 50% duty cycle.

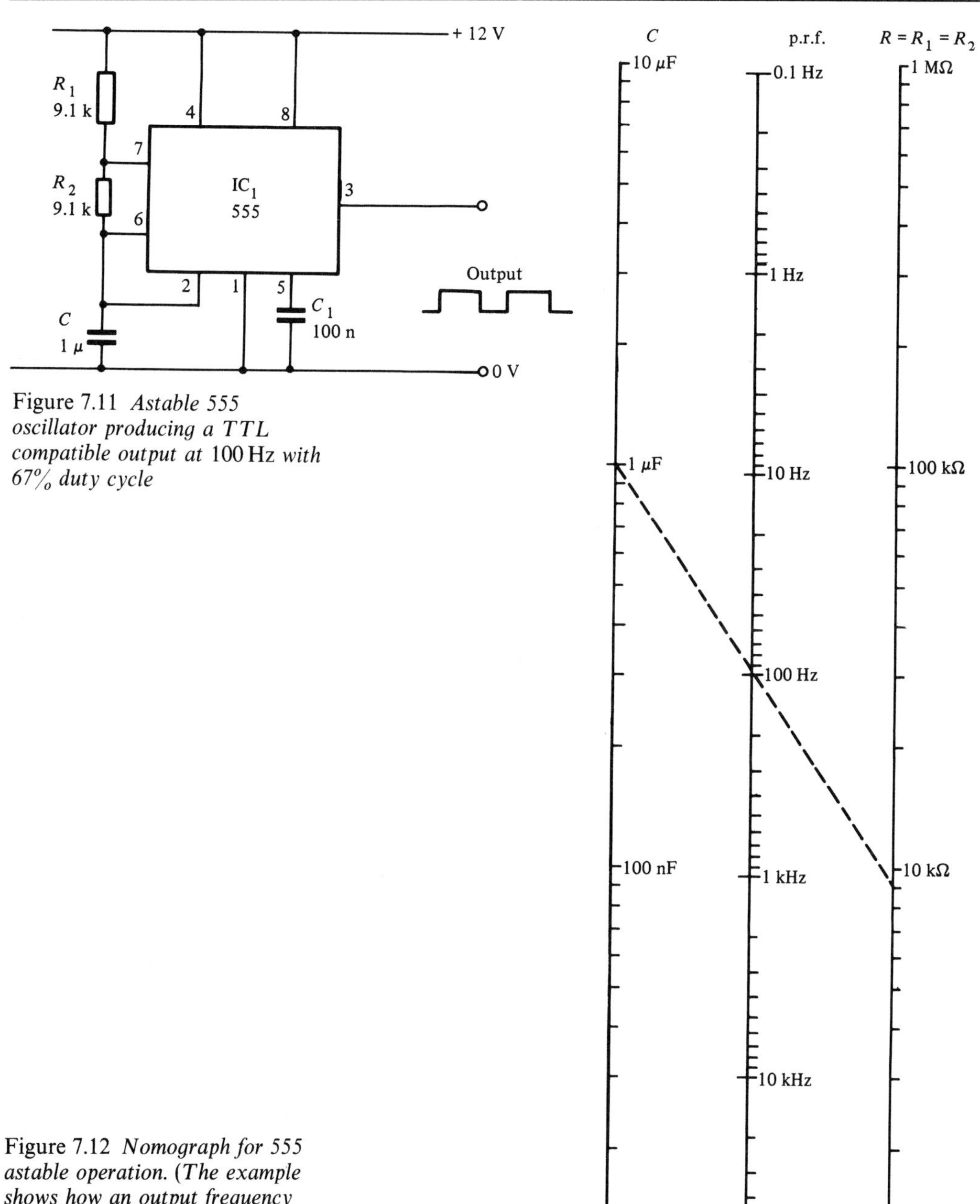

Figure 7.11 *Astable 555 oscillator producing a TTL compatible output at* 100 Hz *with* 67% *duty cycle*

Figure 7.12 *Nomograph for 555 astable operation. (The example shows how an output frequency of* 100 Hz *can be achieved using* $C = 1\,\mu$ *and* $R = R_1 = R_2 = 9.1\text{k}$, *see Figure 7.11)*

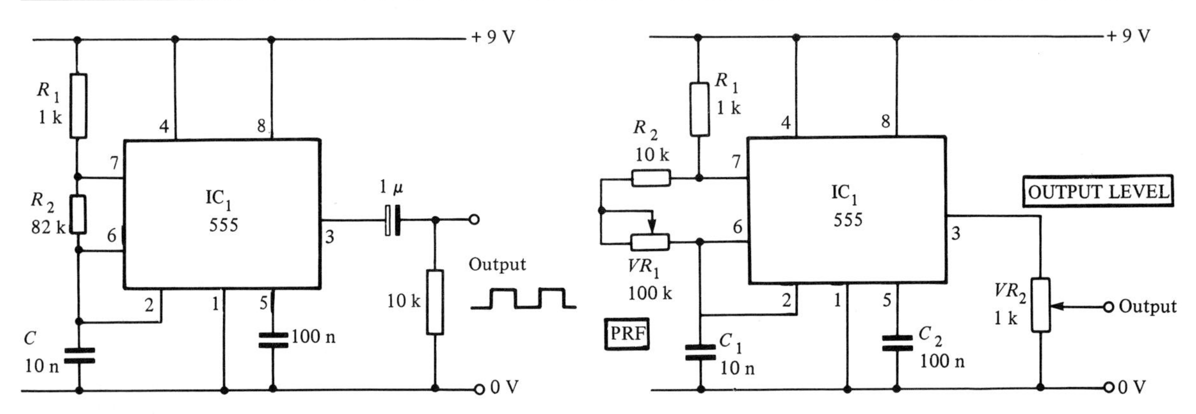

Figure 7.13 *Test oscillator producing 1 kHz output with near 50% duty cycle*

Figure 7.14 *Variable frequency/variable amplitude square wave oscillator*

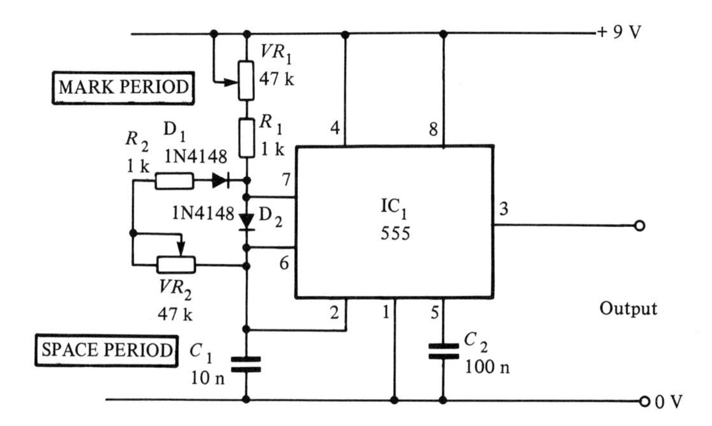

Figure 7.15 *555 astable with independently adjustable mark and space periods*

Provided that R_2 is very much larger than R_1, this circuit provides a reasonably precise square wave output at a p.r.f. of 1 kHz. The circuit can be modified for other p.r.f.s using the nomograph (in which $R = R_2$) shown in Figure 7.8. The periodic time (*not* the pulse width) is determined from this nomograph and the p.r.f. then calculated by taking the reciprocal of the periodic time.

Figure 7.14 shows a variable square wave signal source which offers an approximate 10:1 p.r.f. adjustment range from approximately 700 Hz to just over 7 kHz. At 700 Hz the ‘best case’ duty cycle is 50.2% while at 7 kHz the duty cycle is worsened to about 52.4%.

Figure 7.15 shows how the mark and space periods of an astable timer may be independently varied over a range of approximately 40:1. VR_1 and VR_2 respectively govern the mark and space periods. The p.r.f. of this arrangement will, of course, be governed by the settings by both controls, VR_1 and VR_2. The maximum p.r.f. (with both controls set to minimum resistance) is approximately 66 kHz while the minimum p.r.f. (both controls set to maximum resistance) is approximately 1.4 kHz. Provided both the

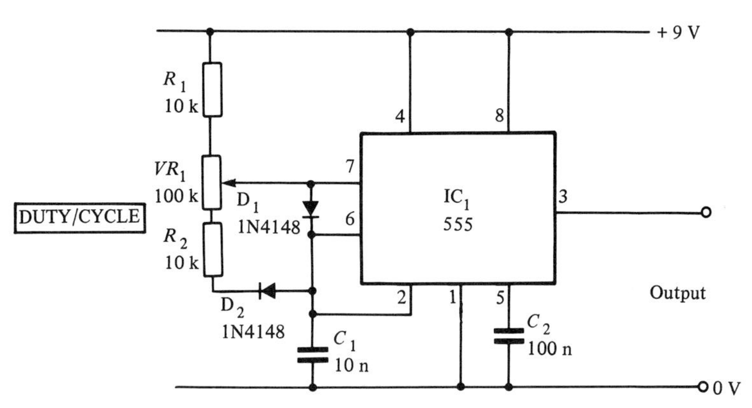

Figure 7.16 *555 astable with fixed p.r.f. and adjustable duty cycle*

variable resistors obey the same law, a square wave output will be produced when the controls have identical settings.

In some applications it may be necessary to vary the duty cycle of a 555 timer without affecting the p.r.f. of its output. Figure 7.16 shows an astable circuit which operates with a p.r.f. of 1.2 kHz and which offers adjustment of duty cycle over the range 10% to 90%.

Figure 7.17 shows how an inexpensive yet versatile pulse generator can be constructed using an astable pulse generator to trigger a monostable circuit. This arrangement provides periods adjustable in four decade ranges from 1.4 s to 140 μs and pulse widths adjustable in four ranges from 0.7 s to 70 μs. The output is adjustable from 0 V to 10 V.

The 'reset' input of the 555 timer can be used to enable or disable normal astable operation. This facility provides us with a method of gating an astable timer. Figure 7.18 shows a simple audible alarm oscillator which operates whenever a 'high' (TTL or CMOS logic 1) is received at the reset input (pin-4 of a standard 555). If a pulsed, rather than continuous output is required, a second astable timer can be added, as shown in Figure 7.19. This circuit is an ideal application for a 556 device.

Figure 7.17 *Versatile pulse generator based on two 555 timers*

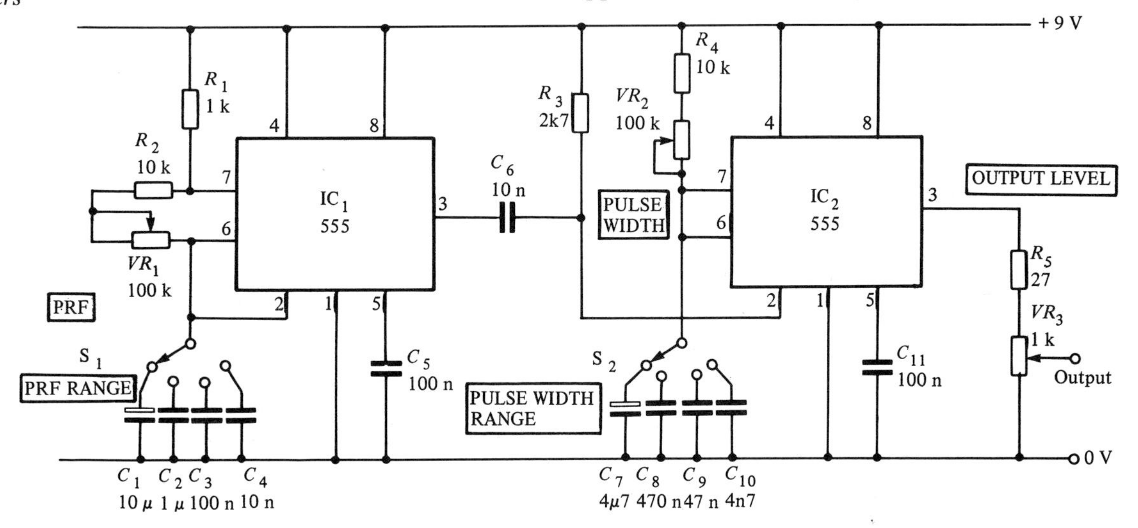

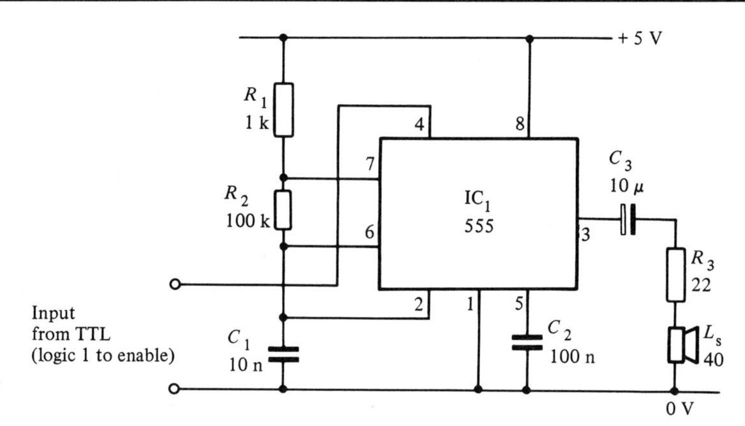

Figure 7.18 *Simple gated audible tone generator*

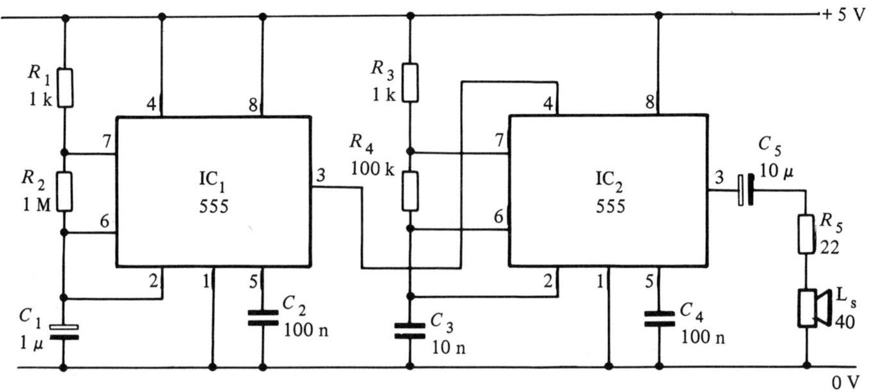

Figure 7.19 *Pulsed alarm tone generator*

Precision timers

Unfortunately, when timing periods in excess of a few tens of seconds are required, the humble 555 timer is somewhat limited in accuracy. This limitation is directly attributable to the poor tolerance and excessive leakage currents associated with large value electrolytic capacitors.

An obvious solution to the problem is that of using a short period time standard (one that can be generated with a high degree of accuracy) and dividing this using a chain of binary dividers to produce the desired output period. This technique is adopted in several 'precision' timers.

The ZN1034 is one such device which is capable of providing precise timing intervals ranging from 50 ms to several days. The device contains an internal oscillator (the frequency of which is determined by a single external resistor and capacitor) and the output of this oscillator is then fed to an internal 12-stage binary divider. The output of the last stage of this binary divider chain changes state after 4095 complete cycles of the fundamental oscillator signal.

The ZN1034 is housed in a 14-pin DIL package and requires a nominal 5 V supply (maintained within standard TTL limits) and consumes a modest 5 mA quiescent supply current. The device can deliver output currents of up to 25 mA and offers a typical temperature stability of 0.01%/°C coupled with a repeat timing accuracy of 0.01%.

The basic configuration of a ZN1034 timer is shown in Figure 7.20. The monostable timing period is triggered by a negative going edge applied to the trigger input (pin-1). The output of this circuit, like that of the previous 555 timer circuits, remains high during the timing period which is determined by the value of an external C-R time constant such that:

$$t = 2735\, C\, R$$

Where t is in seconds, C is in farads, and R is in ohms.

Values of C and R for typical time periods are given in Table 7.1:

Table 7.1

Monostable period	*C*	*R*
1 second	10 n	39 k
1 minute	100 n	22 k
5 minutes	1 μ	100 k
1 hour	1 μ	1.2 M
1 day	10 μ	3.3 M
1 week	100 μ	2.2 M

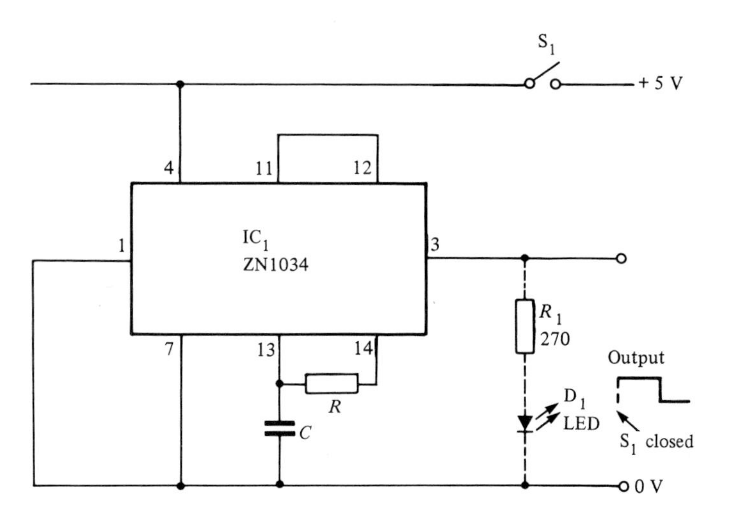

Figure 7.20 *Basic ZN1034 configuration*

The recommended range for ZN1034 timing components is:

$R = 5\,\mathrm{k}\Omega$ to $5\,\mathrm{M}\Omega$
$C = 3.3\,\mathrm{nF}$ to $1000\,\mu\mathrm{F}$

A useful feature of the ZN1034 is that it provides complementary outputs. The output at pin-3 goes 'high' during the timing period while that available at pin-2 goes 'low' during the timing period. Figure 7.21 illustrates this relationship.

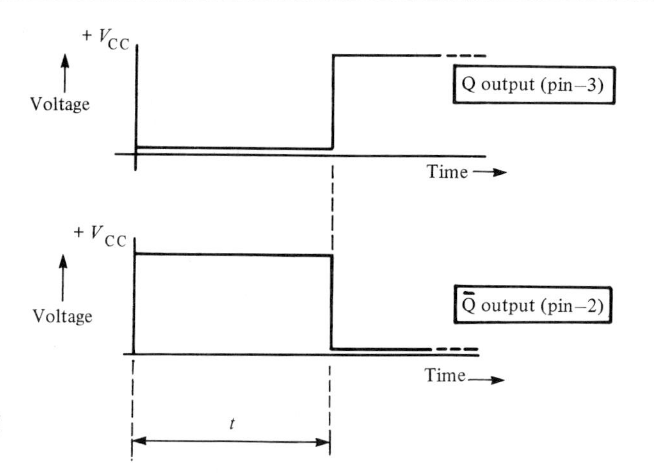

Figure 7.21 *Relationship between the two outputs provided by the ZN1034*

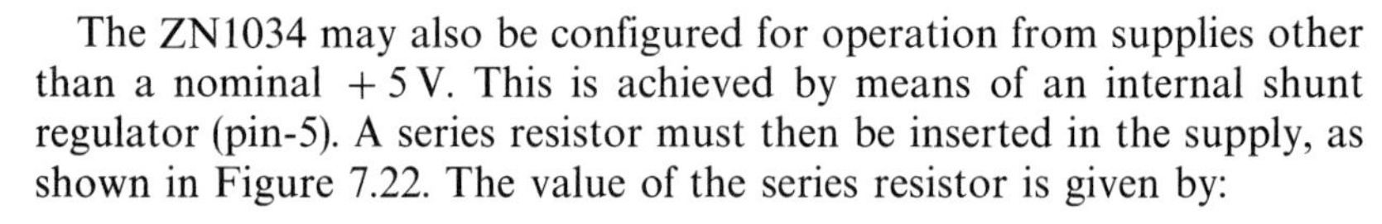

The ZN1034 may also be configured for operation from supplies other than a nominal +5 V. This is achieved by means of an internal shunt regulator (pin-5). A series resistor must then be inserted in the supply, as shown in Figure 7.22. The value of the series resistor is given by:

$$R_s = \frac{V_s - 5}{I_L + 7} \text{k}\Omega$$

Where V_s is the supply rail voltage and I_L is the load current in mA.

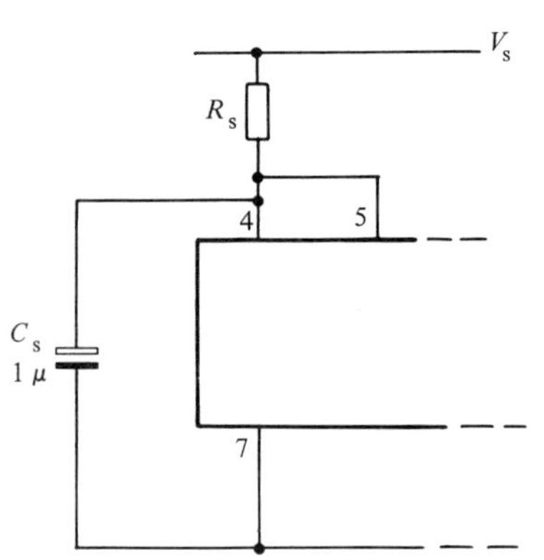

Figure 7.22 *Using the ZN1034's internal regulator*

If desired, and at the expense of a marginal worsening in temperature stability, the ZN1034 timer curcuit may be calibrated using a good quality pre-set resistor (of between 47 k and 220 k maximum resistance) connected between pins 11 and 12.

The output current capability of the ZN1034 timer may easily be increased with the aid of an external transistor, as shown in Figure 7.23. The relay coil (which should have a resistance of at least 110 Ω) is energized throughout the timing period.

When the ZN1034 trigger input (pin-1) is linked to 0 V, the device is triggered whenever the supply is applied. This is a convenient arrangement in many cases. However there may be a number of applications in which the trigger pulse will be derived from some form of transducer or logic arrangement. Figure 7.24 shows various options.

A delay circuit based on the ZN1034 timer is shown in Figure 7.25. A single on/off (start/reset) switch is used to control the circuit. The thyristor is held in the non-conducting state during the monostable period but is triggered into conduction as soon as the timing period ends. The thyristor then remains in the conducting state, passing current through the load, until S_1 is switched to the off (reset), position.

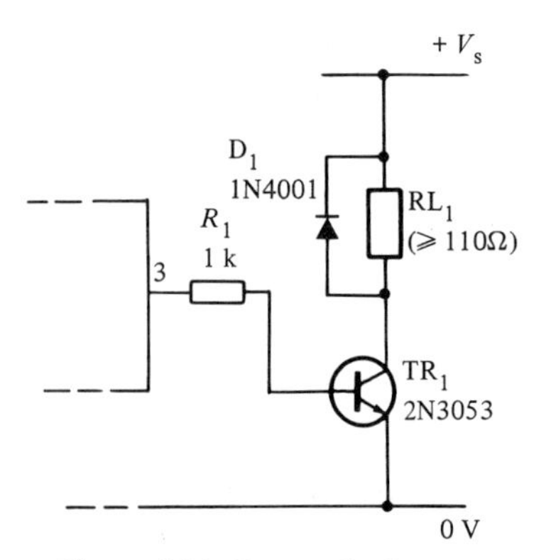

Figure 7.23 *External relay driver for the ZN1034*

Programmable timers

The 2240 is a development of the basic 555 timer which incorporates an internal timebase oscillator together with an 8-stage binary divider chain

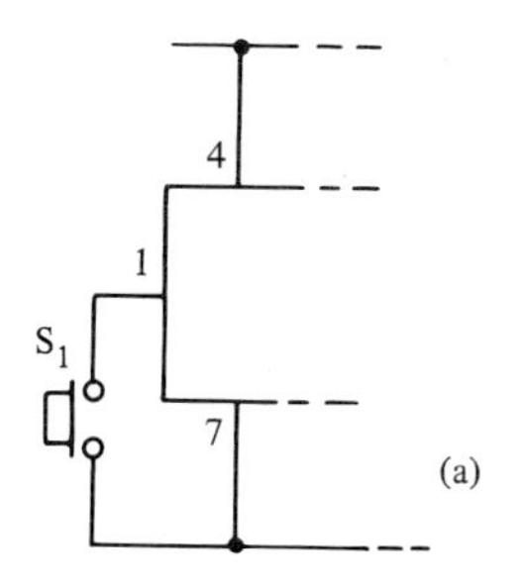

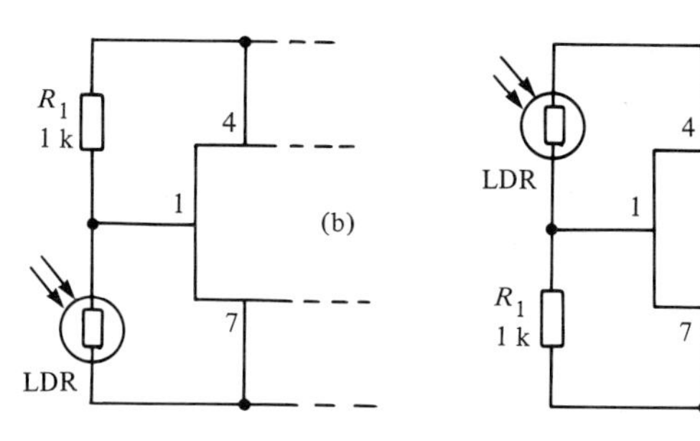

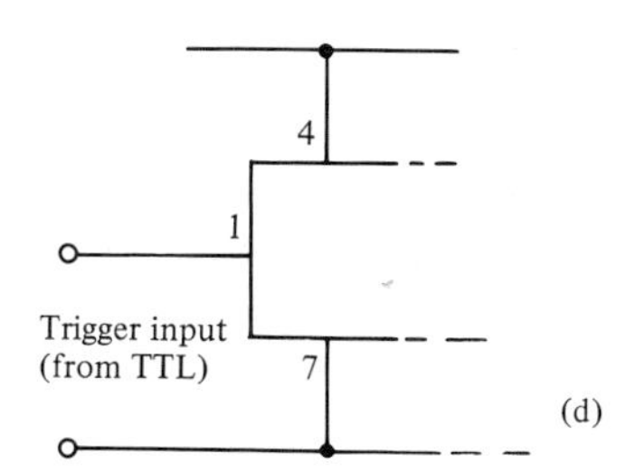

Figure 7.24 *Methods of triggering the ZN1034*

(a) Using a normally open push-button
(b) Using a light dependent resistor (device triggers on rising light level)
(c) Using a light dependent resistor (device triggers on falling light level)
(d) Using TTL compatible logic (the trigger input is derived directly from a logic gate output)

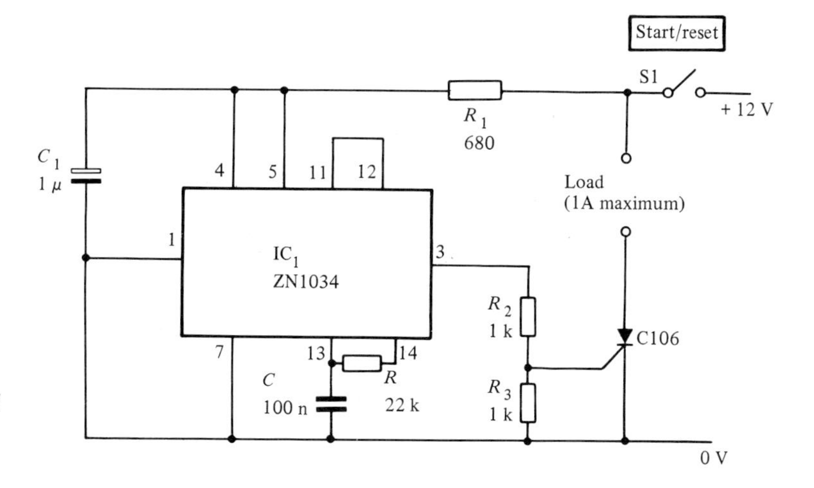

Figure 7.25 *ZN1034 delay circuit (the load should be rated at 12 V 1 A maximum)*

housed in a 16-pin DIL package. The device permits access to the divider chain (such that the actual divisor can be programmed in the range 1 to 255) and is activated by the application of a positive going trigger pulse applied to pin-11. This trigger pulse starts the timebase oscillator, enables the divider chain, and sets all divider outputs (pins 1 to 8) to the 'low' state. Thereafter, a binary counting sequence is generated on the output lines.

The divider chain outputs are, in fact, open collector transistors and these can therefore be connected together in a wired-AND configuration. If several output are linked together, it will be necessary for *all* outputs to go 'high' in order that the collective output should go 'high'. If any one or more of the outputs should be 'low' then the collective output will be 'low'. This arrangement provides a neat method of resetting the timer when the count reaches a particular state determined by the eight divider outputs.

The fundamental timebase period of the 2240 (available at pin-14) operates with a periodic time given by:

$$t = CR \text{ seconds}$$

Where C is in farads and R is in ohms.

Figure 7.26 shows a basic 2240 timer arrangement in which the output period is selected using an 8-way DIP switch bank. The switches have the following binary weightings:

S_1	S_2	S_3	S_4	S_5	S_6	S_7	S_8
1	2	4	8	16	32	64	128

Thus, if S_2, S_3 and S_7 are all placed in the 'on' position (while the other switches remain 'off') the resulting divisor (n) is $(2 + 4 + 64)$ or 70. If the

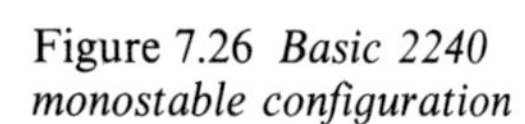

Figure 7.26 *Basic 2240 monostable configuration*

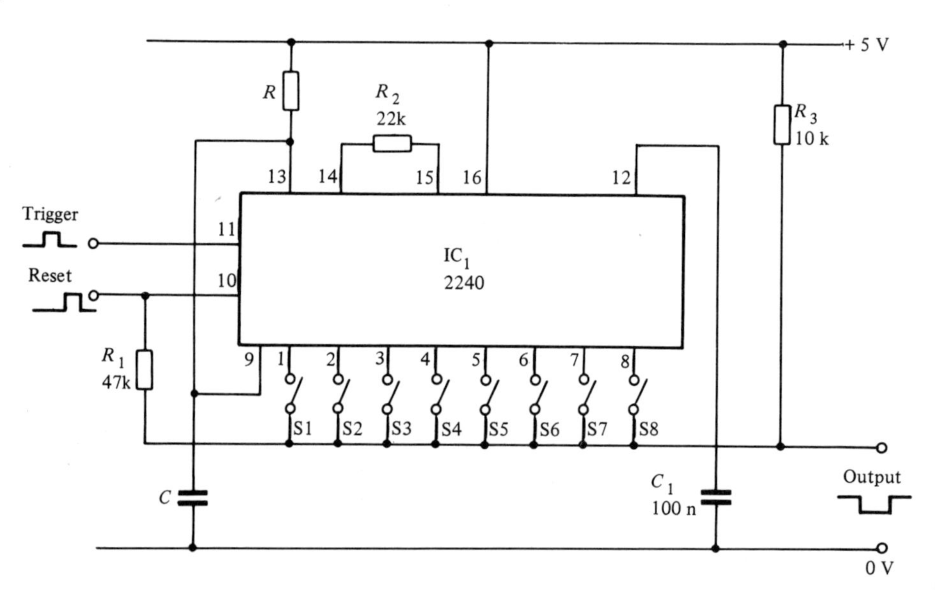

fundamental timebase period is 1 ms, the output period (at pin-15) will then be 70 ms.

The timing sequence is completed or aborted when a positive going reset pulse appears at pin-10. Thereafter the timebase and counter sections are disabled and all eight divider outputs go 'high'.

Unlike the ZN1034 (which is designed primarily for use with a conventional TTL supply rail) the 2240 will operate over a range of supply voltages from 4 V to 15 V with a typical supply current of 15 mA at 15 V. The device is capable of providing reliable output periods which range from 10 μs to well over 10 hours.

The recommended range for the timing components associated with the 2240 is:

$R = 1\,\text{k}\Omega$ to $10\,\text{M}\Omega$
$C = 10\,\text{nF}$ to $1000\,\mu\text{F}$

The 2240 can be configured for astable operation by connecting the 'reset' input to 0 V. Figure 7.27 shows a typical astable arrangement capable of generating 255 different complex bit patterns. Each setting of the DIP switches (S_1 to S_8) produces a different bit pattern; the minimum pulse width appearing in the output pulse train being determined by the duration of the least significant bit selected.

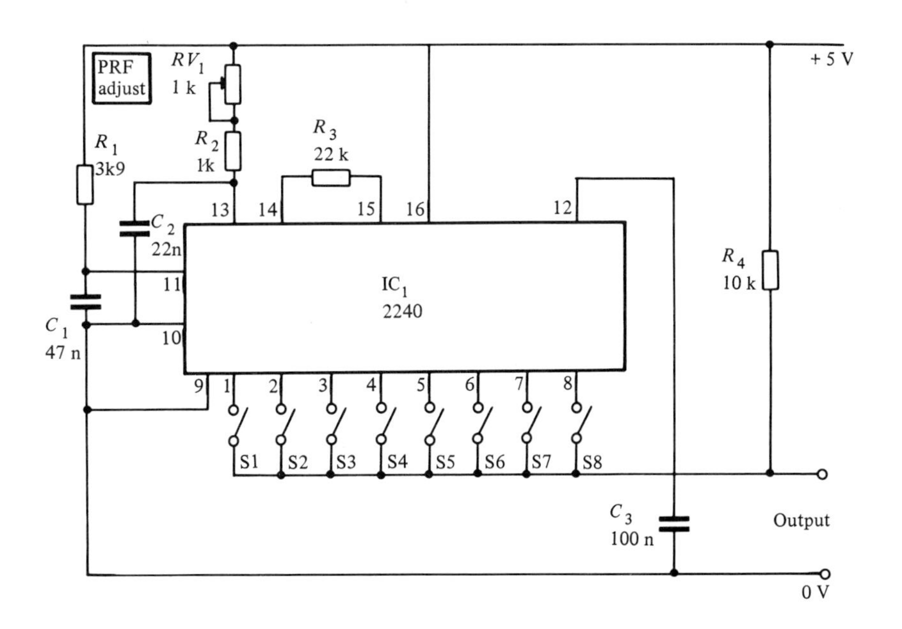

Figure 7.27 *Complex astable pulse generator using a 2240*

If square waves, rather than complex bit patterns, are required just one of the eight switches should be placed in the 'on' position at any particular time. The circuit can then be used as reasonably accurate frequency standard. VR_1 should be adjusted so that a precise 10 kHz square wave output when S_1 is 'on' and all other switches are 'off'. Thereafter, a

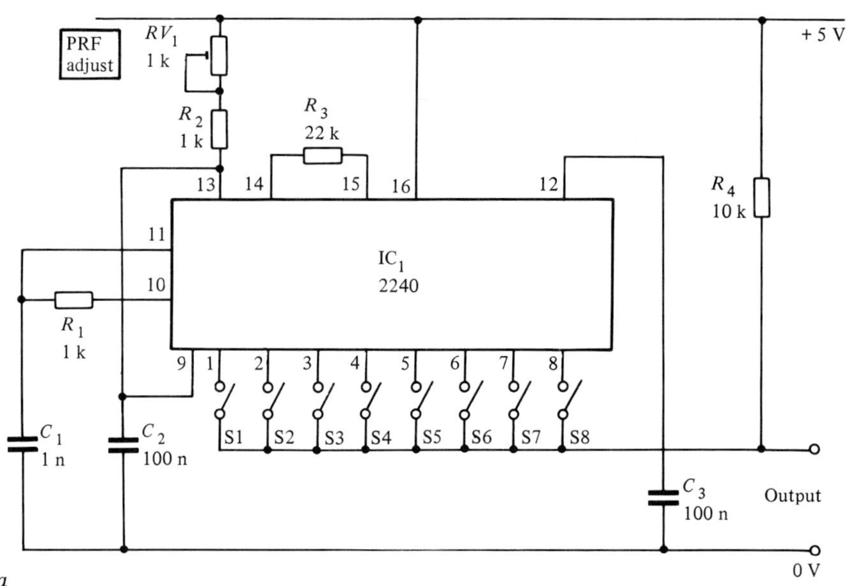

Figure 7.28 *Using the 2240 as a frequency synthesizer*

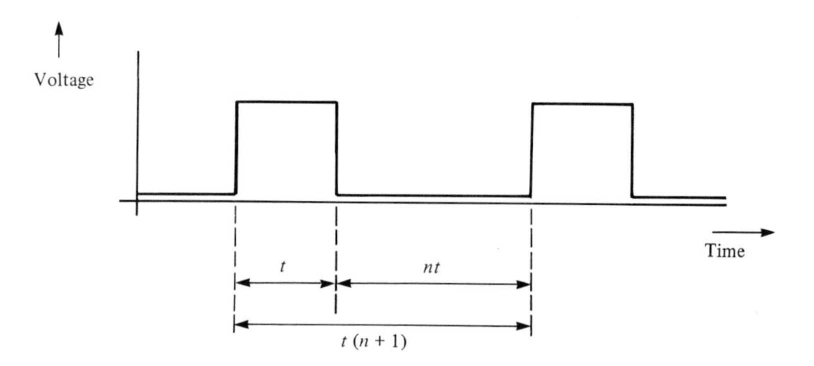

Figure 7.29 *Output waveform produced by the circuit shown in Figure 7.28*

symmetrical square wave output will be produced when just one of the switches is selected according to Table 7.2:

Table 7.2

Switch selected ('on')	*Divisor (n)*	*Output frequency (Hz)*
1	1	10 k
2	2	5 k
3	4	2.5 k
4	8	1.25 k
5	16	625
6	32	312.5
7	64	156.25
8	128	78.125

Finally, Figure 7.28 shows a simple frequency synthesizer based on the 2240. The output of the circuit comprises a train of pulses each having a width equal to t (the fundamental timebase period) and a periodic time equal to $t \times (n + 1)$, as shown in Figure 7.29.

8 Computer interfacing

In recent years, the availability of microcomputers has added a new dimension to electronics in which the computer can be used as a means of controlling electronic hardware. The result is an electronic system which is inherently flexible; significant changes can be made simply by modifying the control program.

As an example, consider the case of an environmental control system. Here we may be concerned with accepting information from a number of transducers and generating the necessary control signals to operate pumps, fans, and heaters. To enable us to monitor the system, we would probably require a display of actual temperatures, as well as status of the system as a whole. As a further refinement, we would probably wish to incorporate some form of clock so that the system can cope with different times of day and days of the week.

The traditional approach to solving such a problem would involve a fairly complex logic system coupled with display drivers, some form of keypad, a digital clock, several relay drivers, and one or more analogue to digital converters. A far better method, requiring only the last two items of hardware, would be based on a microcomputer or microprocessor-based controller. Not only would such a system be capable of fulfilling all of the functions of its conventional counterpart but it would also provide us with a far more sophisticated means of processing our data coupled with the ability to store it and examine it at a later date or even transmit it to a remote supervisory computer installation.

The advent of powerful low-cost single-chip microcomputers has further underlined the cost-effectiveness of the microcomputer as the prime mover in a huge variety of electronic control systems. The time saved on hardware development can usefully be devoted to the software aspects of a project and future modifications can simply involve the substitution of firmware (ROM based software).

Characteristics of 'user ports'

Happily, most of today's microcomputers are well equipped with external ports. Of these, several may be dedicated to such functions as driving a printer or modem. However, more often than not, at least one parallel port is available as a 'user port'. User ports are invariably 'byte wide' (i.e. they comprise eight individual input/output lines) and are implemented by means of one of a number of VLSI programmable parallel I/O devices. Such devices may provide as many as twenty-four separate I/O lines and the following are common examples:

6520	Peripheral interface adaptor (PIA).
6521	Peripheral interface adaptor (PIA) – similar to the 6520.
6522	Versatile interface adaptor (VIA).
6820	Peripheral interface adaptor (PIA) – equivalent to the 6520.
6821	Peripheral interface adaptor (PIA) – equivalent to the 6521.
8255	Programmable parallel interface (PPI).
Z80-PIO	Programmable parallel input/output (PIO).

Programmable parallel I/O devices can normally be configured (under software control) in one of several modes:

(a) All eight lines configured as inputs.
(b) All eight lines configured as outputs.
(c) Lines individually configured as inputs or outputs.

In addition, extra lines are usually provided for 'handshaking'. This is the apt name given to process by which control signals (such as interrupt request) are exchanged between the microcomputer and peripheral hardware.

The nomenclature used to describe port lines and their function tends to vary from chip to chip but there is a reasonable degree of commonality and the following covers the majority of devices listed previously:

PA0 to PA7	Port A I/O lines; 0 corresponds to the least significant bit (LSB) while 7 corresponds to the most significant bit (MSB).
CA_1 to CA_2	Handshaking lines for Port A; CA_1 is an interrupt input while CA_2 can be used as both an interrupt input and peripheral control output.
PB0 to PB7	Port B I/O lines; 0 corresponds to the least significant bit (LSB) while 7 corresponds to the most significant bit (MSB).
CB_1 to CB_2	Handshaking lines for Port B; CB_1 is an interrupt input while CB_2 can be used as both an interrupt input and peripheral control output.

The electrical characteristics of an I/O port tend to vary from chip to chip. However, signals are invariably TTL compatible and, provided one follows the general rules for interfacing TTL gates described in Chapter 6, few problems should be encountered. Despite this, and due to limitations associated with the active 'pull-up' devices employed in programmable parallel I/O chips, difficulties arise when interfacing to CMOS logic. Here it is necessary to employ an additional LS-TTL buffer stage preceding the CMOS stage (again, following the rules described in Chapter 6).

Several programmable parallel I/O devices have port output lines (usually the B group) which are able to source sufficient current to permit the direct connection of the base of an n-p-n transistor. In such cases, and to minimize loading, the external transistor should operate with a base current of less than 1 mA or so. Furthermore, since the high level output voltage may fall to around 1.5 V when a port line is sourcing an appreciable current, Darlington transistors are much preferred.

Adding a 'user port'

In some cases a microcomputer may be limited in its port capability (or an existing parallel port may be dedicated to some other peripheral hardware (such as a printer). It then becomes necessary to add some extra I/O capability. This may be achieved using either a fully-blown programmable parallel I/O device or, alternatively, a single byte (or less) of input or output may be added quite easily.

In either case it will, of course, be necessary to have a reasonably intimate knowledge of both the existing hardware configuration and of the allocation of memory in the system. It will also be necessary to devise an appropriate method of address decoding such that the interface is only enabled (and data latched to or read from the outside world) when the interface is actually being addressed.

We shall now briefly examine the principles of interfacing (without the use of a complex programmable I/O device) to two popular microprocessors; the Z80 and the 6502. Details for the Z80 will also apply to the 8080 and 8085 while those given for the 6502 will also apply to 6800 and 6809 based systems.

Z80 based systems

The Z80 provides for 256 I/O ports. These ports are addressed using the lower byte (lines A0 to A7 inclusive) of the address bus. The Z80 provides a number of instructions for dealing with the I/O ports, the most commonly used of which take the assembly language mnemonic forms, IN A, (n) and OUT (n), A. The first of these transfers a byte from port address *n* into the accumulator, *A*. The second reverses this process and transfers a byte from the accumulator to port address *n*. This makes coding Z80 I/O a relatively painless task.

For those who would prefer to use a high level language, many versions of BASIC for Z80 based microcomputers incorporate commands which are directly equivalent to the assembly language mnemonics, IN and OUT, thus facilitating direct access to port I/O from within BASIC.

Three representative address decoders for use with Z80 based microcomputers are shown in Figure 8.1. To minimize loading on the address bus, the gates employed in the address decoder should be either LS-TTL or CMOS types. The arrangement of 8.1(a) provides an active low chip enable signal when an I/O port address of FF hexadecimal appears on the lower byte of the address bus. The lower byte of the Z80's address bus (comprising lines A0 to A7) will all go simultaneously high when the Z80 executes instructions of the form, OUT (FFH), A or IN A, (FFH).

Different addresses can be easily accommodated by changing the decoding logic. Figures 8.1(b) and 8.1(c) show address decoding of hexadecimal port addresses 7F and FE respectively.

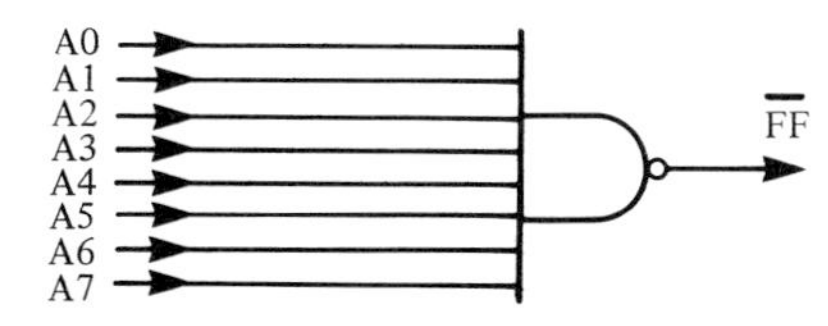

(a) Output goes low when A0 to A7 are all high

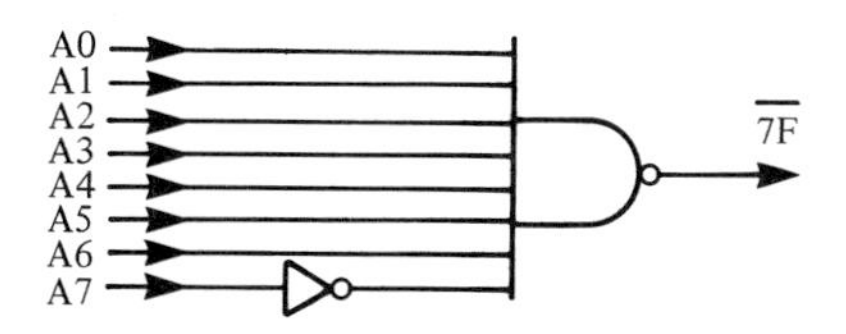

(b) Output goes low when A0 to A6 are high and A7 is low

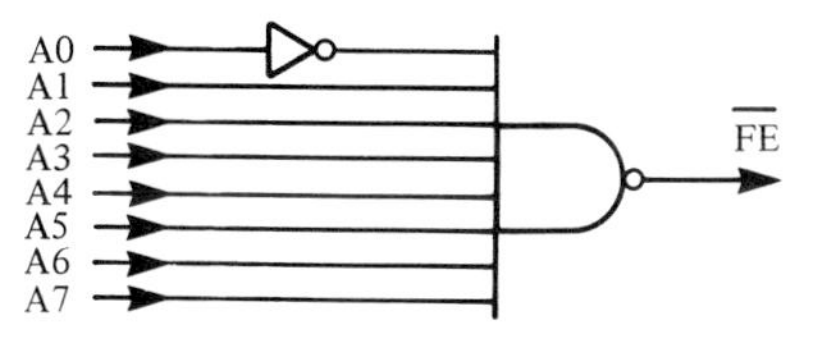

(c) Output goes low when A1 to A7 are high and A0 is low

Figure 8.1 *Simple address decoders*

Decoding the port address is, however, not quite the end of the story. We also need to take into account the state of the input/output request ($\overline{IORQ}$) line in order to distinguish between normal memory operations and those which are specifically directed at the I/O ports. Furthermore, we also need to establish whether an input or output operation is to be performed. This is accomplished by sensing the state of the read ($\overline{RD}$) or write ($\overline{WR}$) lines.

Figures 8.2(a) and 8.2(b) respectively show how active low chip enable signals can be derived for input and output from a Z80 port address of FF hexadecimal. In the former case, the state of the $\overline{IORQ}$ and $\overline{RD}$ lines is combined with the address decoding while, in the latter case, the $\overline{IORQ}$ signal is combined with the $\overline{WR}$ line.

Decoding logic can often be more readily understood using Boolean algebra. There are numerous texts covering this subject but, for the benefit of the newcomer, the + sign is used to indicate the OR function while the sign indicates the AND function. A $^{-}$ is used to denote inversion (i.e. the NOT function).

The Boolean expression for the output of the decoding logic of Figure 8.2(a) is:

$$\overline{ENABLE} = \overline{FF} + \overline{IORQ} + \overline{RD}$$

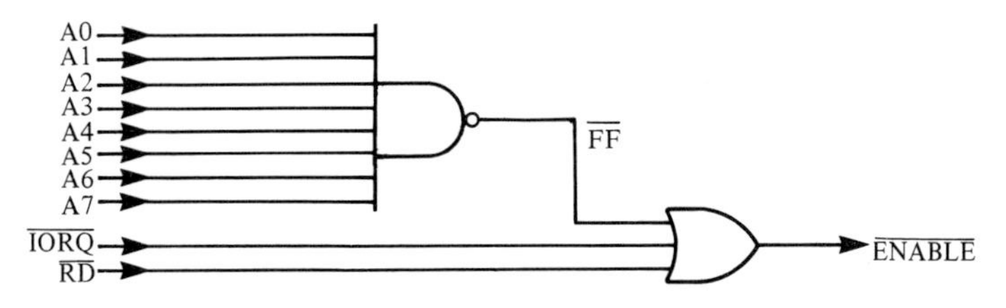

(a) Output goes low when A0 to A7 are high and both $\overline{\text{IORQ}}$ and $\overline{\text{RD}}$ are low

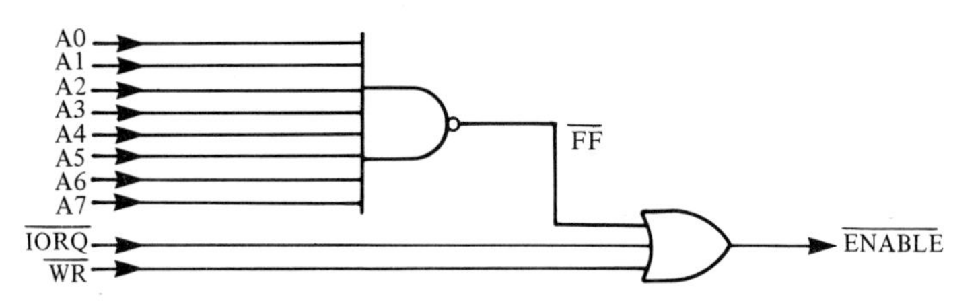

(b) Output goes low when A0 to A7 are high and both $\overline{\text{IORQ}}$ and $\overline{\text{WR}}$ are low

Figure 8.2 *Method of combining address decoding with I/O and read/write control signals*

From De Morgan's theorem this is equivalent to:

$$\overline{\text{ENABLE}} = \overline{\text{FF}\,.\,\text{IORQ}\,.\,\text{RD}}$$

Similarly, the Boolean expression for the output of the decoding logic of Figure 8.2(b) is:

$$\overline{\text{ENABLE}} = \overline{\text{FF}} + \overline{\text{IORQ}} + \overline{\text{WR}}$$

From de Morgan's theorem this is equivalent to:

$$\overline{\text{ENABLE}} = \overline{\text{FF}\,.\,\text{IORQ}\,.\,\text{WR}}$$

If an active high (rather than active low) enable signal is required, the three-input OR gate can simply be replaced by a three-input NOR.

Figure 8.3 shows a how a four-bit output port can be added to a Z80 based microcomputer. This circuit is based on a 74LS175 quad data latch which is clocked whenever all eight address lines are simultaneously high and the $\overline{\text{WR}}$ and $\overline{\text{IORQ}}$ control lines go low. The state of the lower four data lines (D0 to D3) is thus placed in the latch whenever the microprocessor executes an instruction of the form OUT(FFH), A. The accumulator must, of course, be previously loaded with the data to be transferred to the port.

The four $\bar{\text{Q}}$ outputs of the data latch can usefully be employed to drive LED indicators or be used to provide complementary outputs. Additional output current drive capability may be provided by means of the methods described on page 185.

Figure 8.4 shows how a four-bit input port can be added to a Z80 based

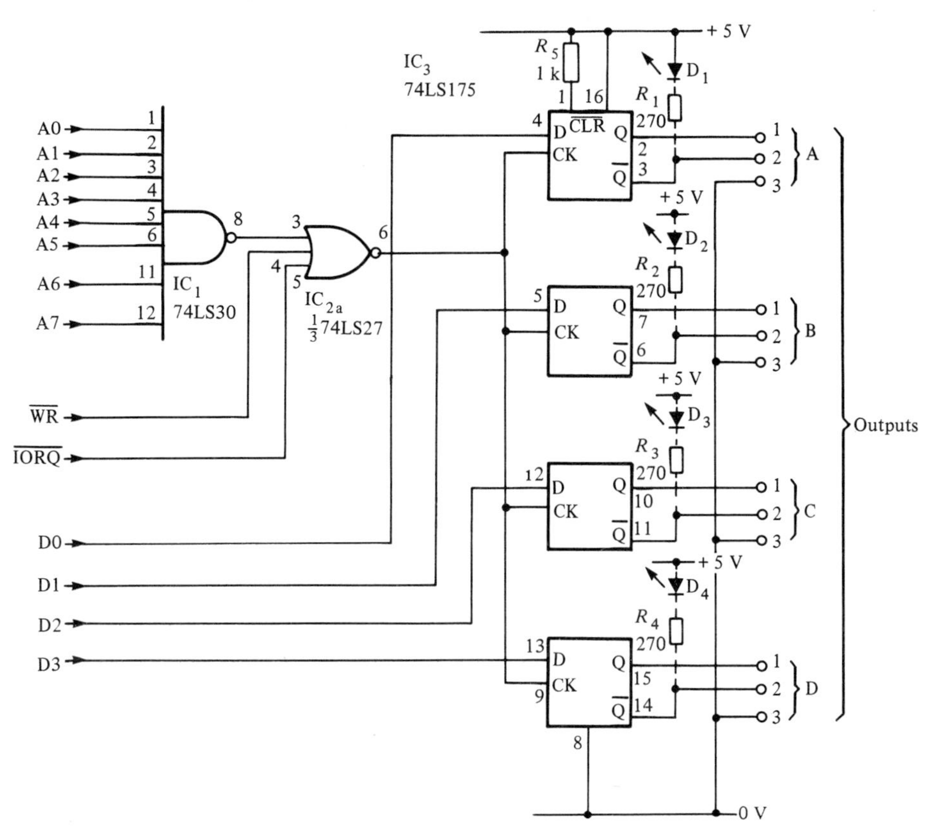

Figure 8.3 *Simple 4-bit output port*

microcomputer. This arrangement is based on a 74LS126 quad tri-state buffer which is enabled whenever all eight address lines are simultaneously high and the $\overline{\text{RD}}$ and $\overline{\text{IORQ}}$ control lines go low. The state of the four input lines is thus transferred to the lower four data lines (D0 to D3) whenever the microprocessor executes an instruction of the form, IN A, (FFH).

Each input is fitted with a pull-up resistor such that an unconnected input will return a logic 1. The state of a switch may be detected by simply connecting it between an input and the 0 V line. An open switch will result in a logic 1 state while a closed switch will generate a logic 0. Under no circumstances should the input voltage be allowed to fall outside the range 0 V to +5 V. Circuits for attaching temperature and light level sensors are described on page 185.

Figure 8.5 shows an alternative four-bit input arrangement based on a 74LS153 quad data selector (multiplexer). Here the state of the four input lines is transferred to the least significant data line (D0) whenever the processor reads from the port addresses given in Table 8.1:

Figure 8.4 *Simple 4-bit input port*

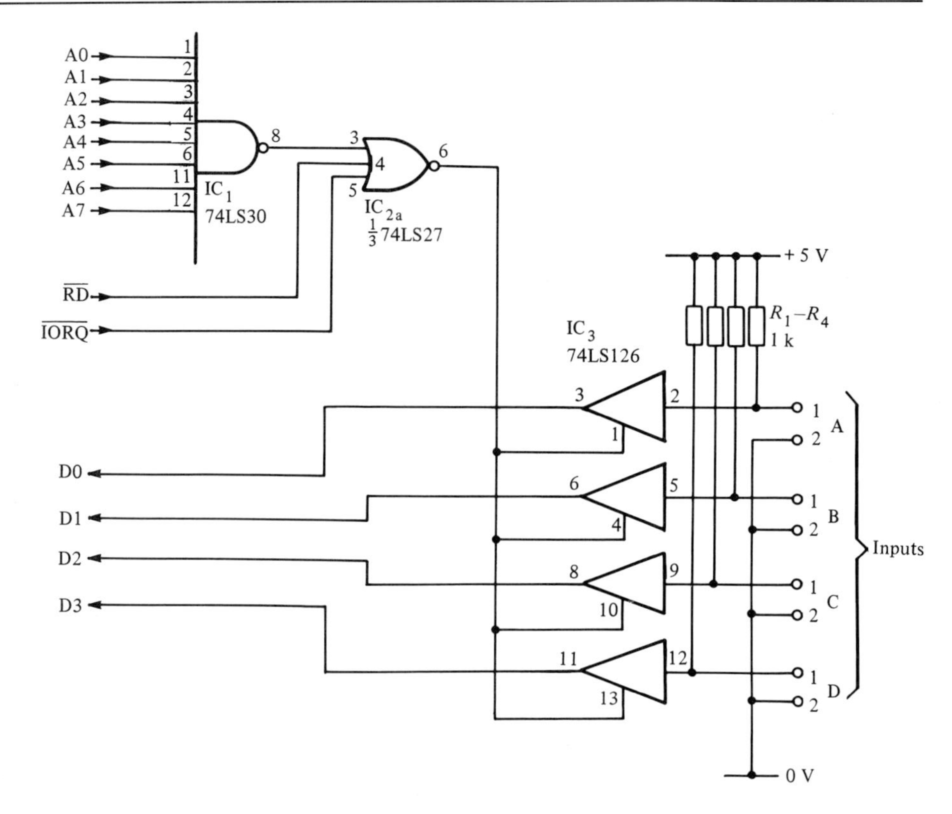

Figure 8.5 *Alternative form of 4-bit input port*

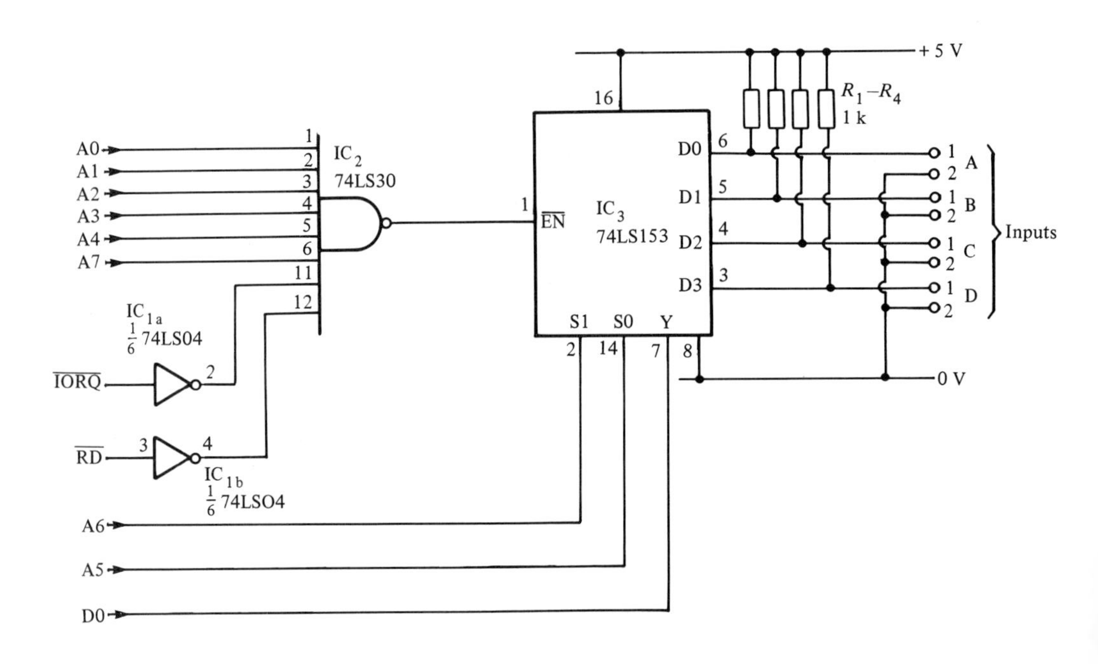

Table 8.1

Port address	*Input selected*
9F	A
BF	B
DF	C
FF	D

6502 based systems

Since the 6502 does not have the IN and OUT instructions possessed by the Z80, it is necessary to 'memory map' the I/O rather than treat it as a port. This simply means finding a memory address which is unpopulated with RAM, ROM or any other I/O device. Assuming that such an address can be found, suitable decoding logic can be devised. Unless one can make use of partial address decoding which may already be present within the microcomputer system, this well generally involve the full width of the address bus (i.e. all sixteen lines, A0 to A15).

Figure 8.6 shows decoding logic which will cope with most requirements. The inputs to this arrangement are arranged in two groups: those which must go low and those which must go high in order to provide an active low chip enable signal. Where a particular input is not required, it should be connected, depending upon the group to which it belongs, to either logic 0 or to logic 1 (via a pull-up resistor). To minimize loading on the address bus, the gates employed in the address decoder should be either LS-TTL or CMOS types.

Figure 8.7 shows an example of this technique in which an active low

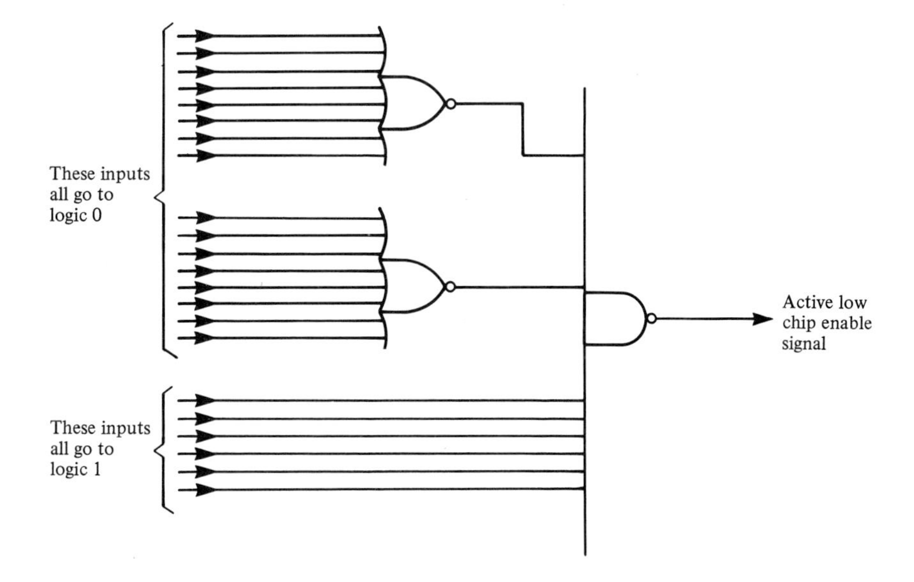

Figure 8.6 *General form of address decoder*

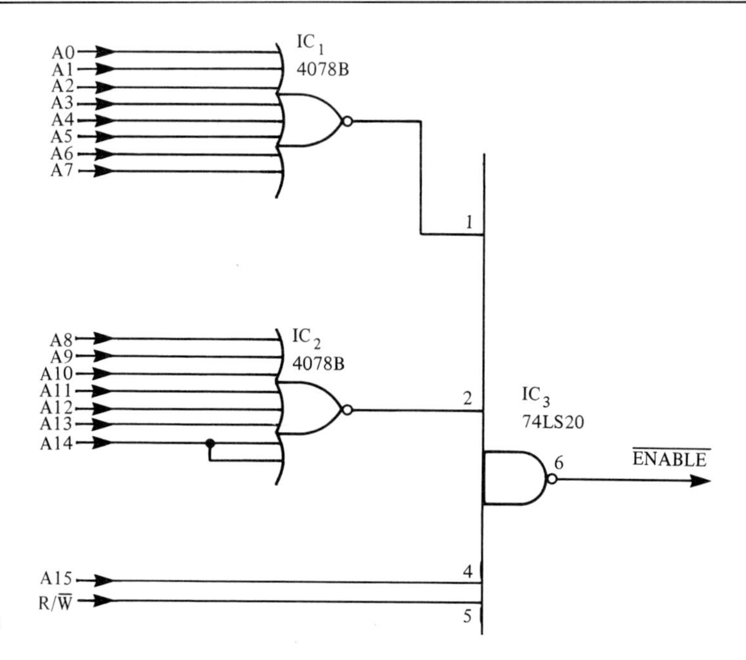

Figure 8.7 *Typical address decoder for a 6502 based system*

chip enable signal is generated when the processor reads data from a hexadecimal memory address of 8000. When this happens, the following bit pattern is generated:

$R\overline{W}$	*A15*	*A14*	*A13*	*A12*	*A11*	*A10*	*A9*	*A8*	*A7*	*A6*	*A5*	*A4*	*A3*	*A2*	*A1*	*A0*
1	1	0	0	0	0	0	0	0	0	0	0	0	0	0	0	0

Since only two inputs (A15 and $R/\overline{W}$) are high when the port is being read, we can make use of a four-input rather than an eight-input NAND.

Byte-wide input and output

A simple, general purpose byte-wide input and output interface can consist of nothing more than an octal input latch, an octal tristate buffer, and appropriate decoding logic, as shown in Figure 8.8. Before considering the adoption of such an arrangement, however, it would be wise to examine the alternative of adding one of the programmable I/O devices mentioned earlier. This may well prove to be not only more flexible but may well also prove to be simpler to construct.

For many applications (such as analogue to digital and digital to analogue conversion) there is an even simpler solution. Most modern peripheral chips are directly bus compatible and thus will not require the services of a buffer or data latch. They can simply be connected direct to the system data bus and have their $\overline{\text{ENABLE}}$ or $\overline{\text{CHIP SELECT}}$ inputs supplied from the address decoder.

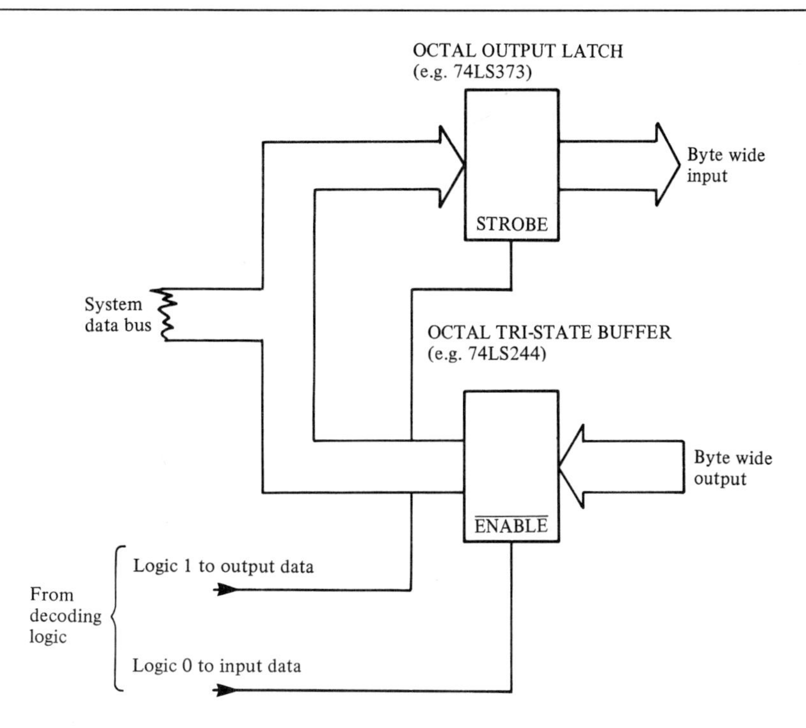

Figure 8.8 *Basic arrangement for implementing byte-wide I/O*

The UCN5801A is an example of one such microcomputer bus compatible device. This can form the basis of a very effective output driver. This versatile device is BI-MOS octal latch/driver consisting of eight CMOS data latches with CLEAR, STROBE and output ENABLE control lines coupled to eight bipolar Darlington driver transistors, as shown in Figure 8.9. This configuration provides an extremely low power latch with a very high output current capability.

All inputs of the UCN5801A are CMOS, NMOS and PMOS compatible (thus permitting direct connection to either an 8-bit data bus or eight of the port output lines of a programmable parallel I/O device). The outputs of the UCN5801A are all open-collectors, the positive supply voltage for which may be up to 50 V. Each Darlington output device is rated at 500 mA maximum. However, if that should prove insufficient for a particular application then several output lines may be paralleled together subject, of course, to the limits imposed by the rated load current of the high voltage supply.

Figure 8.10 shows a typical arrangement of the UCN5801 in which the high voltage supply is + 12 V. The state of the bus is latched to the output of IC_1 whenever the STROBE input is taken high. A logic 0 present on a particular data line will turn the corresponding Darlington output device 'off' while a logic 1 will turn it 'on'. It should also be noted that the output stages are protected against the effects of an inductive load by means of internal diodes. These are commoned at pin-12 and this should thus be returned to the positive load supply.

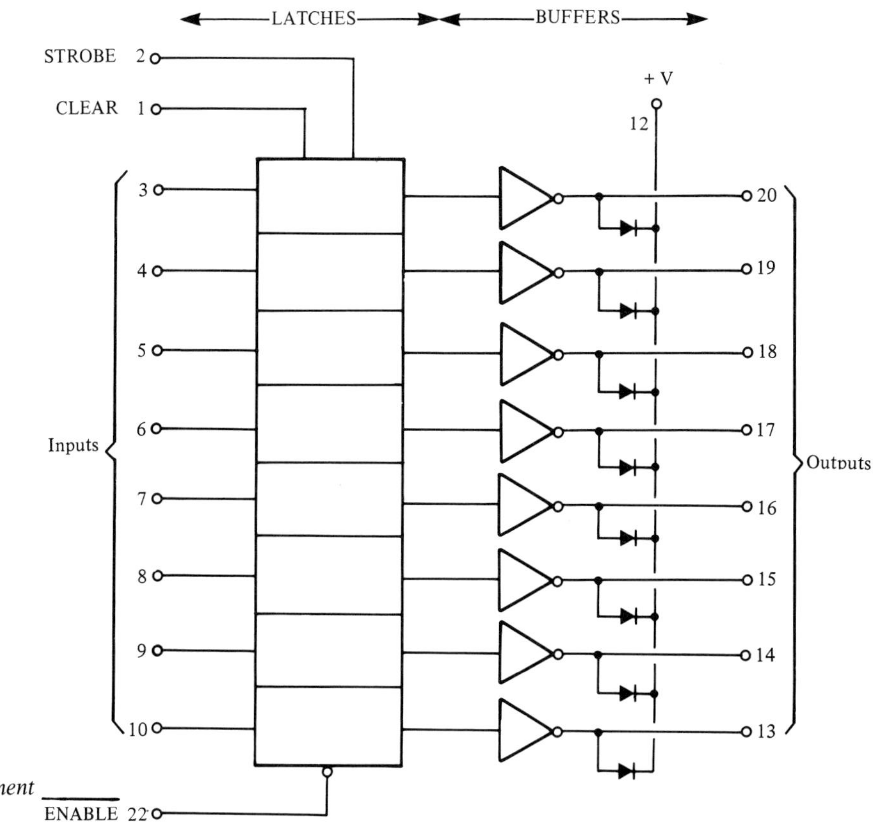

Figure 8.9 *Internal arrangement of the UCN5801A*

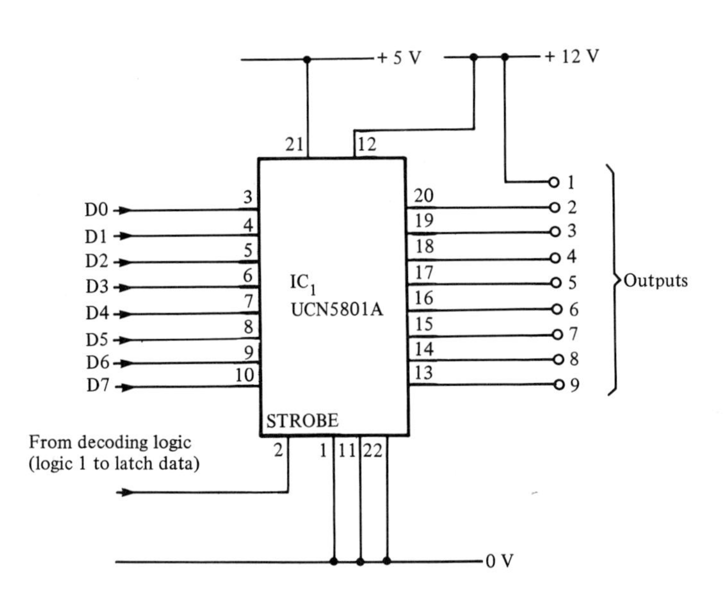

Figure 8.10 *Typical output driver based on the UCN5801A*

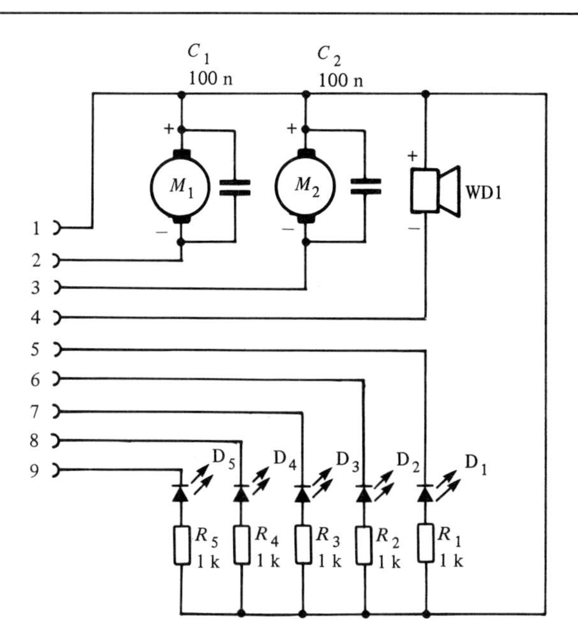

Figure 8.11 *Typical arrangement of output loads for the circuit of Figure 8.10*

Figure 8.11 shows how the circuit of Figure 8.10 can be used to drive two 12 V motors, a warning device, and five LEDs. The control byte sent to the device would then take the following form:

Device controlled	D5	D4	D3	D2	D1	WD1	M2	M1
Data line	D7	D6	D5	D4	D3	D2	D1	D0

As an example, suppose we wish to simultaneously operate M1 and D1, the corresponding output bit pattern would be, 00001001 (equivalent to a hexadecimal value of 09). Alternatively, to operate D5, D4, and WD1 simultaneously we would need to output a bit pattern of 11000100 (C4 hexadecimal).

Output drivers

Due to the limited output current and voltage capability of most I/O ports, external circuitry will normally be required to drive a load. Figure 8.12 shows some typical arrangements for operating various types of load. Figure 8.12(a) shows how a transistor can be used to operate a low-power relay. Where the relay requires an appreciable operating current (say, 100 mA or more) a VMOS FET should be used, as shown in Figure 8.12(b). Such devices offer very low values of 'on' resistance coupled with a very high 'off' resistance and, unlike conventional bipolar transistors, impose a negligible load on the I/O port.

Figure 8.12(c) shows a filament lamp driver (simple LED driver circuits

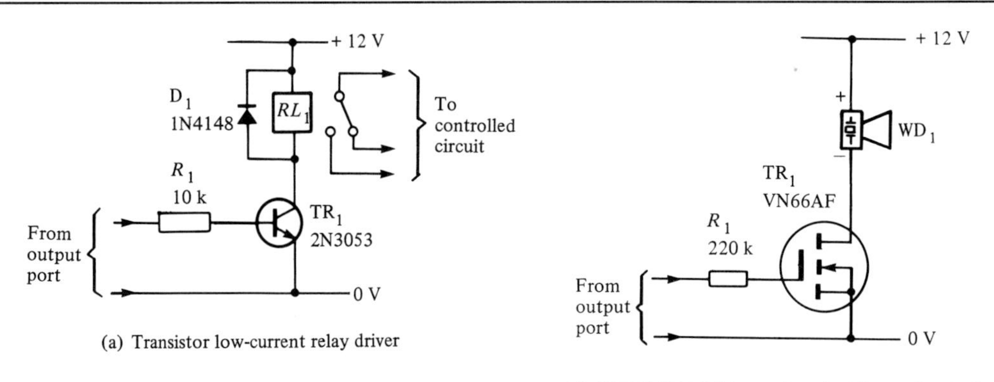

(a) Transistor low-current relay driver

(d) VMOS FET driving a piezoelectric warning device

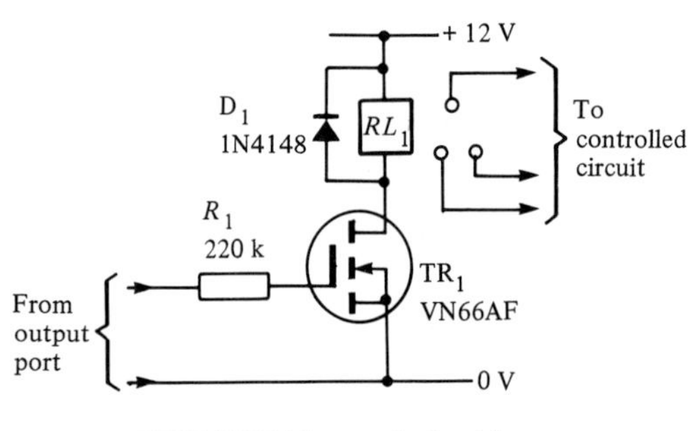

(b) VMOS FET high-current relay driver

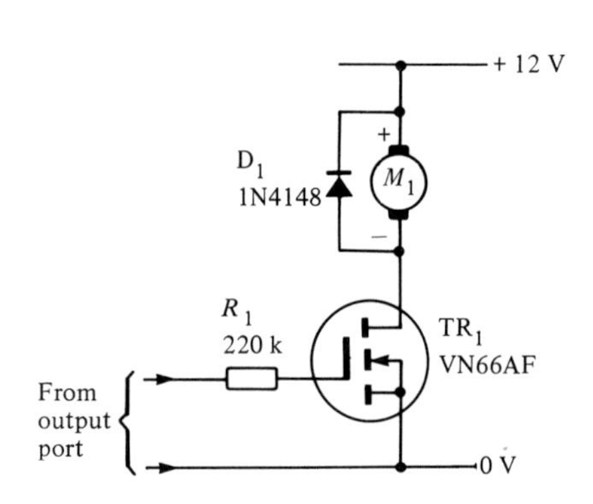

(e) VMOS FET d.c. low-voltage d.c. motor driver

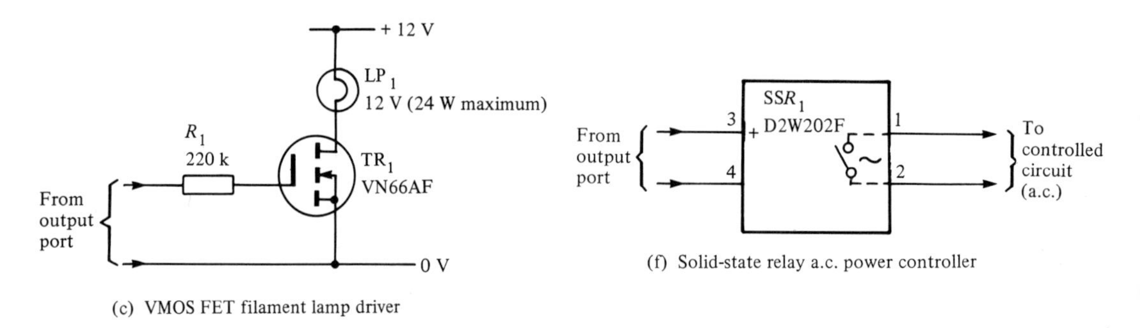

(c) VMOS FET filament lamp driver

(f) Solid-state relay a.c. power controller

Figure 8.12 *Typical output drivers*

appear in Chapters 2 and 6) while Figure 8.12(d) shows an arrangement for driving a piezoelectric warning device.

A VMOS FET low-voltage d.c. motor driver circuit is shown in Figure 8.12(e). A high from the output port operates the motor which should operate with a stalled current of less than 1.5 A.

Figure 8.12 (f) shows an arrangement for operating a mains connected load using a solid-state relay. Such devices are internally optically coupled and provide a very high degree of isolation between the port and the load. The D2W202F from International Rectifier has a typical input resistance of 1.5 kΩ and will thus interface directly with most TTL devices and Port B output lines. Controlled voltages can range from 60 V to 280 V a.c. at currents of up to 2 A. Maximum 'off' state leakage current is 5 mA and the device provides an isolation of up to 2.5 kV.

Input conditioning

The input signals required by microcomputer control systems are unfortunately rarely TTL compatible. Hence, in the majority of applications it is necessary to include additional circuitry between the sensor and input port.

Where switches are to be interfaced, a number of circuits have already been described in Chapter 6. Similarly, where sensors producing low levels of analogue signal are to be employed, the comparator arrangements described in Chapter 7 can be used. In both cases, the objective is that of making the port input signal TTL compatible.

There are, however, a few situations in which minimal input signal conditioning is required. These include simple temperature and light level sensing along the lines shown in Figure 8.13.

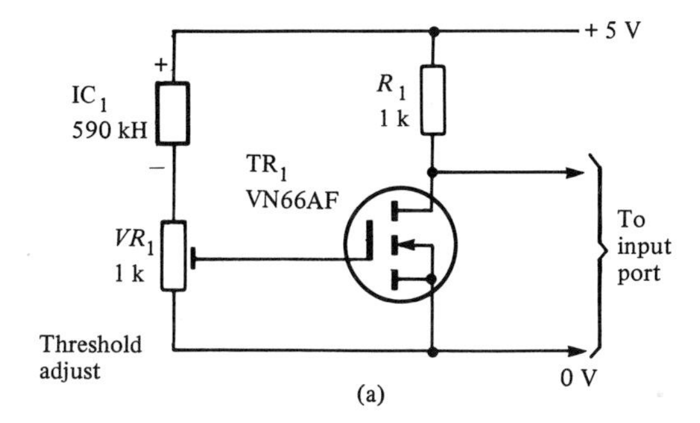

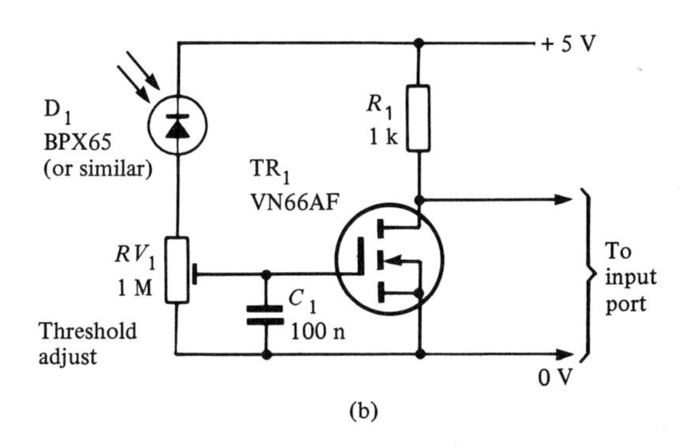

Figure 8.13 (*a*) *Simple interface for a temperature sensor*
(*b*) *Simple interface for a light level sensor*

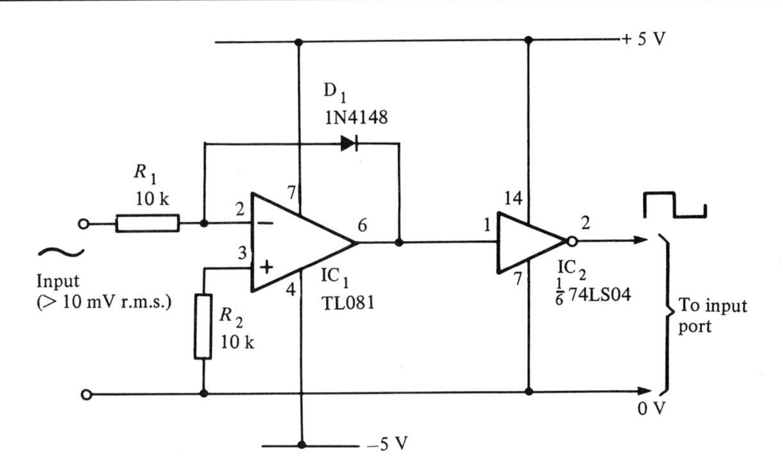

Figure 8.14 *Interface for low-level audio and mains failure detection*

Figure 8.13(a) shows how light level can be sensed using a photodiode sensor. This arrangement generates a logic 0 input whenever the light level exceeds the threshold setting, and vice versa.

Figure 8.13(b) shows how temperature can be sensed using a 590 kH sensor. This arrangement generates a logic 0 input whenever the temperature level exceeds the threshold setting, and vice versa.

Finally, Figure 8.14 shows how an external a.c. source can be coupled to an input port. This arrangement produces TTL compatible input pulses having 50% duty cycle. The circuit requires an input of greater than 10 mV r.m.s. for frequencies up to 10 kHz and greater than 100 mV r.m.s. for frequencies up to 100 kHz. The obvious application for such an arrangement is detection of audio frequency signals but, with its input derived from the low voltage secondary of a mains transformer (via a 10:1 potential divider), it can also function as a mains failure detector.

Driving stepper motors

The complex task of interfacing a stepper motor to a microcomputer system can be much simplified by using a dedicated stepper motor driver chip such as the SAA1027. This device includes all necessary logic to drive a stepper motor as well as output drivers for each of the four phases. The chip operates from a nominal + 12 V supply rail but its inputs are not directly TTL compatible and thus transistor or open-collector drivers will normally be required.

Figure 8.15 shows a typical stepper motor interface based on the SAA1027. The motor is a commonly available four-phase two-stator type having the following characteristics:

Supply voltage:	12 V
Resistance per phase:	47 Ω
Inductance per phase:	400 mH
Maximum working torque:	50 mNm
Step rotation:	7.5°/step

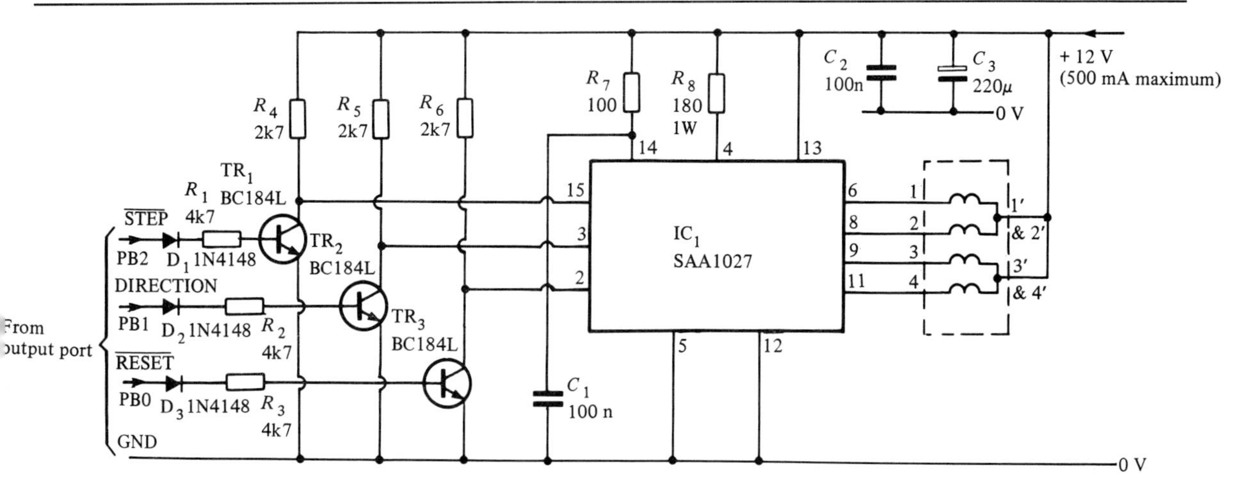

Figure 8.15 *Stepper motor interface using an SAA1027 driver*

The pin connections of the motor are shown in Figure 8.16.

The stepper motor interface requires only three port output lines which operate on the following basis:

(a) The STEP input is pulsed low to produce a step rotation.
(b) The DIRECTION input determines the sense of rotation. A low on the DIRECTION input selects clockwise rotation. Conversely, a high on the DIRECTION input selects anticlockwise rotation.
(c) The RESET input can be taken low to reset the driver.

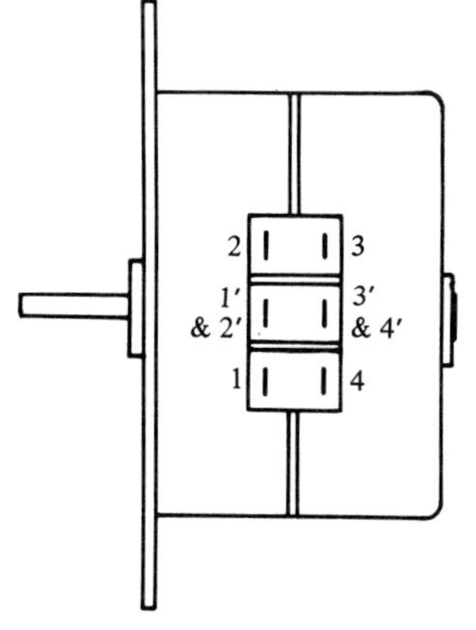

Figure 8.16 *Stepper motor connections*

The software routines for driving the stepper motor are quite straightforward. As an example, assume that the interface is connected to the Port B lines of a Z80-PIO as follows:

Table 8.2

Function	*Port Line*
RESET	PB0
DIRECTION	PB1
STEP	PB2

If the hexadecimal port address is 7E (data register) and 7 (control register); the following code will initialize the PIO into mode 3 such that the three least significant lines are outputs and the remainder are inputs:

```
LD A, FFH     :Select mode three
OUT (7FH), A :operation
LD A, F8H     :PB0 to PB2 will be outputs
OUT (7FH), A :the remainder will be inputs
```

To reset the driver the following lines of code will be required:

```
XOR A          :Turn all bits ‘off’
OUT (7EH), A :to reset
```

To select clockwise rotation:

```
LD A,0          : Select clockwise
OUT (7EH), A : rotation
```

To select anticlockwise rotation:

```
LD A,2          : Select anticlockwise
OUT (7EH), A : rotation
```

To produce a 7.5° anticlockwise rotation:

```
LD A, FEH      : First take PB2
OUT (7EH), A : high,
LD A, FBH      : then take it
OUT (7EH), A : low
```

The last segment of code can be enclosed within a conditional loop in order to produce continuous rotation. In practice, bit masks should be used in order to preserve the state of unchanged bits. This technique is adequately documented in most assembly language textbooks.

Digital to analogue conversion

In many control applications it is necessary to provide an analogue, rather than digital, output. This can be achieved by means of a digital to analogue converter (DAC). The principle of operation of two types of DAC is illustrated in Figure 8.17.

Figures 8.17(a) and 8.17(b) respectively show how binary weighted and *R*-2*R* resistor networks can be used in conjunction with a four-bit latch and a summing operational amplifier. This circuit is capable of producing 16 (2^4) different analogue output levels (including zero). In practice, four-bit DAC offer insufficient resolution and it is invariably better to make use of a dedicated DAC chip rather than attempt to construct such an arrangement using individual digital and analogue devices in conjunction with conventional discrete resistors.

A representative 8-bit DAC is shown in Figure 8.18. The circuit is based on the popular Ferranti ZN428E which incorporates an internal *R*-2*R* network and bus-compatible 8-bit data latch. The chip also contains its own precision 2.5 V voltage reference. The analogue output appearing at pin-3 of IC_1 is fed to an adjustable gain non-inverting operational amplifier stage, IC_2.

The circuit is calibrated by first outputing a zero byte to the interface and adjusting RV_1 for an output of exactly 0 V. A second byte of FF (hex) is then output and RV_2 adjusted for the desired full-scale output. If, for example, this is set to 2.55 V, the DAC will produce 256 discrete output voltages ranging from 0 V to 2.55 V in 10 mV steps.

As an example of the use of the DAC let us assume that we need to generate a stepped-ramp waveform and that the decoding logic produces an active low enable signal corresponding to a hexadecimal port address

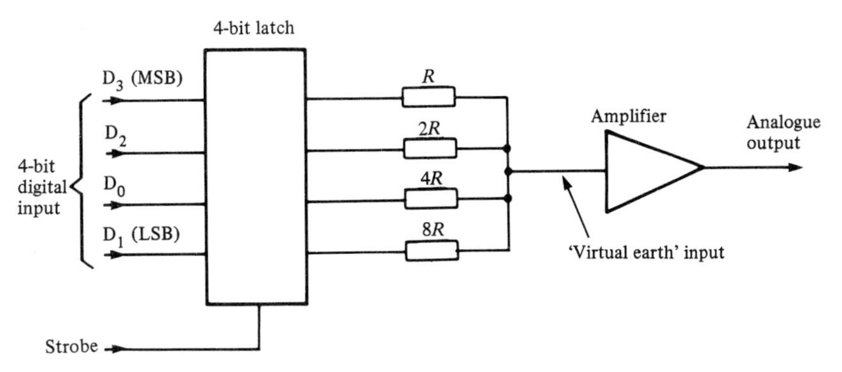

(a) Using a binary weighted ladder network

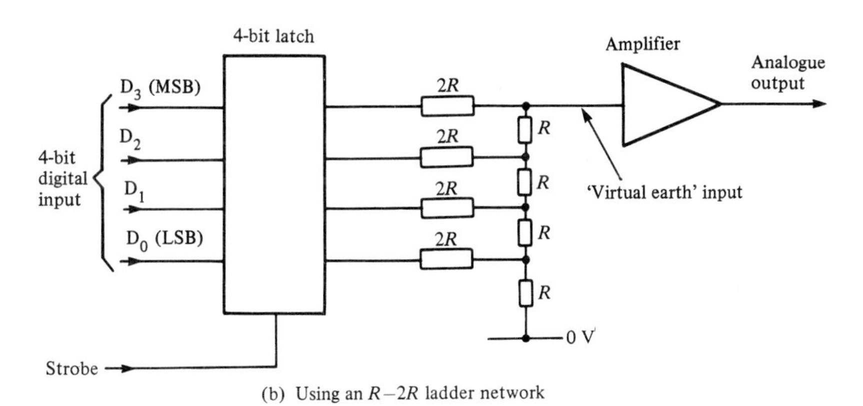

(b) Using an $R-2R$ ladder network

Figure 8.17 *Basic arrangement of 4-bit DAC*

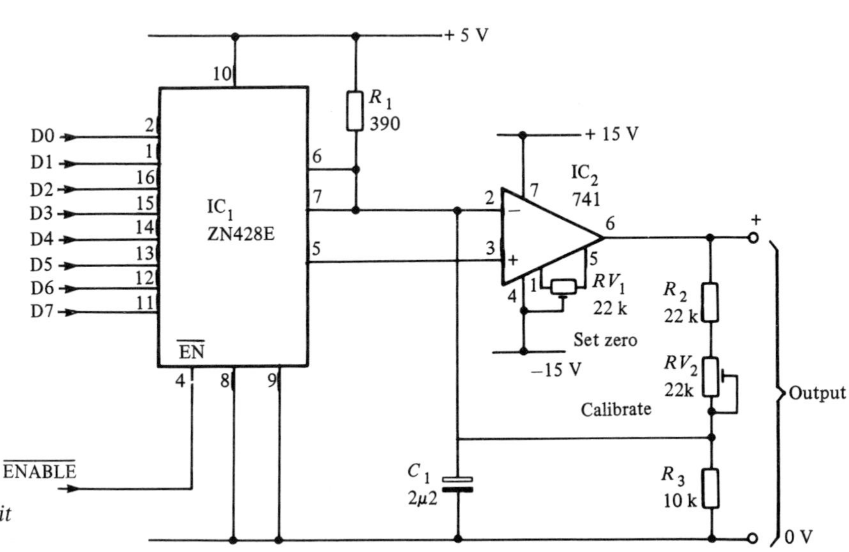

Figure 8.18 *Representative 8-bit DAC using a ZN428E*

of FFH. The following Z80 code could be used to generate a low-frequency stepped ramp waveform:

```
LOOP1  LD A, FFH       :Generate all 256 voltage levels
       LD C, 0         :Starting at zero
LOOP2  OUT (FFH), C    :Send it to the DAC
       INC C           :Count up in C
       DEC A           :Count down in A
       CP 0            :Finished one cycle?
       JR NZ, LOOP2    :No – keep going!
       JP LOOP1        :Yes – now start another!
```

In practice, appropriate delays can be added to produce any desired output frequency up to a maximum which is ultimately determined by the microprocessor's clock frequency.

In applications where an appreciable value of output current is required, an additional output buffer may be added. Figure 8.19 shows how this can be achieved using a 759 power operational amplifier. The stage operates with a unity gain non-inverting buffer. Where a larger output is required, or where the output polarity is to be reversed, the circuit can be readily modified using the general information given for operational amplifiers in Chapter 5.

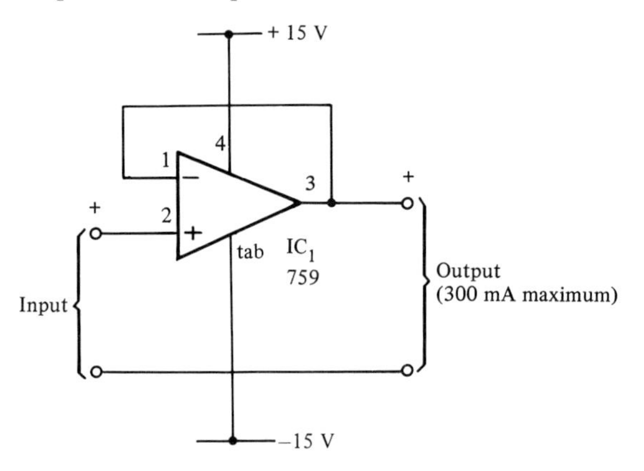

Figure 8.19 *Unity gain buffer for use with the arrangement shown in Figure 8.18*

Analogue to digital conversion

The output of many transducers exists in analogue rather than digital form. In such cases, and assuming that we require more than just a simple on/off form of control, it is necessary to incorporate a means of analogue to digital conversion within an interface. This can readily be accomplished using one of several commonly available dedicated analogue to digital converter (ADC) chips.

A simple ADC arrangement is shown in Figure 8.20. This circuit uses an 8-bit CMOS ADC, the ADC0804, which incorporates a tristate output latch thus permitting direct connection to a microcomputer bus. The ADC0804 requires an external voltage reference which is provided by a

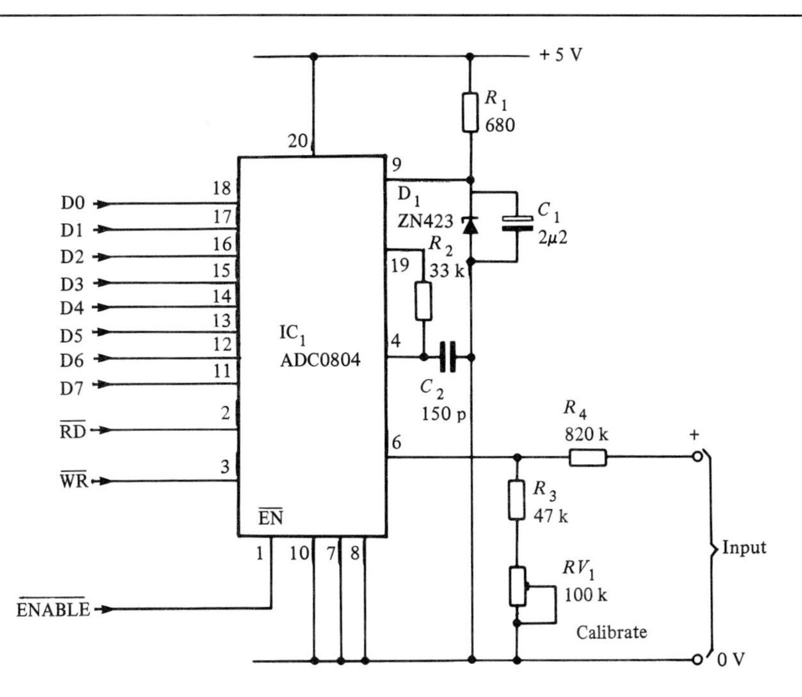

Figure 8.20 *Simple 8-bit ADC using an ADC0804*

precision band-gap device, D_1. The frequency of the ADC0804's internal clock is determined by R_2 and C_2.

Scaling of the input voltage is provided by a simple potential divider arrangement comprising RV_1, R_3 and R_4. The input resistance of this circuit is nominally 1 MΩ and RV_1 should be adjusted so that a full-scale output results from an input of 25.5 V. In condition the converter operates in steps of 100 mV.

A typical Z80 assembly language routine for starting conversion and returning the converted value (in the *C* register) would be as follows:

```
GETIN   LD A, FFH       : Start conversion
        OUT (BFH), A    : by setting all bits high
        IN A, (BFH)     : Read the value and
        LD C, A         : preserve it in the C reg.
        RET             : Go back!
```

The interface of Figure 8.20 suffers from a number of limitations, not the least of which is that it only provides for a single analogue input channel. The obvious solution to this problem is that of using a number of CMOS analogue switches connected to the input of an ADC0804. Individual analogue inputs could then be connected when required or, alternatively, the software could be arranged so that the input channels are successively sampled. A four-channel ADC would require four analogue switches while an eight-channel unit would need eight analogue switches, and so on.

In either case an output latch would be necessary in order to provide

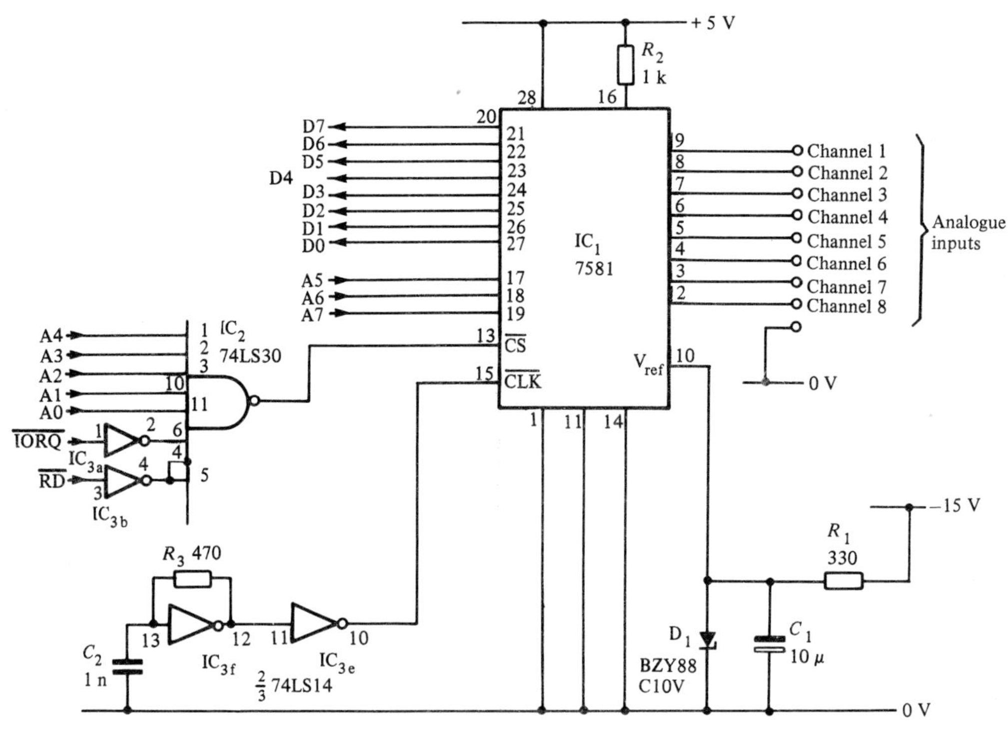

Figure 8.21 *Eight channel 8-bit ADC using a 7581*

temporary storage for the data applied to the control inputs of the analogue switches. Unfortunately, such an arrangement would probably necessitate the use of more than six DIL integrated circuits and construction could be something of a headache. A far better solution would be that of making use of a multi-channel ADC such as the 7581.

The 7561 is an 8-bit 8-channel ADC which incorporates its own internal 8 × 8 dual-port RAM. The device employs successive approximation techniques and results are stored internally until required. Conversion of a single channel takes 80 clock periods with a complete scan through all eight channels taking 640 clock cycles.

When channel conversion is complete, the successive approximation register's contents are transferred into the appropriate internal RAM location. The contents of this RAM can later be examined by placing the appropriate binary address pattern on the A0, A1, and A2 lines while, at the same time, taking the $\overline{CS}$ line low.

To ensure that memory updates only occur when the host micro-computer is not addressing the converter's memory, automatic interleaved direct memory access (DMA) is provided by on-chip logic.

The circuit of a 7581-based ADC is shown in Figure 8.21. Appropriate address decoding must be provided for the $\overline{CS}$ line (in this case it is provided by IC_2 and IC_3 which ensures that the output of the 7581 is only placed on the data bus when address lines A0 to A4 are all high with the $\overline{RD}$ and $\overline{IORQ}$ simultaneously low). The remaining three address lines (A5 to A7) used for conventional I/O port addressing are taken to the 7581's address inputs. This arrangement results in the following address allocation:

Table 8.3

	Address		
Channel no.	*Binary*	*Decimal*	*Hex*
1	00011111	31	1F
2	00111111	63	3F
3	01011111	95	5F
4	01111111	127	7F
5	10011111	159	9F
6	10111111	191	BF
7	11011111	223	DF
8	11111111	255	FF

IC_{3f} and IC_{3e} act as a simple buffered Schmitt oscillator (see Chapter 6 for further information) which provides a square wave to the 7581's clock input at a frequency of approximately 2 MHz (the precise frequency of this signal is unimportant). The negative reference voltage required by the 7581 is regulated by a shunt zener regulator, D_1.

The 7581-based eight channel analogue to digital converter is delightfully simple to use. The converter continuously samples each of the eight input channels without the need for any start conversion commands from the host CPU. Since the outboard clock runs at approximately 2MHz and the 7581 takes 80 clock cycles to convert each channel input, the total time taken to update the 7581's internal RAM is around 320 μs. In order to read the state of a particular channel it is only necessary to include an instruction of the form IN A, (*n*) where *n* is the port address. The value returned in the accumulator will be in the range 0 to FF hexadecimal (the maximum value corresponding to a full-scale value equal, but of opposite polarity, to the reference voltage).

9 Tools and test equipment

This section deals with the tools and test equipment necessary for the successful design, manufacture, and construction of electronic circuits. Simple circuits, such as audio frequency amplifiers and power supplies, can be constructed using nothing more than a few basic tools and some very simple items of test equipment. More complex circuitry naturally requires the services of more complex production equipment as well as a range of more sophisticated test equipment. This fact, however, should not serve to daunt the newcomer; the acquisition of the 'right tools and test equipment for the job' can be a gradual process.

Provided one is not too ambitious, it is not necessary to have access to a vast range of tools and test equipment at the outset. Readers should, therefore, not be deterred from 'having a go' – some of the best, and most innovative items of equipment having been constructed by enthusiastic amateurs on the 'kitchen table'! Nowadays, this is the exception rather than the rule. Electronic design is now almost entirely the province of the professional engineer working from a well equipped design laboratory. Don't forget, though, that everybody has to start from somewhere – the very best of facilities will be of little use of the user is lacking in flair and imagination.

Tools

Provided they are used properly and are well cared for, good quality tools can be expected to last a lifetime. There is little sense in purchasing inferior tools that will need replacing every few years and thus it is wise to purchase the highest quality items that you can afford. Many of the basic tools for electronics will probably already be available and it should therefore not be necessary to devote a vast capital outlay to the purchase of such items.

Fortunately, relatively few basic tools are absolutely essential in order to construct and test the majority of simple electronic projects. A set of small hand tools (long nosed pliers, side cutters, cross point and flat bladed screwdrivers of various sizes) is a good starting point. Added to this initial purchase should be a good quality, preferably low-voltage temperature controlled, soldering iron, with a selection of bits. If a temperature controlled iron is considered prohibitively expensive, then a good quality mains operated iron rated at between 15 W and 25 W should suffice for most applications. Readers should, however, check that spare bits, elements, and accessories are available for any iron that they purchase – some of the cheaper soldering irons fail badly on this point.

In addition to a soldering iron, a desoldering pump is also a worthwhile purchase. Such an item need only be a manually operated suction type and need not cost more than £5 or so. This item will, however, prove to be invaluable for desoldering a variety of devices (including integrated circuits).

A set of two or three trimming tools is also a worthwhile investment. These differ from conventional screwdrivers in that they have non-metallic shafts and bodies and are designed to facilitate the adjustment of ferrite cores (some of which have hexagonal slots) and miniature preset capacitors and potentiometers. These components can be easily damaged when using conventional screwdrivers, so beware.

As a guide, and for those who prefer to make a systematic collection of the tools needed, 'basic' and 'extended' lists of tools are given. The basic list represents the minimum complement of tools considered necessary by the author for even the most basic of electronic construction and testing. The extended list includes items that, while not required for everyday use, will be found invaluable in the longer term.

Basic list of tools

Small pair of side cutters
Small pair of snipe-nose (half-round) pliers
Pair of combination pliers and cutters
Set of flat-blade screwdrivers (small, medium, and large)
Set of cross-point screwdrivers (small, medium, and large)
Set of trimming tools
Small hand drill
Set of drill bits
Desoldering pump

And either a temperature controlled soldering iron with selection of bits or a miniature soldering iron rated at between 15 W and 24 W.

Extended list of tools

As basic list plus the following additional items:
Spot face cutter
Small pair of flat-nose pliers
Pair of wire strippers
Set of jeweller's screwdrivers
Set of small files
Small hacksaw
Small bench vice
Trimming vice
Pair of tweezers
Magnifying glass or bench magnifier

Portable soldering iron (butane or rechargeable type)
p.c.b. drill and drill-stand
Set of p.c.b. drill bits
Metal rule
Scribe
Centre punch
Set of chassis cutters
Set of files (including 'round' and 'flat' types)

Test equipment

Test equipment is important not only as a means of testing equipment to specification but also as an aid to calibration, trouble shooting and general fault-finding. Currently, a vast range of electronic test equipment is available which varies not only in sophistication but also in price. At one extreme are such low-cost everyday items as the humble 'non-electronic' analogue multimeter while at the other can be found sophisticated microprocessor controlled equipment which provide semi-automatic testing and calibration facilities at prices which run well into four figures.

As in the previous section, 'basic' and 'extended' lists have been provided for guidance. The basic list includes those items which, in the author's experience, are most regularly used. It is not necessary to have access to *all* of these items to accomplish any particular task. Indeed, the only *essential* item is a good quality analogue or digital multimeter. Items on the 'basic' list are essential if a broad range of work is to be undertaken. This should not, however, deter the enthusiast from making a start. Test equipment can be acquired over a period of time, starting, of course, with a good multirange meter. Additional items, including those on the extended list can be added when funds become available. Furthermore, it is best not to be in too much of a hurry to extend the range of test equipment available. Readers will soon get to know the instruments from which they derive most benefit and this will undoubtedly point the way to future purchases.

Basic list of test equipment

Multirange meter (good quality analogue or digital type)
Oscilloscope (preferably dual-beam type) with '$\times$ 10' probe
Audio frequency signal generator (with sine and square wave outputs)
Logic probe
Variable low-voltage d.c. power supply
Selection of test leads and prods

Extended list of test equipment

As basic list plus:

Digital counter/frequency meter

Transistor tester
Logic pulser
Pulse generator
Function generator
Audio frequency power meter
Radio frequency signal generator
Radio frequency power meter
Distortion test set
Selection of connectors and adaptors
Selection of IC test clips
Integrated circuit tester

With the exception of d.c. power supplies which are described in detail in Chapter 3, we shall now consider each of the basic items of test gear in turn. Desirable performance specifications will be given for each instrument and typical applications discussed. Full operating principles have not been included as these are beyond the scope of this book. Where further information is required, readers are advised to consult an appropriate text on test equipment design.

Multirange meters

The multirange meter is undoubtedly the most often used test instrument in any electronic laboratory of workship. The instrument should provide d.c. and a.c. voltage and current ranges together with resistance ranges. The choice of analogue or digital types is very much a question of personal preference and either type should suffice for basic construction, testing and fault-finding. Additional ranges which permit continuity testing, diode and transistor measurements, etc. are useful, but not essential. In any event, readers are advised to 'shop around'; a useful starting point being the advertisements which appear in various technical magazines.

Analogue versus digital

Analogue instruments employ conventional moving coil meters and the display takes the form of a pointer moving across a calibrated scale. This arrangement is not so convenient in use as that employed in digital instruments. It does, however, offer some advantages, not the least of which lies in the fact that it is very difficult to make adjustments using a digital readout to monitor continuously varying circuit conditions. In such applications, the analogue meter is vastly superior; the absolute reading produced by such an instrument being less significant than the direction in which the pointer is moving.

The principal disadvantage of many analogue meters is the rather cramped, and somewhat confusing, scale calibration. To determine the exact reading requires first an estimation of the pointer's position and then the application of some mental arithmetic based on the range switch setting. Digital meters, on the other hand, are usually extremely easy to read and have displays which are clear, unambiguous, and capable of providing a resolution which is far greater than that which can be obtained with an analogue instrument. See Plate 1.

Another very significant difference between conventional analogue and digital meters is the input resistance which they present to the circuit under investigation when making voltage measurements. The resistance of a reasonable quality non-electronic analogue multimeter can be as low as 50 kΩ on the 2.5 V range. With digital instruments the corresponding resistance is typically 10 MΩ on the 2 V range. The digital instrument is thus vastly superior when accurate readings are to be taken on circuits which are susceptible to loading.

It is well worth putting this into context with a simple example. Let us suppose that two typical multimeters (one digital and one analogue) are used to measure the voltage produced by the two simple potential dividers shown in Figure 9.1. To make the arithmetic simple, we shall assume that the analogue meter has a sensitivity of 10 kΩ/V and that it is used on the 10 V d.c. range. Its internal resistance (i.e. the resistance which it presents to the outside world) will thus be $10\,k\Omega/V \times 10\,V = 100\,k\Omega$. Now let us assume that the digital instrument has an internal resistance of 10 MΩ (constant on all ranges).

Since both potential dividers use equal value resistors, the true value of the voltage produced should be 5 V (i.e. exactly half the input voltage) and an 'ideal' meter should indicate this value. The actual readings are given in Table 9.1:

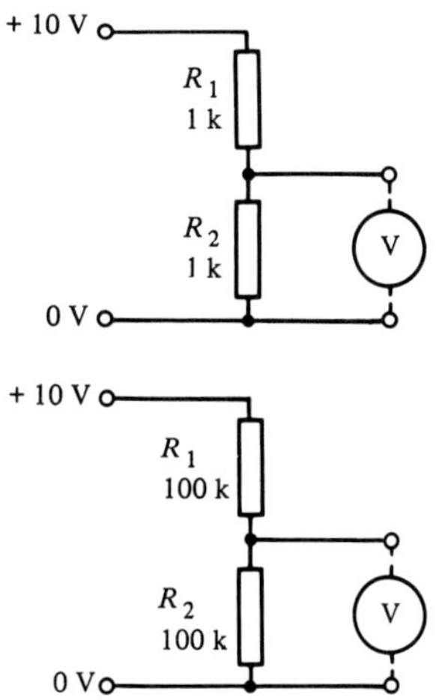

Figure 9.1 *Potential dividers to illustrate the effects of voltmeter loading*

Table 9.1

Meter	*Potential divider*	
	(*a*)	(*b*)
Typical analogue type (10 kΩ/V)	4.9 V	3.3 V
Typical digital type (10 MΩ)	4.997 V	4.974 V

The staggering difference in the case of circuit (b) illustrates the effect of voltmeter loading on a high resistance circuit. An appreciable current is drawn away from the circuit when the relatively low internal resistance analogue meter is used and this is clearly a very undesirable effect.

Digital instruments are often somewhat limited in terms of their frequency response on the a.c. current and voltage ranges. This is often restricted to the low end of the audio frequency spectrum and many analogue instruments offer far superior response in this respect.

One of the principal limitations of the 'non-electronic' analogue meter concerns the resistance ranges. These are distinctly non-linear and it is extremely difficult to obtain any sort of meaningful absolute indication of resistance for high values of resistance (say, above 100 kΩ). If you require a meter which can provide accurate measurements of resistance (to within, say, ±1%) it is important to opt for a digital type or an 'electronic' analogue meter.

Another important difference between the two types of meter relates to their mechanical ruggedness. With one or two notable exceptions (e.g. the traditional AVO meter) analogue instruments do not have a very good track record in this respect. The reason for this is simply that a moving coil meter is a rather delicate device. Furthermore, the more sensitive the instrument the more susceptible it will be to damage due to mishandling. A drop from bench surface to the floor will almost certainly do the meter movement no good at all. Digital instruments are much superior in this respect; they can readily withstand the knocks which arise from everyday use in a busy workshop.

So that comparisons can be made, the following specifications are typical of medium cost analogue and digital instruments:

Analogue type

Voltage ranges:	
d.c.	1 V, 3 V, 10 V, 30 V, 100 V, 300 V, and 1 kV (full-scale).
a.c.	3 V, 10 V, 30 V, 100 V, 300 V, and 1 kV (r.m.s. full-scale).
Sensitivity:	
d.c.	20 kΩ/V.
a.c.	5 kΩ/V.
Current ranges:	
d.c.	300 μA, 3 mA, 30 mA, 300 mA, and 3 A (full-scale).
a.c.	3 mA, 30 mA, 300 mA, and 3 A (r.m.s. full-scale, 20 Hz to 10 kHz).
Resistance:	50 Ω, 500 Ω, 5 kΩ, and 50 kΩ (mid-scale).
Accuracy:	
d.c. voltage and current ranges	±1% of full-scale.
a.c. voltage and current ranges	±2.5% of full-scale.
resistance	±10% typical at mid-scale.

Digital type

Voltage ranges:	
d.c.	200 mV, 2 V, 20 V, 200 V, and 1 kV (full-scale).
a.c.	200 mV, 2 V, 20 V, 200 V, and 750 V (r.m.s. full-scale, 40 Hz to 1 kHz).
Input resistance:	
d.c.	10 MΩ on all ranges.
Input impedance:	
a.c.	10 MΩ shunted by 100 pF on all ranges

Current ranges:	
d.c.	200 μA, 2 mA, 20 mA, 200 mA, and 2 A (full-scale).
Current ranges:	
a.c.	200 μA, 2 mA, 20 mA, 200 mA, and 2 A (r.m.s. full-scale, 40 Hz to 1 kHz).
Resistance:	200 Ω, 2 kΩ, 20 kΩ, 200 kΩ, 2 MΩ and 20 MΩ (full-scale).
Accuracy:	
d.c. voltage and current ranges	±(0.5% reading +1 digit)
a.c. voltage and current ranges	±(0.75% reading +5 digits)
resistance	±(0.3% reading +1 digit)

Choosing a digital multimeter

There is such a wide range of digital multimeters currently available that to say one is spoilt for choice is something of an understatement. However, the majority of such instruments conform to a fairly standard specification (see above). There are, however, one or two important differences of which the potential buyer should be fully aware.

Due to their modest power requirements, portable digital meters invariably use liquid crystal displays (LCD) displays. These, unfortunately, can be somewhat difficult to read under certain lighting conditions, the viewing angle is limited and the display response can be rather slow. Having said all that, LCD instruments are a 'must' for portable instruments since a comparable LED display has an excessive power requirement. On the other hand, where one is looking for a bench instrument which is to be operated from the a.c. mains supply, an LED display has much to recommend it.

It is important to check that the instrument which you purchase has an effective means of protection in the event of misconnection or incorrect range selection. None of us are perfect in this respect and it is well worth checking that your prized possession is protected against human failure.

Protection circuitry takes various forms and digital multimeters are invariably separately protected on the current and voltage ranges. In the former case, most instruments employ either a non-linear 'varistor' element or a low-capacitance spark-pap, both of which conduct heavily whenever the input voltage exceeds a certain critical value. On the current ranges, a quick-blow fuse and pair of anti-parallel silicon high-current fast-switching diodes are usually fitted. Finally, it is important to check that the fuse is fitted in a reasonably accessible position; at least one popular instrument has to be partially dismantled in order to replace the current protection fuse.

Choosing an analogue multimeter

Having already mentioned the consequences of using a meter which has a relatively low internal resistance, it should be obvious that one of the prime requirements of an analogue multimeter is that it should have a high sensitivity. The sensitivity of an analogue multimeter can be expressed in two ways: one of these involves specifying the full-scale deflection current of the moving coil meter (typically 50 μA, or less) while the other involves quoting an 'ohms per volt' rating. This latter specification is, in effect, the resistance presented by the meter when switched to the 1 V d.c. range.

The 'ohms per volt' (Ω/V) rating is inversely proportional to the basic full-scale sensitivity of the meter movement and, to determine the resistance of the meter on any particular range, it is only necessary to multiply the range setting by the 'ohms per volt' rating. Some common ratings are given below:

Meter f.s.d.	*Ohms per volt*
10 μA	100 kΩ/V
20 μA	50 kΩ/V
50 μA	20 kΩ/V
100 μA	10 kΩ/V
200 μA	5 kΩ/V

In the author's experience, values of less than 20 kΩ/V are unsatisfactory and it is well worth spending a little more in order to obtain an instrument which has a higher sensitivity.

Having avoided instruments which are very likely to cause problems due to loading, it is worth checking the extent of the various ranges provided. In general, it is a good idea to choose an instrument with as many functions and ranges as possible. Do not worry, however, if the meter does not have a 'decibel' or 'capacitance' range. These are, in the author's experience, of very limited use. Check, also, that the range switching is logical and can be easily accomplished with one hand. Avoid instruments with a multiplicity of input sockets; the process of swapping leads from socket to socket can be extremely frustrating.

Another important point relates to the meter scale. It should not be cramped, neither should it be so small that it cannot be read from a reasonable distance. The 'ohms' scale is likely to be the worst offender as far as this is concerned so do check that this particular scale has markings which can be easily read over at least the range 10 Ω to 10 kΩ.

The accuracy of an analogue meter is unlikely to be much better than about ±2% on the d.c. ranges and ±4% on the a.c. ranges. However, there is little point in worrying about this; meter loading will undoubtedly contribute to further (and in many cases larger) errors.

Finally, since meter movements can be extremely expensive (and uneconomic to repair in many cases) it is vitally important to check that an instrument has some form of overload protection. The simplest form of protection takes the form of an anti-parallel diode limiter shown in Figure 9.2. More sophisticated (and more effective) forms of protection are fitted to more expensive instruments.

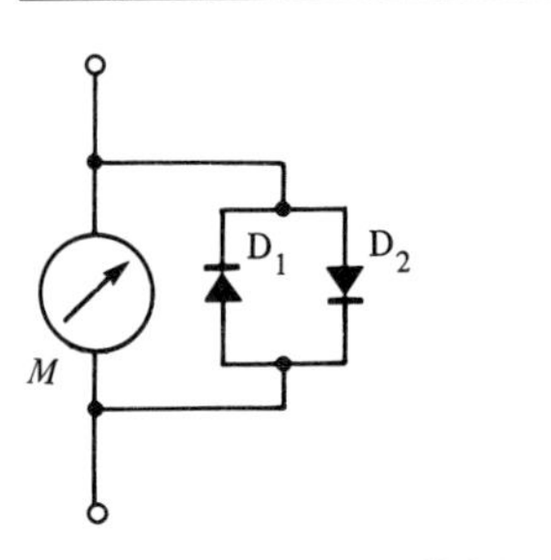

Figure 9.2 *Anti-parallel diode limiter used to protect a meter movement*

Audio frequency signal generators

Audio frequency (a.f.) signal generators generally provide sinusoidal and square wave outputs over a frequency range extending from at least 20 Hz to 20 kHz and can be invaluable for general purpose testing and calibration.

The specification of a typical audio frequency signal generator is given below:

Frequency range:	10 Hz to 100 kHz in four decade ranges.
Waveform:	Sine and square.
Output impedance:	600 Ω unbalanced.
Output voltage:	10 mV, 100 mV, 1 V and 10 V maximum (peak-peak).
Accuracy:	±2.5% of actual frequency. ±4% of full-scale output voltage.
Sine wave distortion:	Less than 0.05%.
Square wave rise-time:	Better than 150 ns.

Choosing an audio frequency signal generator

An a.f. generator for general purpose use should provide a sine wave output over the frequency range extending from 20 Hz to 20 kHz and ideally from below 10 Hz to above 100 kHz. The output level should be adjustable from a minimum of 10 mV (or less) to a maximum of 10 V (or more). Calibration of the output should be provided in either r.m.s. or peak-peak values (to convert from one to the other it is only necessary to multiply the r.m.s. values by 2.828 in order to determine the corresponding peak-peak value or divide peak-peak values by 2.828 in order to find the r.m.s. quantity). If an output meter is not fitted, the device should have a suitably calibrated output scale. It is also important to ensure that the instrument has a known output impedance. Often the manufacturer's specification will state that the output is calibrated into a given impedance (usually 600 Ω). This, however, does not necessarily imply that the output impedance of the generator is 600 Ω since, in practice, it may vary from a few Ω to over 1 kΩ. A low impedance generator is quite useful where it is necessary to test low impedance circuits. However, it would be eminently possible to add an external unity voltage gain current amplifier to the output (using, for example, the circuits given in Figures 4.34, 4.35, 4.36 or 8.19) where an item of equipment is deficient in this respect.

It is also worth looking for an instrument that provides a square, as well as sinusoidal, output. The instrument can then double as a signal source for testing pulse and logic circuits. Where such a facility is lacking, a simple TTL compatible square wave output facility can be added quite simply using the circuit shown in Figure 9.3. This circuit configuration is known as a Schmitt trigger and it is capable of producing a square wave output having very fast rise and fall times when fed with a sinusoidal input.

Finally, a word of caution is necessary. Some low-cost audio frequency signal generators are based on function generator chips (e.g. 8038) which synthesize sinusoidal signals rather than generate them using a low-

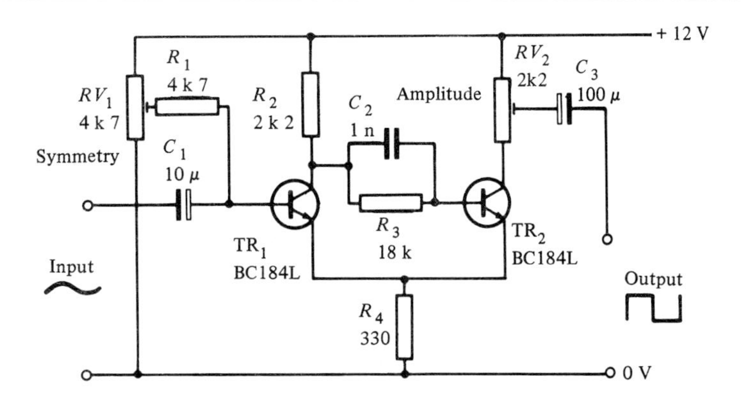

Figure 9.3 *Schmitt trigger sine-square converter*

distortion oscillator based on, for example, a Wien bridge network. Function generator chips generate quasi-sinusoidal output which, though perfectly acceptable for general use, will almost certainly have such high levels of harmonic distortion present that they will preclude its use for meaningful audio frequency measurements. If you intend to specialize in high quality audio applications, it is important to avoid such an instrument at all costs.

Oscilloscopes

An oscilloscope is an extremely versatile item of test equipment which can be used in a wide variety of measuring and test applications, the most important of which is the display of time related waveforms. Such an item probably respresents the single most costly item of test equipment in many workshops and laboratories and it is therefore important that full benefit is derived from it. See Plates 2 and 3.

The oscilloscope display is provided by a cathode ray tube (CRT) which has a typical screen area of 80 mm × 100 mm. The CRT is fitted with a graticule which either can be integral with the tube face or can take the form of a separate translucent sheet. The graticule is ruled with a 1 cm grid to which further bold lines are usually added to mark the major axes on the central viewing area. Accurate voltage and time measurements can be made with reference to the graticule, applying a scale factor from the appropriate range switch. A word of caution is appropriate at this point; before attempting to take measurements from the graticule it is essential to check that the variable front panel controls are set in the calibrate (CAL) position. Failure to observe this simple precaution may result in readings which are grossly inaccurate.

We shall now continue by briefly examining the function of each of the more commonly available oscilloscope controls. It is, however, worth mentioning that there is a considerable variation not only in the range of controls provided but also in the nomenclature used by oscilloscope manufacturers to describe them. The list which follows, therefore, is necessarily general and contains details which may not be applicable in all cases.

Cathode ray tube display

Focus: Provides a correctly focused display on the CRT screen.

Intensity: Adjusts the brightness of the display.

Astigmatism: Provides a uniformly defined display over the entire screen area and in both planes. This control is normally used in conjunction with the focus and intensity controls.

Trace rotation: Permits accurate alignment of the display with respect to the graticule.

Scale illumination: Control the brightness of the graticule lines.

Horizontal deflection system

Timebase (time/cm): Adjusts the timebase range and determines the horizontal scale. This control usually takes the form of a multi-position rotary switch and an additional continuously variable control (often concentric with the timebase selector) is often provided. The 'CAL' position is usually at one, or other, extreme setting of this control.

Stability: Adjusts the timebase so that a stable (synchronized) display is obtained.

Trigger level: Selects the particular level on the triggering signal at which the timebase sweep commences.

Trigger slope: This control usually takes the form of a switch which determines whether triggering occurs on the positive or negative edge of the triggering signal.

Vertical deflection system

Vertical attenuator (volts/cm): Adjusts the vertical size of the trace and sets the vertical voltage scale. This control is invariably a multi-position rotary switch. However, an additional variable control is sometimes also provided. Often this control is concentric with the main control and the 'CAL' position is usually at one or other extrème setting of the control.

Vertical position: Positions the trace along the vertical axis of the CRT.

AC-Gnd-DC: All modern oscilloscopes employ d.c. coupling throughout the vertical amplifier. Hence a shift along the vertical axis will occur whenever a direct voltage is present at the input. When investigating waveforms in a circuit one often encounters a.c. signals superimposed on d.c. levels; the letter may be removed by inserting a capacitor in series with the input. With the 'AC-Gnd-DC' switch in the 'AC' position the capacitor is inserted at the input, whereas in the 'DC' position the capacitor is shorted. If 'Gnd' is selected the vertical input is taken to common (0 V) and the input terminal is left floating. In order to measure the d.c. level of an input signal, the 'AC-Gnd-DC' switch is first placed in the 'Gnd' position and the 'vertical position' is then adjusted so that the trace is coincident with the central horizontal axis. The switch should then be placed in the 'DC' position and the shift along the vertical axis measured in order to quantify the d.c. level.

Chopped-alternate: This control, which is only used in dual-beam oscilloscopes, provides selection of the beam splitting mode. In the 'chopped' position, the trace displays a small portion of one vertical channel waveform followed by an equally small portion of the other. The traces are, in effect, sampled at a relatively fast rate, the result being two apparently continuous traces. In the 'alternate' position, a complete horizontal sweep is devoted to each channel on an alternate basis.

A typical oscilloscope specification is as follows:

Vertical sensitivity:	10 mV/cm to 10 V/cm in decade ranges.
Vertical bandwidth:	d.c. to 30 MHz (10 Hz to 30 MHz when a.c. coupled).
Vertical rise time:	12.5 ns
Input impedance:	1 MΩ in parallel with 30 pF.
Modes:	Channel 1/Channel 2/Chopped/Alternate
Timebase:	1 μs/cm to 1 s/cm (plus variable control).
Calibrator:	1 V peak-peak square wave at 1 kHz.
Trigger sensitivity:	Better than 5 mV (or 0.5 div.) over the range 20 Hz to 20 MHz.
CRT viewing area:	8 cm × 10 cm.

See Plates 4 to 7.

Choosing an oscilloscope

When such a comparatively large capital outlay is involved, it is vitally important to choose an instrument which will satisfy one's long-term needs. Fortunately, there is no shortage of suitable instruments from which to select, indeed the 'low-end' of the market is particularly well catered for with well over a dozen different general purpose oscilloscopes currently available at prices which start below £400. Apart from the cost (which will undoubtedly be a determining factor in many cases) the following points are worth bearing in mind:

(a) Dual-beam oscilloscopes are only marginally more expensive than their single-beam counterparts (the additional cost of the beam switching circuitry only amounts to around 10% of the cost of the equipment) and it is well worth paying a little more for a dual beam instrument.

(b) Purchase an instrument which has as wide a bandwidth as possible. Unless you are only likely to be concerned with audio and low-frequency applications, 30 MHz should be considered the minimum acceptable.

(c) Check that the front panel controls are logically laid out and are easily accessible. Avoid equipment which is fitted with every conceivable control including the 'kitchen sink' – an oscilloscope should be as simple to use as possible.

(d) Examine the display produced at various settings of the brightness and focus controls. Check, in particular, that it is possible to produce a finely focused trace at the maximum setting of the brightness control – many of the cheaper oscilloscopes fail badly on this point.

(e) Carefully examine the graticule. Check that it is clearly visible (better quality instruments will include some form of graticule illumination). Check also that the graticule is marked in centimetre divisions – some manufacturers 'cheat' on this in order to make the display look larger than it is.

(f) If you get the chance, connect a sinusoidal signal to the vertical input and check that the display suffers from no astigmatism as the trace is adjusted to fill the complete viewing area. Check also that the display does not change in size as the brightness control is advanced. If any change is detected, this can usually be attributed to poor regulation in the EHT power supply. Finally, reduce the input level and notice the point at which the instrument fails to trigger – a good oscilloscope should trigger properly for input levels of less than 5 mV.

(g) Some modern oscilloscopes use switched mode power supplies in order to reduce weight, cost and size. Unfortunately, switched mode power supplies can radiate significant levels of noise. If you suspect that the instrument in question uses such a power supply, set the vertical amplifier for maximum gain and then carefully examine the trace for noise. Trigger circuits may also be susceptible to noise radiated by switched mode power supplies, so beware.

(h) One of the best indicators of the quality of any item of test equipment is the standard of internal construction and, while it is not always possible to dismantle an instrument prior to purchase (unless one has a very understanding salesperson), a chance to 'peek' inside should not be missed. 'Afterthoughts' on the part of the equipment designer and 'bodges' during manufacture will be easy to spot and equipment with such visible defects should be avoided.

Oscilloscope probes

An important requirement of an oscilloscope is that it should faithfully reproduce pulses of fast duration and should not significantly capacitively load the circuit to which it is connected. Unfortunately, the input capacitance of an oscilloscope (which itself is not insignificant at typically around 50 pF) appears in parallel with the capacitance of the coaxial input cable, as shown in the equivalent circuit of Figure 9.4.

The capacitance of the coaxial cable is often in the region of 150 pF or so, and hence the total shunt capacitance presented to the circuit under investigation is somewhere in the region of 200 pF or so. At low frequencies, this capacitance is negligible. However, at frequencies in excess of a few kHz, it can pose a serious problem. At frequencies around 1 MHz, for example, a capacitance of 200 pF exhibits a reactance of as little as 800 Ω. Hence fast rise-time pulses are likely to be very significantly distorted by the presence of the oscilloscope and its input cable.

Figure 9.4 *Oscilloscope probe-cable-input equivalent circuit*

Fortunately, this problem can be very easily resolved with the aid of a compensated oscilloscope probe. The most common type of oscilloscope probe provides ten-times attenuation, and is usually marked '× 10'. The effect of such a probe is to raise the input resistance by a factor of 10 and reduce the input capacitance by a similar amount. A typical 'x 10' probe exhibits an input resistance together with a capacitance of 16 pF and is provided with a variety of interchangeable probe tips to facilitate connection to various types of circuitry.

Logic probes

Testing and fault-finding on digital circuits is greatly simplified with the aid of a logic probe. This invaluable low-cost test instrument comprises a hand-held probe fitted with two, or more, light emitting diodes (LED) which indicate the logical state of its probe tip.

Whereas a pulse of relatively long duration, say, one second or more, can be readily detected using a logic probe which only provides logic 0 and logic 1 indications, a short duration pulse (of, say, one tenth of a second or less) can only be detected when the probe incorporates circuitry which stretches the pulse so that a third LED remains illuminated for sufficient time to be seen.

Logic probes normally derive their power supply from the circuit under test and are invariably connected by means of a short length of twin flex fitted with insulated crocodile clips. While almost any convenient connecting point may be used, the leads of an electrolytic decoupling capacitor or the output terminals of a regulator both make ideal connecting points which can be readily identified.

Typical specifications for a logic probe are also follows:

Input resistance at probe tip:	400 kΩ approximately.
Threshold voltages:	
Logic 1 (TTL)	2.4 V
(CMOS)	70% of supply.
Logic 0 (TTL)	1.2 V.
(CMOS)	30% of supply.
Stretched pulse duration:	100 ms.
Minimum detectable pulse width:	20 ns.
Maximum input signal frequency (50 % duty cycle):	30 MHz.
Power supply:	
TTL	4.5 V to 5.5 at less than 30 mA.
CMOS	3 V to 15 V at less than 60 mA.

Choosing a logic probe

Several types of logic probe are commonly available ranging in price from around £10 to well over £50. As with most items of test equipment you get what you pay for; the more expensive instruments usually offer a performance which is significantly better than their cheaper counterparts. In particular, a distinction can be made between logic probes which are optimized for one type of logic or the other (i.e. either TTL or CMOS but not necessarily both). There are others, generally more expensive types, which cater for both types of logic. Some of these units accept a compromise on the logic levels used by the two major logic families while others are switchable into one mode or the other. If you intend to regularly use both types of logic, it is important to have an instrument of this type.

Logic probes should be expected to cope with signals having a pulse repetition frequency of at least 10 MHz and preferably extending to 30 MHz or so. This represents the normal upper limit for TTL devices though some 'top of the range' probes can be used at frequencies up to 50 MHz. Another feature found in more expensive logic probes is the pulse stretching mentioned earlier. This feature can be absolutely invaluable for detecting narrow pulses which would otherwise go unnoticed and is well worth the extra outlay.

Finally, it is important to check that any prospective purchase incorporates adequate protection against inadvertent misconnection of the probe tip. In this respect it is advisable to look for an instrument which can survive connection to a rail which has at least ± 100 V d.c. present. Some better quality instruments provide protection against inadvertent connection to voltage sources of up to 250 V r.m.s. and will thus safeguard the instrument from direct connection to an a.c. mains supply.

Storage

Having assembled a collection of tools, test leads, test equipment, as well as the necessary components and materials used for electronic circuit construction, it is well worth giving some consideration to the method of storage employed for each.

Tools and test leads should ideally not be stored in drawers (which appears to be the most generally used method of housing such items). Tools should be openly accessible and visible and undoubtedly the best method of storing them is with the use of a 'tool board'. This is simply a plane chipboard or plywood panel which is securely fastened to the wall of the workshop or laboratory. A total area of about 1 square metre will satisfy most requirements. Tools are then laid out on this surface and retained by means of spring clips. If desired, the profile of each tool may be outlined on the board; this will not only identify the correct position of each item on the board but will also act as a reminder that the particular instrument is not in its proper place.

When stored in a drawer, test leads seem to develop an instant affinity for one another. The result is a tangled mass of assorted cables and wires which can often take considerable time to sort out. Again an open storage method has much to recommend it. This may take the form of a ready-made test lead rack. One such proprietary system comprises a metal rail which accommodates a number of moulded hooks. Each hook then retains one test lead or cable. A cheaper alternative is the use of a short length of timber on to which is secured a number of cup hooks. Each cup hook supports an individual test lead or cable and it is a good idea to label each hook so that leads can be returned to their proper place. The user soon gets to know the whereabouts of a particular item and it may thus be located with the minimum of delay.

Components should be stored in one or more multiple drawer units. These can be purchased as complete units housed in an exterior metal steel frame suitable for free-standing or wall-mounting or, alternatively, may be assembled from modular interlocking drawers. Drawers are usually provided with internal dividers and with slots to accommodate labels. An alternative, though somewhat more bulky storage system (more suited to small scale production than prototype work) involves the use of polypropylene bins which may either be stacked several deep or mounted on a metal panel. These are ideal for large quantities of components or for more bulky parts (such as electrolytic capacitors, transformers, etc.). Readers may wish to consider a mixture of both storage systems, which is the arrangement much preferred by the author.

Test equipment should preferably not be stored away; if it is earning its keep, it should be in a permanent state of readiness for use, not gathering dust in some out of the way place! The moral, of course, is that the work area should be arranged so that all regularly used test equipment is within easy reach. Similar items of test gear can be conveniently placed on an 'equipment shelf' above the normal work area. Where necessary, proprietary steel shelving can be used to accommodate the bulkier and less often used items of equipment.

10 Circuit construction

This chapter deals with the techniques used to produce a working circuit from an initial 'paper design'. We begin with the stages involved in the layout and manufacture of stripboards and printed circuit boards. Consideration is given to the correct choice of enclosure and connectors as this is crucial in making the finished equipment both functional and attractive. The chapter also deals with the selection of heatsinks, soldering and desoldering techniques, and basic fault-finding.

Matrix boards and stripboards

Matrix boards and stripboards are ideal for initial breadboarding and prototype construction. The distinction between matrix boards and stripboards is simply that the former has no copper tracks and the user has to make extensive use of press-fit terminal pins which are used for component connection. Extensive inter-wiring is then necessary to link terminal pins together. This may be carried out using sleeved tinned copper wire (of appropriate gauge) or short lengths of PVC-insulated 'hook-up' wire. Like their matrix board counterparts, stripboards are also pierced with a matrix of holes which, again, are almost invariably placed on a 0.1 in pitch. The important difference, however, is that stripboards have copper strips bounded to one surface which link together rows of holes along the complete length of the board. The result, therefore, is something of a compromise between a 'naked' matrix board and a true printed circuit. Compared with the matrix board, the stripboard has the advantage that relatively few wire links are required and that components can be mounted and soldered directly to the copper strips without the need for terminal pins.

Conventional types of stripboard (those with parallel runs of strips throughout the entire board surface) are generally unsuitable for relatively complex circuitry of the type associated with microprocessor systems. Fortunately, several manufacturers have responded with special purpose stripboards. These have strips arranged in groups which not only permit the mounting of DIL integrated circuits (including the larger 28-pin and 40-pin types) but are also available in a standard range of 'card' sizes (both single and double-sided and with and without plated through holes). These boards also have sensibly arranged strips for supply distribution together with edge connectors (either direct or indirect type) popularly used for microprocessor bus systems.

Stripboard layout techniques

The following steps are required when laying out a circuit for stripboard construction:

(a) Carefully examine a copy of the circuit diagram. Mark all components to be mounted 'off-board' and identify (using appropriate letters and/or numbers, e.g. SK_1 pin-2) all points at which an 'off-board' connection is to be made.
(b) Identify any multiple connections required between integrated circuits or between integrated circuits and connectors. Arrange such components in physical proximity and with such orientation that will effectively minimize the number of links required.
(c) Identify components which require special attention (such as those which require heatsinks or have special screening requirements). Ensure that such components are positioned sensibly bearing in mind their particular needs.
(d) Keep inputs and outputs at opposite ends of the stripboard. This not only helps maintain a logical circuit layout (progressing from input to output) but, in high gain circuits, it may also be instrumental in preventing instability due to unwanted feedback.
(e) Use standard sizes of stripboard wherever possible. Where boards have to be cut to size, it is usually more efficient to align the strips along the major axis of the board.
(f) Consider the means of mounting the stripboard. If it is to be secured using bolts and threaded spacers (or equivalent) it will be necessary to allow adequate clearance around the mounting holes.
(g) Produce a rough layout for the stripboard first using paper ruled with squares, the corners of the squares representing the holes in the stripboard. This process can be carried out 'actual size' using 0.1 in graph paper or suitably enlarged by means of an appropriate choice of paper. For preference, it is wise to choose paper with a feint blue or green grid as this will subsequently disappear after photocopying leaving you with a 'clean' layout.
(h) Identify all conductors which will be handling high currents (i.e. those in excess of 1 A) and use adjacent strips connected in parallel at various points along the length of the board.
(i) Identify the strips which will be used to convey the supply rails; as far as possible these should be continuous from one end of the board to the other. It is often convenient to use adjacent strips for supply and 0 V since decoupling capacitors can easily be distributed at strategic points. Ideally, such capacitors should be positioned in close proximity to the positive supply input pin to all integrated circuits which are likely to demand sudden transient currents (e.g. 555 timers, comparators, IC power amplifiers).
(j) In high frequency circuits, link all unused strips to 0 V at regular points. This promotes stability by ensuring that the 0 V rail is effective as a common rail.

(k) Minimize, as far as possible, the number of links required. These should be made on the upper (component) side of the stripboard. Only in exceptional cases should links be made on the underside (foil side) of the board.

(l) Experiment with positioning of integrated circuits (it is good practice, though not essential, to align them all in the same direction). In some cases, logic gates may be exchanged from package to package (or within the same package) in order to minimise strip usage and links. (If you have to resort to this dodge, do not forget to amend the circuit diagram.)

When the stripboard layout is complete, it is important to carefully check it against the circuit diagram. Not only can this save considerable frustration at a later stage but it can be instrumental in preventing some costly mistakes. In particular, one should follow the positive supply and 0 V strips and check that all chips and other devices have supplies. The technique employed by the author involves the use of coloured pencils which are used to trace the circuit and stripboard layout; relating each line in the circuit diagram to a physical interconnection on the stripboard. Colours are used as follows:

Positive supply rails	Red
Negative supply rails	Black
Common 0 V rail	Green
Analogue signals	Yellow
Digital signals	Brown
Off-board connections	Mauve

Assembly of the stripboard is happily a quite straightforward process. The sequence used for stripboard assembly will normally involve mounting IC sockets first followed by transistors, diodes, resistors, capacitors and other

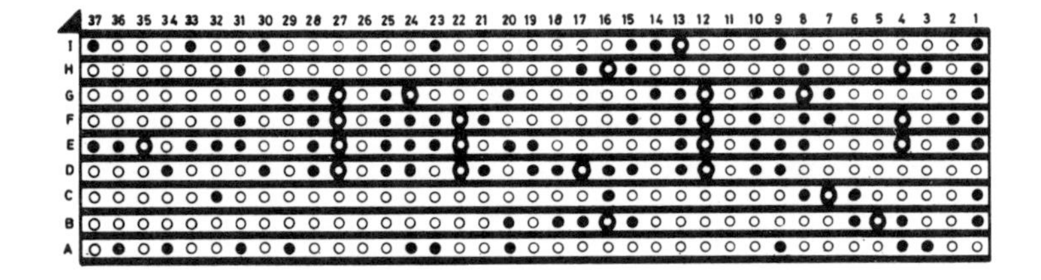

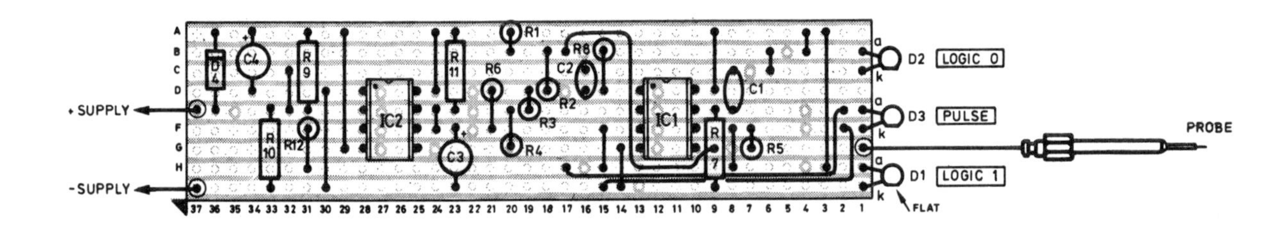

Figure 10.1 *Typical stripboard layout and matching component overlay* (reproduced by kind permission of *Everyday Electronics*)

passive components. Finally, terminal pins and links should be fitted before making the track breaks. On completion, the board should be carefully checked, paying particular attention to all polarized components (e.g. diodes, transistors and electrolytic capacitors). Figure 10.1 shows a typical stripboard layout together with matching component overlay.

Printed circuits

Printed circuit boards (PCB) comprise copper tracks bonded to an epoxy glass or synthetic resin bounded paper (SRBP) board. The result is a neat and professional looking circuit which is ideal for prototype as well as production quantities. Printed circuits can easily be duplicated or modified from original master artwork and the production techniques are quite simple and should thus not deter the enthusiast working from home.

Printed circuit board layout techniques

The following steps should be followed when laying out a printed circuit board:

(a) Carefully examine a copy of the circuit diagram. Mark all components which are to be mounted 'off-board' and, using appropriate letters and/or numbers e.g. SK_1 pin-2), identify all points at which an 'off-board' connection is to be made.
(b) Identify any multiple connections (e.g. bus lines) required between integrated circuits or between integrated circuits and edge connectors. Arrange such components in physical proximity and with such orientation that will effectively minimize the number of links required.
(c) Identify components which require special attention (such as those which require heatsinks or have special screening requirements). Ensure that such components are positioned sensibly bearing in mind their particular needs.
(d) Keep inputs and outputs at opposite ends of the PCB wherever possible. This not only helps maintain a logical circuit layout (progressing from input) but, in high gain circuits, it may also be instrumental in preventing instability due to unwanted feedback.
(e) Use the minimum board area consistent with a layout which is uncramped. In practice, and to prevent wastage, you should aim to utilize as high a proportion of the PCB surface area as possible. In the initial stages, however, it is wise to allow some room for manoeuvre as there will doubtless be subsequent modifications to the design.
(f) Unless the design makes extensive use of PCB edge connectors, try to ensure that a common 0 V foil is run all round the periphery of the PCB. This has a number of advantages not the least of which is the fact that it will be then be relatively simple to find a route to the 0 V rail from almost anywhere on the board.
(g) Consider the means of mounting the PCB and, if it is to be secured using bolts and threaded spacers (or equivalent) you should ascertain

the number and location of the holes required. You may also wish to ensure that the holes are coincident with the 0 V foil, alternatively where the 0 V rail is not to be taken to chassis ground, it will be necessary to ensure that the PCB mounting holes occur in an area of the PCB which is clear of foil.

(h) Commence the PCB design in rough first using paper ruled with squares. This process can be carried out 'actual size' using 0.1 in graph paper or suitably enlarged by means of an appropriate choice of paper. For preference, it is wise to choose paper with a feint blue or green grid as this will subsequently disappear after photocopying leaving you with a 'clean' layout.

(i) Using the square grid as a guide, try to arrange all components so that they are mounted on a standard 0.1 in pitch matrix. This may complicate things a little but is important if you should subsequently wish to convert to computer-aided PCB design or automated manufacture.

(j) Arrange straight runs of track so that they align with one dimension of the board or the other. Try to avoid haphazard track layout.

(k) Identify all conductors which will be handling high currents (i.e. those in excess of 500 mA) and ensure that tracks have adequate widths. The following should act as a rough guide:

Current (d.c. or r.m.s. a.c)	*Minimum track width*
less than 500 mA	0.6 mm
500 mA to 1.5 A	1.6 mm
1.5 A to 3 A	3.0 mm
3 A to 6 A	6.0 mm

Note that, as a general rule of thumb, the width of the 0 V track should be at least TWICE that used for any other track.

(l) Identify all conductors which will be handling high voltages (i.e. those in excess of 150 V d.c. or 100 V r.m.s. a.c.) and ensure that these are adequately spaced from other tracks. The following should act as a rough guide:

Voltage between adjacent conductors (d.c. or peak a.c.)	*Minimum track spacing*
less than 150 V	0.6 mm
150 V to 300 V	1.6 mm
300 V to 600 V	2.5 mm
600 V to 900 V	4.0 mm

(m) Identify the point at which the principal supply rail is to be connected. Employ extra wide track widths (for both the 0 V and supply rail) in this area and check that suitable decoupling capacitors are placed as close as possible to the supply. Check that other decoupling capacitors are distributed at strategic points around the board. These should be positioned in close proximity to the positive supply input pin to all integrated circuits which are likely to demand sudden transient currents (e.g. 555 timers, comparators, IC power amplifiers).

(n) Fill unused areas of PCB with 'land' (areas of foil which should be linked to 0 V). This helps ensure that the 0 V rail is effective as a common rail, minimizes use of the 'etchant', helps to conduct heat away from heat producing components, and furthermore, is essential in promoting stability in high-frequency applications.

(o) Lay out the 0 V and positive supply rails first. Then turn your attention to linking to the pads or edge connectors used for connecting the off-board components. Minimize, as far as possible, the number of links required. *Do not* use links in the 0 V rail and avoid using them in the positive supply rail.

(p) Experiment with positioning of integrated circuits (it is good practice, though not essential, to align them all in the same direction). In some cases, logic gates may be exchanged from package to package (or within the same package) in order to minimize track runs and links. (If you have to resort to this dodge, do not forget to amend the circuit diagram.)

(q) Be aware of the pin spacing used by components and try to keep this consistent throughout. With the exception of the larger wirewound resistors (which should be mounted on ceramic stand-off pillars) axial lead components should be mounted flat against the PCB (with their leads bent at right angles). Axial lead components should *not* be mounted vertically.

(r) Do not forget that tracks may be conveniently routed beneath other components. Supply rails in particular can be routed between opposite rows of pads of DIL integrated circuits; this permits very effective supply distribution and decoupling.

(s) Minimize track runs as far as possible and maintain constant spacing between parallel runs of track. Corners should be radiused and acute internal and external angles should be avoided. In exceptional circumstances, it may be necessary to run a track between adjacent pads of a DIL integrated circuit. In such cases, the track should not be a common 0 V path, neither should it be a supply rail.

Figure 10.2 shows examples of good and bad practice associated with PCB layout while Figure 10.3 shows an example of a PCB layout and matching component overlay which embodies most of the techniques and principles discussed.

As with stripboard layouts, it is well worth devoting some time to checking the final draft PCB layout before starting on the master artwork. This can be instrumental in saving much agony and heartache at a later stage. The same procedure should be adopted as given on page 214 (i.e. simultaneously tracing the circuit diagram and PCB layout). The next stage depends upon the actual PCB production. Three methods are commonly used for prototype and small-scale production. These may be summarized as follows:

(a) Drawing the track layout directly on the copper surface of the board using a special pen filled with etch resist ink. The track layout should, of course, conform as closely as possible with the draft layout.

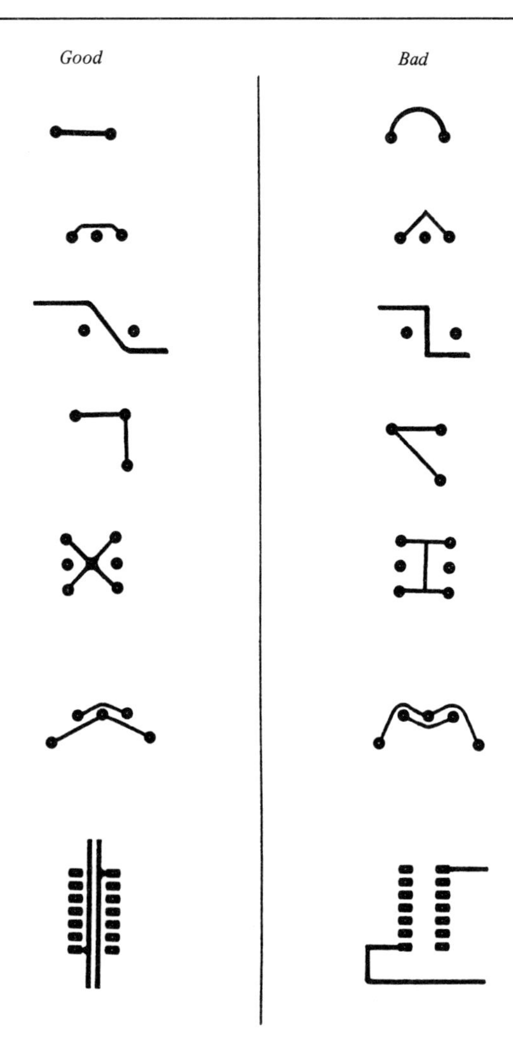

Figure 10.2 *Examples of good and bad practice in PCB layout*

(b) Laying down etch resist transfers of tracks and pads on to the copper surface of the PCB following the same layout as the draft but appropriately scaled.
(c) Producing a transparency (using artwork transfers of tracks and pads) conforming to the draft layout and then applying photographic techniques.

Methods (a) and (b) have the obvious limitation that they are a strictly 'one-off' process. Method (a) is also extremely crude and only applicable to very simple boards. Method (c) is by far the most superior and allows one to re-use or modify the master artwork transparency and produce as many further boards as are required. The disadvantage of the method is that it is slightly more expensive in terms of materials (specially coated copper board is required) and requires some form of ultra-violet exposure unit. This device normally comprises a light-tight enclosure into the base

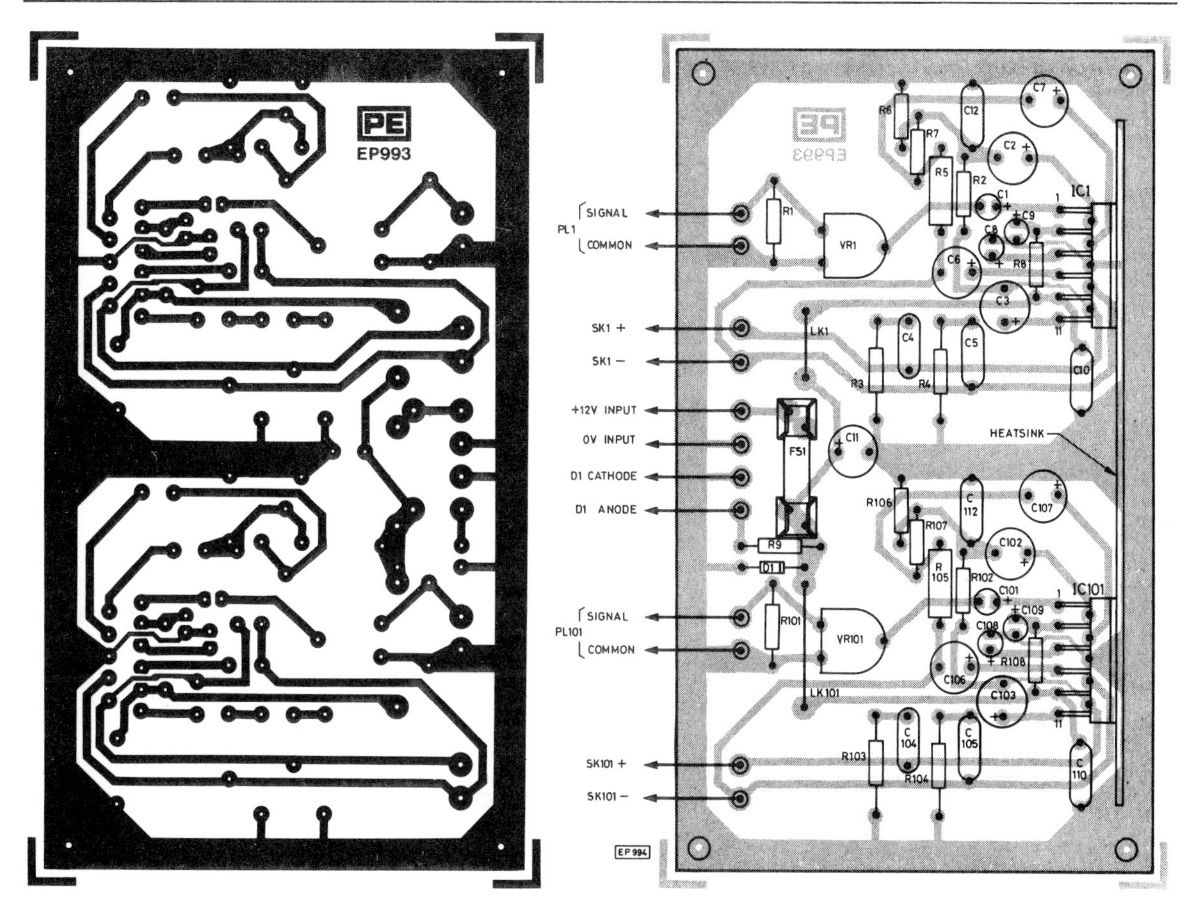

Figure 10.3 *Typical PCB layout and matching component overlay* (Reproduced by kind permission of *Practical Electronics*)

of which one or more ultra-violet tubes are fitted. Smaller units are available which permit exposure of boards measuring 250 mm × 150 mm while the larger units are suitable for boards of up to 500 mm × 350 mm. The more expensive exposure units are fitted with timers which can be set to determine the actual exposure time. Low cost units do not have such a facility and the operator has to refer to a clock or wristwatch in order to determine the exposure time.

In use, the 1:1 master artwork (in the form of opaque transfers and tape on translucent polyester drafting film) is placed on the glass screen immediately in front of the ultra-violet tubes (taking care to ensure that it is placed so that the component side is uppermost). The opaque plastic film is then removed from the photo-resist board (previously cut roughly to size) and the board is then placed on top of the film (coated side down). The lid of the exposure unit is then closed and the timer set (usually for around four minutes but see individual manufacturer's recommendations. The inside of the lid is lined with foam which exerts an even pressure over the board such that it is held firmly in place during the exposure process.

It should be noted that, as with all photographic materials, sensitised copper board has a finite shelf-life. Furthermore, boards should ideally be stored in a cool place at a temperature of between 2 °C and 10 °C. Shelf-life at 20 °C will only be around twelve months and thus boards should be used reasonably promptly after purchase.

Note that it is not absolutely essential to have access to an ultra-violet light box as we all have at least occasional access to an entirely free source of ultra-violet light. Provided one is prepared to wait for a sunny day and prepared to experiment a little, the exposure process can be carried out in ordinary sunlight. As a guide, and with the full sun present overhead, exposure will take around fifteen minutes. Alternatively, one can make use of a sun-ray lamp. Again, some experimentation will be required in order to get the exposure right. With a lamp placed approximately 300 mm from the sun ray source (and arranged so that the whole board surface is evenly illuminated) an exposure time of around four minutes will be required. Note that, if you use this technique, it is important to follow the sun-ray lamp manufacturer's instructions concerning eye protection. A pair of goggles or dark sunglasses can be used to protect the eyes during the exposure process. However one should *never* look directly at the ultra-violet light source even when it is 'warming up'.

Finally, one can easily manufacture one's own light-box (using low-power ultra-violet tubes) or make use of a standard ultra-violet bulb (of between 150 W and 300 W) suspended above the work area. If this technique is used, the bulb should be hung approximately 400 mm above

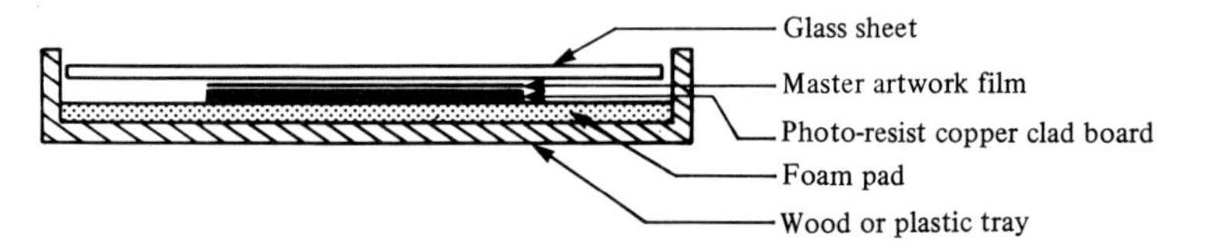

Figure 10.4 *Exposure frame suitable for photo-etch PCB production*

the table on which the sensitized board, artwork and glass sheet have been placed. A typical exposure time for a 300 W bulb is in the range of ten to fifteen minutes. A pair of dark sunglasses can again be used to protect the eyes during the process. Where one is using an 'alternative' technique, a frame should be constructed along the lines shown in Figure 10.4. This can be used to hold the transparency in contact with the coated copper board during the exposure process.

Whichever method of exposure is used some experimentation may be required in order to determine the optimum exposure time. After this time has elapsed, the board should be removed and immersed in a solution of sodium hydroxide which acts as a developer. The solution should be freshly made and the normal concentration required is obtained by mixing approximately 500 ml of tap water (at 20 °C) with one tablespoon of sodium hydroxide crystals. A photographic developing tray (or similar shallow plastic container) should be used to hold the developer. Note that care should be taken when handling the developer solution and the use of plastic or rubber gloves is strongly recommended. This process should be carried out immediately after exposure and care should be taken not to allow the board to be further exposed under room lights.

The board should be gently agitated while immersed in the developer and the ensuing process of development should be carefully watched. The board should be left for a sufficiently long period for the entire surface to be developed correctly but not so long that the tracks lift. Development times will depend upon the temperature and concentration of the developer and on the age of the sensitized board. Normal development times are in the region of 30 to 90 seconds and after this period the developed image of the track layout (an etch-resist positive) should be seen.

After developing the board it should be carefully washed under a running tap. It is advisable not to rub or touch the board (to avoid scratching the surface) and the jet of water should be sufficient to remove all traces of the developer. Finally, the board should be placed in the etchant which is a ferric chloride solution ($FeCl_3$). For obvious reasons, ferric chloride is normally provided in crystalline form (though at least one major supplier is prepared to supply it on a 'mail order basis' in concentrated liquid form) and should be added to tap water (at 20 °C) following the instructions provided by the supplier. If no instructions are given, the normal quantities involved are 750 ml of water to 500 g of ferric chloride crystals. Etching times will also be very much dependent upon temperature and concentration but, for a fresh solution warmed to around 40 °C the time taken should typically be ten to fifteen minutes. During this time the board should be regularly agitated and checked to ascertain the state of etching. The board should be removed as soon as all areas not protected by resist have been cleared of copper; failure to observe this precaution will result in 'undercutting' of the resist and consequent thinning of tracks and pads. Where thermostatically controlled tanks are used, times of five minutes or less can be achieved when using fresh solution.

It should go without saying that great care should be exercised when handling ferric chloride. Plastic or rubber gloves should be worn and care must be taken to avoid spills and splashes. After cooling, the ferric chloride solution may be stored (using a sealed plastic container) for future use. In general, 750 ml of solution can be used to etch around six to ten boards of average size; the etching process taking longer as the solution nears the end of its working life. Finally, the exhausted solution must be disposed of with care – *it should not be poured into an ordinary mains drainage system.*

Having completed the etching process, the next stage involves thoroughly washing, cleaning, and drying the printed circuit board. After this, the board will be ready for drilling. Drilling will normally involve the services of a 0.6 mm or 1 mm twist drill bit for standard component leads and IC pins. Larger drill bits may be required for the leads fitted to some larger components (e.g. power diodes) and mounting holes. Drilling is greatly simplified if a special PCB drill and matching stand can be enlisted. Alternatively, provided it has a bench stand, a standard electric drill can be used. Problems sometimes arise when a standard drill or hand drill is unable to adequately grip a miniature twist drill bit. In such cases one should make use of a miniature pin chuck or a drill fitted with an enlarged shank (usually of 2.4 mm diameter).

Connectors

Various forms of connectors may be required in any particular item of electronic equipment. These may be categorized in various ways but the following should serve as a guide:

(a) *Mains connectors.* These connectors are intended for use with an a.c. mains supply. Standard 3-pole IEC types should be employed wherever possible.

(b) *Single-pole connectors.* These are available in ranges having diameters of 4 mm, 3 mm, 2 mm and 1 mm and are ideal for use with test-leads (i.e. as input and output connectors in test equipment) and for low-voltage power supplies. Plugs and sockets are available in various colours to permit identification.

(c) *Multi-pole connectors.* A huge range of multi-pole connectors is currently available and the following types are worthy of special note:

   (i) DIN standard connectors of the type commonly used in audio equipment.
   (ii) DIN 41612 indirect stripboard and PCB edge connectors with 32, 64, or 96-ways.
   (iii) DIN 41617 low-cost indirect edge connectors (see Chapter 11).
   (iv) D-connectors. These are available with 9, 15, 25, and 37-ways and are popularly used in microcomputer applications. (The 25-way D-connector being that conventionally associated with the RS-232C standard.)
   (v) IEEE-488 connectors. These 14, 24, 36 and 50-way connectors are commonly used with equipment which uses the popular IEEE-488 instrument bus system.

   (Note that insulation displacement connector (IDC) provide a means of terminating multi-way ribbon cables without the need for soldering. Simple tools are used to assemble the connector and strain relief clamp onto the cable, the insulation of which is pierced by the tines of the connector pins.)

(d) *Coaxial connectors.* These connectors are used for screened test leads and also for r.f. equipment. The following three types are most popular:

   (i) BNC. These are available in both 50 Ω and 75 Ω series and are suitable for operation at frequencies up to 4 GHz.
   (ii) PL259/SO239 (popularly known as UHF). These 50 Ω types are suitable for use at frequencies up to 250 MHz.
   (iii) Belling-Lee (popularly known as TV). These low-cost connectors are suitable for use at frequencies up to 800 MHz with an impedance of 75 Ω.

The process of choosing a connector is usually very straightforward. It will first be necessary to ensure that an adequate number of ways are catered for and that the connector is suitably rated as far as current and voltage are concerned (one should consult individual manufacturer's ratings where any doubt arises). It is advisable to maintain compatibility with equipment of similar type and one should avoid using too wide a range of connectors and cables. In addition, a common pin-usage convention should be adopted and strictly adhered to. This will help to avoid problems later on and will make interwiring of equipment and exchange of modules a relatively straightforward process.

Enclosures

An appropriate choice of enclosure is vitally important to the 'packaging' of any electronic equipment. Not only will the enclosure provide protection for the equipment but it should also be attractive and add to the functionality of the equipment. Enclosures can be divided into five main types:

(a) *Instrument cases.* These are ideal for small items of test gear (e.g. meters and signal generators) and are available in a wide variety of styles and sizes. One of the most popular low-cost ranges of instrument cases is that manufactured by BICC-Vero and known as Veroboxes. This enclosure comprises plastic top and bottom sections with anodized aluminium front and back panels. Other ranges of instrument cases feature steel and aluminium construction and are thus eminently suited for larger projects or those fitted with larger mains transformers (such as heavy-duty power supplies).

(b) *Plastic and diecast boxes.* These low-cost enclosures comprise a box with removable lid secured by means of four or more screws. Boxes are available in a range of sizes and the diecast types (which are ideal for use in relatively hostile environments) are available both unpainted and with a textured paint finish.

(c) *Rack systems.* These are designed to accept standard cards and are ideal for modular projects. The outer case comprises an aluminium framework fitted with covers and a series of connectors at the rear from which the modules derive their power and exchange signals. Individual circuit cards (which may be stripboard or PCB) are fitted to a small supporting chassis and anodized aluminium front panel (available in various widths). The card assembly slides into the rack using appropriately positioned clip-in guides. Rack systems are expensive but inherently flexible (see Chapter 11).

(d) *Desk consoles.* These enclosures are ideal for desktop equipment and generally have sloping surfaces which are ideal for mounting keyboards and keypads.

(e) *Special purpose enclosures.* Apart from the general purpose types of enclosure described earlier, a variety of special purpose enclosures are also available. These include such items as clock and calculator housings, enclosures for hand-held controllers, and cases for in-line mains power supplies (including types having integral 13 A plugs).

Having decided upon which of the basic types of enclosure is required, the following questions should be borne in mind when making a final selection of enclosure:

Is the size adequate?
Will the enclosure accommodate the stripboard or PCB circuitry together with the 'off-board' components (including, where appropriate, the mains transformer)?
Will the enclosure be strong enough?
Will other equipment be stacked on top of the enclosure?
Can the mains transformer be supported adequately without deforming the case?
Is the front panel of sufficient size to permit mounting all of the controls and displays?
Is any protection (in the form of handles or a recessed front panel) necessary for the front panel mounted components?
Is there any need for ventilation?
How much heat is likely to be produced within the equipment?
Can heat producing components be mounted on the rear panel?
Will a cooling fan be required?
Is the style of the equipment commensurate with other equipment of its type and within the same range?

When laying out the front panel of the equipment, it is important to bear in mind the basic principles of ergonomic design. All controls should be accessible. They should be logically arranged (grouping related functions together) and clearly labelled. Consideration should be given to the type of controls used (e.g. slider versus rotary potentiometers, push-button versus toggle switches).

It is also important to wire controls such that their action follows the expected outcome. Rotary 'gain' and 'volume' controls, for example, should produce an increase in output when turned in a clockwise direction. Indicators should operate with adequate brightness and should be viewable over an appropriate angle. Indicators and controls should be arranged so that it is possible to ascertain the status of the instrument at a glance. If necessary, a number of opinions should be sought before arriving at a final layout for the front panel; one's own personal preferences are unlikely to coincide exactly with those of the 'end user'.

Soldering and de-soldering techniques

Soldered joints effectively provide for both electrical and mechanical connection of components with pins, tags, stripboards, and PCB. Before a soldering operation is carried out, it is vitally important that all surfaces to be soldered are clean and completely free of grease and/or oxide films. It is also important to have an adequately rated soldering iron (see Chapter 9) which has an appropriate bit. The soldering iron bit is the all-important point of contact between the soldering iron and the joint and it should be kept scrupulously clean (using a damp sponge) and free from oxide. To aid this process, and promote heat transfer generally, the bit should be regularly 'tinned' (i.e. given a surface coating of molten solder).

The selection of a soldering bit is very much a matter of personal

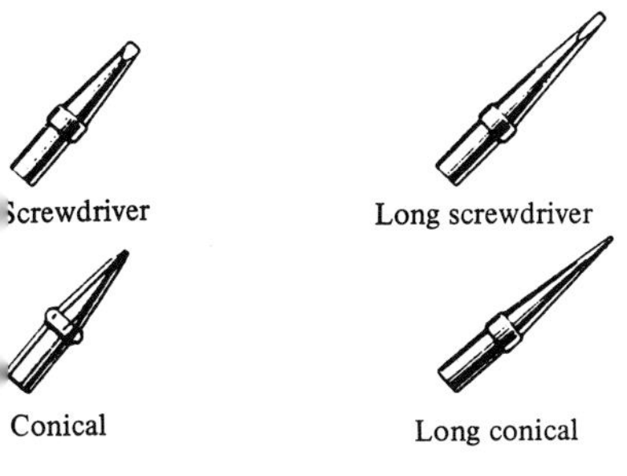

Figure 10.5 *Some typical soldering iron bit profiles*

preference, however, Figure 10.5 shows a number of bit profiles together with suggested applications. The procedure for making soldered joints to terminal pins and PCB pads are shown in Figures 10.6 and 10.7, respectively.

In the case of terminal pins, the component lead or wire, should be wrapped tightly around the pin using at least one turn of wire made using a small pair of long-nosed pliers. If necessary, the wire should be appropriately trimmed using a small pair of side cutters before soldering. Next, the pin and wire should then be simultaneously heated by suitable application of the soldering iron bit and then sufficient solder should be fed on to the pin and wire (*not* via the bit) for it to flow evenly around the joint thus forming a solder 'seal'. The solder should then be left to cool (taking care not to disturb the component or wire during the process).

The finished joint should be carefully inspected and re-made if it suffers from any of the following faults:

(a) Too little solder. The solder has failed to flow around the entire joint and some of the wire turn or pin remains exposed.
(b) Too much solder. The solder has formed into a large 'blob' the majority of which is not in direct contact with either the wire or the pin.
(c) The joint is 'dry'. This usually occurs if either the temperature of the joint was insufficient to permit the solder to flow adequately or if the joint was disturbed during cooling.

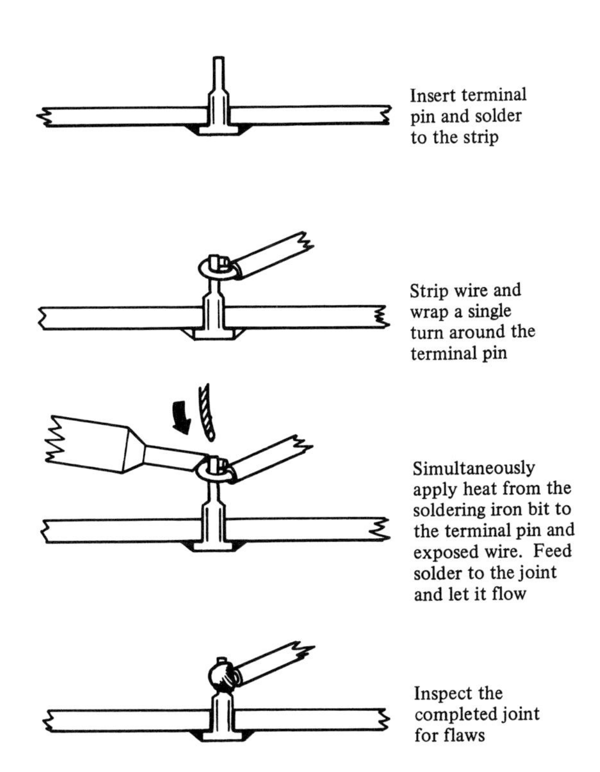

Figure 10.6 *Procedure for making a soldered joint to a stripboard terminal pin*

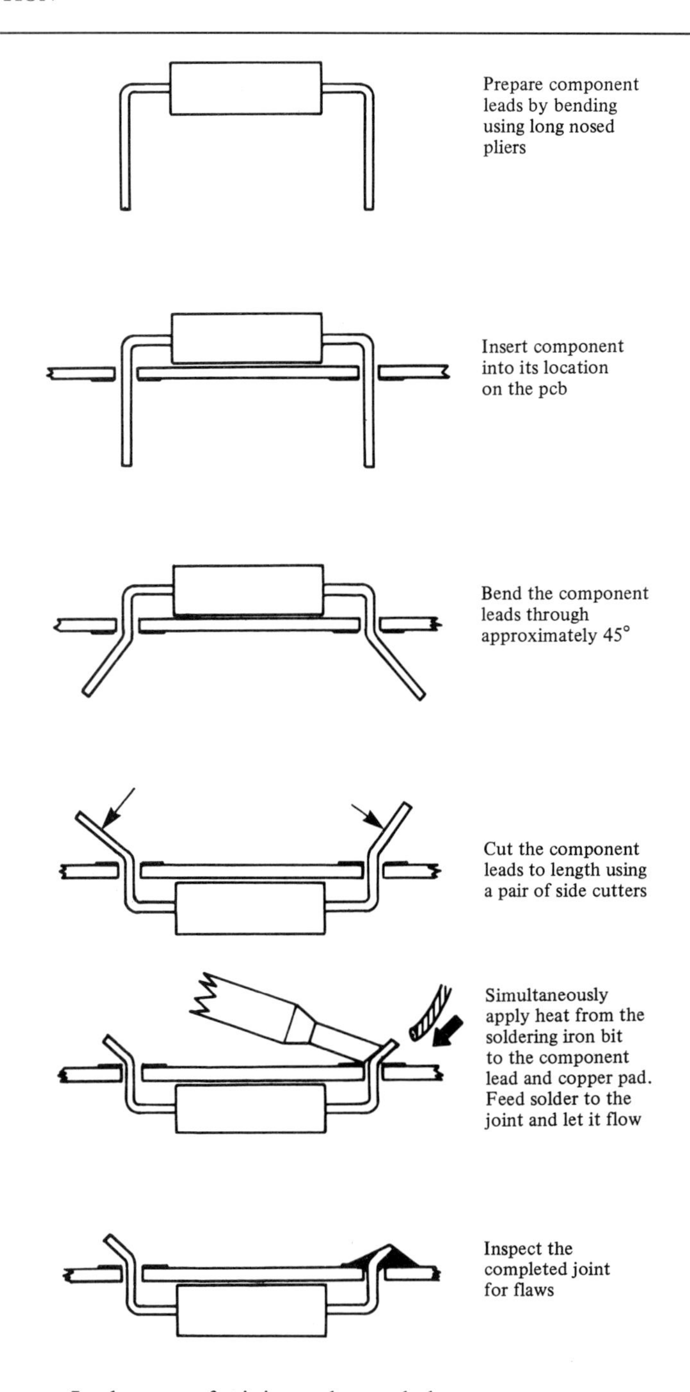

Figure 10.7 *Procedure for making a soldered joint to a PCB pad*

In the case of a joint to be made between a component and a PCB pad, a slightly different technique is used (though the requisites for cleanliness still apply). The component should be fitted to the PCB (bending its leads appropriately at right angles if it is an axial lead component) such that its leads protrude through the PCB to the copper foil side. The leads should be trimmed to within a few millimetres of the copper pad then bent slightly (so that the component does not fall out when the board is inverted) before soldering in place. Opinions differ concerning the angles through which the component leads should be bent. For easy removal, the leads should not be bent at all while, for the best mechanical joint, the leads should be bent through 90°. A good compromise, and that preferred by the author involves bending the leads through about 45°. Care should again be taken to use the minimum of solder consistent with making a sound electrical and mechanical joint.

The finished joint should be carefully inspected and re-made if it suffers from any of the following faults:

(a) Too little solder. The solder has failed to flow around the entire joint and some of the lead or pad remains exposed.
(b) Too much solder. The solder has formed into a large 'blob' the majority of which is not in direct contact with either the lead or the pad.
(c) The joint is 'dry'. This usually occurs of either the temperature of the joint was insufficient to permit the solder to flow adequately or if the joint was disturbed during cooling.

Finally, it is important to realize that good soldering technique usually takes time to develop and the old adage 'practice makes perfect' is very apt in this respect. Do not despair if your first efforts fail to match with those of the professional!

Component removal and replacement

Care must be exercised when removing and replacing printed circuit mounted components. It is first necessary to accurately locate the component to be removed on the upper (component) side of the PCB and then to correctly identify its solder pads on the underside of the PCB. Once located, the component pads should be gently heated using a soldering iron. The soldering iron bit should be regularly cleaned using a damp sponge (a small tin containing such an item is a useful adjunct to any soldering work station). The power rating of the iron should be the minimum consistent with effective removal of the components and should therefore not normally exceed 20 W for types which do not have temperature control since excessive heat can not only damage components but may also destroy the bond between the copper track and the board itself. This latter effect causes lifting of pads and tracks from the surface of the PCB and should be avoided at all costs. See Plates 8 to 14.

Once the solder in the vicinity of the pad has become molten (this usually only takes one or two seconds) and desoldering suction pump should be used to remove the solder. This will often require only one operation of the desoldering pump; however, where a large area of solder is present or where not all of the solder in the vicinity of the pad has become molten, a second application of the tool may be required. If the desoldering tool has to be repeatedly applied this usually indicates that either the tool is clogged with solder or that the soldering iron is not hot enough. With practice only one application of the soldering iron and soldering pump should be required. Once cleared of solder, the component lead should become free and a similar process should be used for any remaining leads or pins. Special tools are available for the rapid removal of multi-pin IC devices and these permit the simultaneous heating of all pins.

When de-soldering is complete, the component should be gently withdrawn from the upper (component) side of the board and the replacement component should be fitted, taking care to observe the correct polarity and orientation where appropriate. The leads should protrude through the PCB to the copper foil side and should be trimmed to within a few millimetres of the copper pad then bent slightly (so that the component does not fall out when the board is inverted) before soldering in place. Care should be taken to use the minimum of solder consistent with making a sound electrical and mechanical joint. As always, cleanliness of the soldering iron bit is extremely important if dry joints are to be avoided.

Finally, a careful visual examination of the joint should be carried out. Any solder splashes or bridges between adjacent tracks should be removed and, if necessary, a sharp pointed instrument should be used to remove any surplus solder or flux which may be present.

In some cases it may be expedient (or essential when the PCB has been mounted in such a position that the copper foil side is not readily accessible) to remove a component by cutting its leads on the upper (component) side of the board. However, care should be taken to ensure that sufficient lead is left to which the replacement component (with its wires or pins suitably trimmed) may be soldered. During this operation, extra care must be taken when the soldering iron bit is placed in close proximity to densely packed components on the upper side of the PCB. Polystyrene capacitors and other plastic encapsulated components are particularly vulnerable in this respect.

Heatsinks

Heatsinks are required for semiconductor devices which dissipate appreciable levels of power. Heatsinks are available in various styles (to permit mounting of different semiconductor case styles) and in various ratings (to cater for different levels of power dissipation). See Plate 15.

The efficiency of a heatsink is quoted in terms of its 'thermal resistance' expressed in °C/W. When a system has reached thermal equilibrium, the power dissipated by the semiconductor device is equal to the ratio of temperature difference (between junction and ambient) to the sum of the thermal resistance present. Hence:

$$\text{Power dissipated, } P_{\mathrm{D}} = \frac{T_{\mathrm{j}} - T_{\mathrm{a}}}{\theta_{\mathrm{jc}} + \theta_{\mathrm{cs}} + \theta_{\mathrm{sa}}} \text{ W}$$

Where T_j is the maximum junction temperature (°C) specified by the manufacturer (usually before derating is applied), T_a is the ambient temperature (°C), and θ_{jc}, θ_{cs} and θ_{sa} are the thermal resistance of junction to case (specified by the manufacturer), case to heatsink, and heatsink to air respectively (measured in °C/W).

The previous formula may be conveniently rearranged so that the value of θ_{sa} may be found:

$$\theta_{sa} = \frac{T_j - T_a}{P_D} - (\theta_{jc} + \theta_{cs})\,°\text{C/W}$$

As an example, consider the case of a semiconductor device which has the following ratings specified by the manufacturer:

Maximum junction temperature: 80 °C.
Thermal resistance (junction to case): 2 °C/W

If the device is to dissipate 10 W, the ambient temperature is 30 °C, and the thermal resistance of the mounting hardware (θ_{cs}) is 1 °C/W, the required thermal resistance of the heatsink will be given by:

$$\theta_{sa} = \frac{80 - 30}{10} - (2 + 1)\,°\text{C/W}$$

Thus $\theta_{sa} = 2\,°\text{C/W}$

Hence a heatsink rated at 2 °C/W (or better) is required.

Fault-finding

Before we outline the basic steps for fault-finding on typical electronic circuits, it is vitally important that readers are aware of the potential hazards associated with equipment which uses high voltages or is operated from the a.c. mains supply.

Whereas many electronic circuits operate from low voltage supplies and can thus be handled quite safely, the high a.c. voltages present in mains operated equipment represent a potentially lethal shock hazard. The following general rules should *always* be followed when handling such equipment:

1 Switch off the mains supply *and* remove the mains power connector whenever *any* of the following tasks are being performed:
   (a) Dismantling the equipment.
   (b) Inspecting fuses.
   (c) Disconnecting or connecting internal modules.
   (d) Desoldering or soldering components.
   (e) Carrying out continuity tests on switches, transformer windings, bridge rectifiers, etc.
2 When measuring a.c. and d.c. voltages present within the power unit take the following precautions:
   (a) Avoid direct contact with incoming mains wiring.

(b) Check that the equipment is properly earthed.
(c) Use insulated test prods.
(d) Select appropriate meter ranges *before* attempting to take any measurements.
(e) If in any doubt about what you are doing, switch off at the mains, disconnect the mains connector and *think*.

Fault-finding on most circuits is a relatively straightforward process. Furthermore, and assuming that the circuit has been correctly assembled and wired, there is usually a limited number of 'stock faults' (such as transistor failure, IC failure) which can occur. To assist in this process, it is a good idea to identify a number of 'test points' at which the voltages present (both a.c. and d.c.) can be used as indicators of the circuit's functioning. Such test points should be identified prior to circuit construction and marked with appropriate terminal pins.

Readers should note that the most rapid method of fault diagnosis is not necessarily that of following voltages or signals from one end to the other. Most textbooks on fault-finding discuss the relative methods of the so-called 'end to end' and 'half split' methods.

Transistor faults

As long as a few basic rules can be remembered, recognizing the correct voltages present at the electrodes of a transistor is a fairly simple process. The potentials applied to the transistor electrodes are instrumental in determining the correct bias conditions for operation as an amplifier, oscillator, or switch. If the transistor is defective, the usual bias voltages will be substantially changed. The functional state of the transistor may thus be quickly determined by measuring the d.c. potentials present at the transistor's electrodes while it is still in circuit.

The potential developed across the forward biased base-emitter junction of a silicon transistor is approximately 600 mV. In the case of an n-p-n transistor, the base will be positive with respect to the emitter whilst, for a p-n-p transistor, the base will be negative with respect to the emitter, as shown in Figure 10.8. The base-emitter voltage drop tends to be larger when the transistor is operated as a saturated switch (and is in the 'on' state) or when it is a power type carrying an appreciable collector current. In these applications, base-emitter voltages of between 0.65 V and 0.7 V are not unusual. Small-signal amplifiers, on the other hand, operate with significantly lower values of collector current and values of base-emitter voltage in the range 0.55 V to 0.65 V are typical.

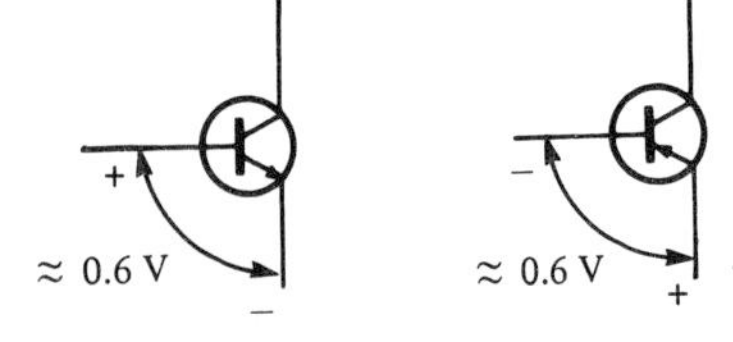

Figure 10.8 *Base-emitter voltages for n-p-n and p-n-p silicon transistors*

A measured base-emitter voltage greatly in excess of 0.6 V is an indication of a defective transistor with an open-circuit base-emitter junction. A measured base-emitter potential very much less than 0.6 V indicates that the transistor is not being supplied with a base bias and, while this may be normal for a switching transistor in the 'off' state, it is indicative of a circuit fault in the case of a linear small-signal amplifier stage. In the case of oscillators and medium/high power r.f. amplifiers operating in class C little or no bias will usually be present and, furthermore, the presence of r.f. drive during measurement can produce some very misleading results. In such cases it is probably worth removing a transistor suspected of being defective and testing it out of circuit.

Unfortunately, it is not so easy to predict the voltage present at the collector-base junction of a transistor. The junction is invariably reverse biased and the potential present will vary considerably depending upon the magnitude of collector current, supply voltage, and circuit conditions. As a rule-of-thumb, small-signal amplifiers using resistive collector loads usually operate with a collector-emitter voltage which is approximately half that of the collector supply. The collector-base voltage will be slightly less than this. Hence, for a stage which is operated from a decoupled supply rail of, say, 8.5 V a reasonable collector-base voltage would be somewhere in the range 3 V to 4 V. Tuned amplifiers having inductive collector loads generally operate with a somewhat higher value of collector-emitter voltage since there is no appreciable direct voltage drop across the load. As a result, the collector-base voltage drop is greater (a typical value being 6 V). Figure 10.9 shows the electrode potentials normally associated with transistors operating as linear small-signal amplifiers.

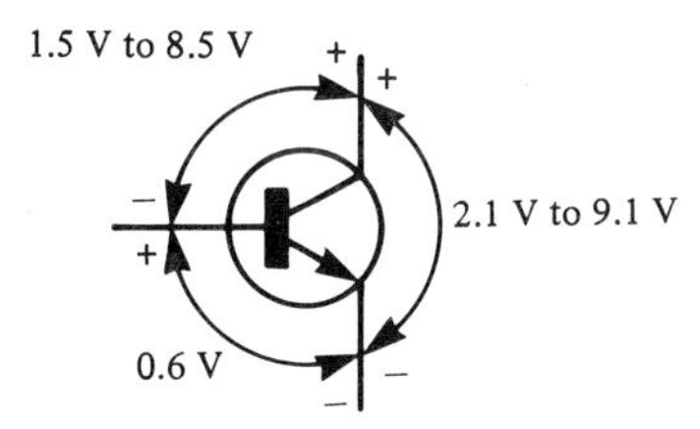

Figure 10.9 *Typical voltages found in a small-signal transistor amplifier (polarities will be reversed in the case of a p-n-p device)*

Where a transistor is operated as a saturated switch, the junction potentials depend upon whether the transistor is in the 'on' or 'off' state. In the 'off' condition, the base-emitter voltage will be very low (typically 0 V) whereas the collector-base voltage will be high and, in many cases, almost equal to the collector supply voltage. In the 'on' state, the base-emitter voltage will be relatively large (typically 0.7 V) and the collector-base voltage will fall to a very low value (typically 0.5 V). It should be noted that, in normal saturated switching, the collector-emitter voltage may fall to as low as 0.2 V and thus the base-emitter voltage will become reversed in polarity (i.e. base positive with respect to collector in the case of an n-p-n transistor). The junction potentials associated with transistors operating as saturated switches are shown in Figure 10.10.

Transistors may fail in a number of ways. Individual junctions may become open-circuit or short-circuit. In some cases the entire device may become short-circuit or exhibit a very low value of internal resistance. The precise nature of the fault will usually depend upon the electrical conditions prevalent at the time of failure. Excessive reverse voltage, for example, is likely to cause a collector-base or base-emitter junction to become open-circuit. Momentary excessive collector current is likely to rupture the base-emitter junction while long-term over-dissipation is likely to cause an internal short circuit or very low resistance condition which may, in some cases, affect all three terminals of the device.

In order to illustrate the effects of various transistor fault conditions consider the circuit of a typical tuned amplifier stage shown in Figure 10.11. The transistor is operated in class-A with conventional base bias potential divider and a tuned transformer collector load. Normal working voltages are shown in the circuit diagram. Note that the junction potentials (0.6 V and 4.7 V for the base-emitter and collector-base junctions respectively)

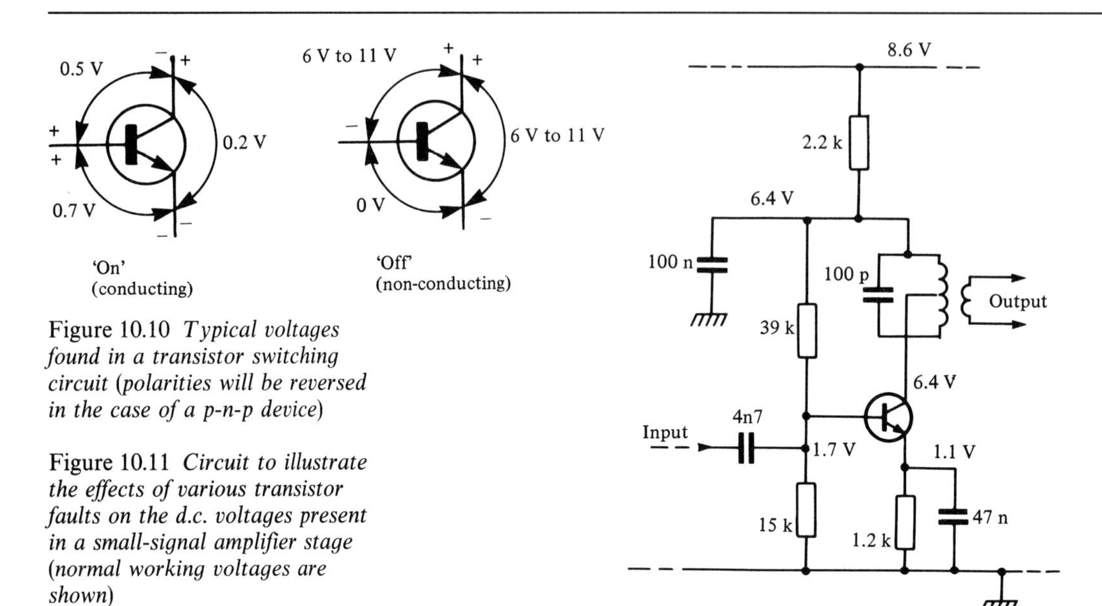

Figure 10.10 *Typical voltages found in a transistor switching circuit (polarities will be reversed in the case of a p-n-p device)*

Figure 10.11 *Circuit to illustrate the effects of various transistor faults on the d.c. voltages present in a small-signal amplifier stage (normal working voltages are shown)*

are in agreement with the voltages given in Figure 10.9. The circuit voltages corresponding to six different transistor fault conditions are shown in Table 10.1:

Table 10.1

Fault number	*Transistor voltages*			*Fault condition*
	e	*b*	*c*	
1	2.7	3.3	3.3	b-c short
2	0.2	0.2	8.2	b-e short
3	3.0	0.8	3.0	c-e short
4	2.9	2.9	2.9	c-b-e short
5	0	2.2	8.3	b-e open
6	0.2	0.8	8.2	c-b open

Each fault will now be discussed individually:

Fault 1

The collector-base short circuit gives rise to identical base and collector voltage. The base-emitter junction is still intact and thus the normal forward voltage drop of 0.6 V is present. In this condition a relatively high value of current is drawn by the stage and thus the supply voltage falls slightly.

Fault 2

The base-emitter short circuit produces identical base and emitter voltages. No collector current flows and the collector voltage rises to almost the full supply voltage. The base and emitter voltages are relatively low since the emitter resistor effectively appears in parallel with the lower section of the base bias potential divider, thus pulling the base voltage down.

Fault 3

Identical voltages at the collector and emitter result from a collector-emitter short-circuit. The emitter voltage rises above the base voltage and the base-emitter junction is well and truly turned off. The short circuit causes a higher value of current to be drawn from the supply and hence the supply voltage falls slightly.

Fault 4

Perhaps the most obvious fault condition is when a short-circuit condition affects all three terminals. The voltages are, naturally, identical and, as with fault 3, more current than usual is taken from the supply and, as a consequence, the supply voltage falls slightly.

Fault 5

With the base-emitter junction open-circuit, the base-emitter voltage rises well above the 0.6 V which would normally be expected. No collector or emitter current is flowing and thus the collector voltage rises, while the emitter voltage falls.

Fault 6

No collector current flows when the collector-base junction is open-circuit and, as with fault 2, the collector voltage rises towards the supply. Note that , since the base-emitter junction is intact, the normal forward bias of 0.6 V is present and this condition distinguishes this fault from fault 2.

Integrated circuit faults

Integrated circuits may fail in various ways. Occasionally, the manifestation of the fault is simply a chip which is chronically overheated – the application of a finger tip to the centre of the plastic package will usually identify such a failure. Any chip which is noticeably hotter than others of a similar type should be considered suspect. Where integrated circuits are fitted in sockets, it will be eminently possible to remove and replace them with known functional devices (but, do remember to switch 'off' and disconnect the supply during the process).

In the case of digital circuitry, the task of identifying a logic gate which is failing to perform its logical function can be accomplished by various means but the simplest and most expedient is with the aid of a logic probe. This invaluable tool comprises a hand-held probe fitted with LED to indicate the logical state of its probe tip. In use, the logic probe is moved from point to point within a logic circuit and the state of each node is noted and compared with the expected logic level. In order to carry out checks on more complex logic arrangements a logic pulser may be used in conjunction with the logic probe. The pulser provides a means of momentarily changing the state of a node (regardless of its actual state) and thus permits, for example, the clocking of a bistable element.

Operational amplifiers can usually be checked using simple d.c. voltage measurements. The d.c. voltages appearing at the inverting and non-inverting inputs should be accurately measured and compared. Where the voltage at the inverting input is more positive with respect to that at the non-inverting input, the output voltage will be high (positive if the operational amplifier is operated from a dual supply rail). Conversely, if the voltage at the inverting input is negative, with respect to that at the non-inverting input, the output voltage will be high (positive if the operational amplifier is operating from a dual supply). Finally, if both inputs are at 0 V and there is virtually no difference in the input voltages, the output should also be close to 0 V. If it is high or low (or sitting at one or other of the supply voltages) the device should be considered suspect.

The detection of faults within other linear integrated circuits can be rather more difficult. However, a good starting point is that of disconnecting the supply and inserting a meter to determine the supply current under quiescent conditions. The value should be compared with that given by the manufacturer as 'typical'. Where there is a substantial deviation from this figure the device (or its immediate circuitry) should be considered suspect.

11 Projects

This final chapter provides outline constructional details of ten simple test equipment projects which not only provide illustrations of the techniques described in previous chapters but also provide a comprehensive range of equipment useful in its own right.

Each project comprises a module which, together with its companion modules, is mounted in a low-cost miniature subrack system (e.g. RS/Electromail 501–323). This versatile system accepts modules available in a variety of different width 'panel sets'. Panels are linked together using a bus system based on 21-way DIN 41617 connectors.

Each of the standard width (6TE, 12TE and 18TE) panel sets specified is supplied complete with front panel, stripboard circuit card, card supports, handles and DIN 41617 plug. Modules may be 'mixed and matched' to suit individual requirements. Note, however, that the a.c. and d.c. power supplies described on pages 236 and 237 will be required as a minimum since other units derive their power supplies from them.

The following terminology is used to describe the signals present on the rack system bus:

Pin number	*Abbreviation*	*Signal/function*
1	d.p.1	Meter decimal point position 1
2	d.p.2	Meter decimal point position 2
3	d.p.3	Meter decimal point position 3
4	d.p.sel	Meter decimal point selector
5	Meter −	Meter low input (−ve)
6	Meter +	Meter high input (+ve)
7	Signal 1	General purpose signal line 1
8	Signal 2	General purpose signal line 2
9	−17 V	−17 V unregulated at up to 1.25 A
10	+17 V	+17 V unregulated at up to 1.25 A
11	r.f.u.	Reserved for future use
12	−12 V	−12 V regulated at up to 500 mA
13	−5 V	−5 V regulated at up to 500 mA
14	+5 V	+5 V regulated at up to 500 mA
15	+12 V	+12 V regulated at up to 500 mA
16	n.c.	Not connected
17	12 V a.c.	12 V a.c. 50 Hz at up to 1.25 A
18	n.c.	Not connected
19	12 V a.c.	12 V a.c. 50 Hz at up to 1.25 A
20	n.c.	Not connected
21	0 V	Common 0 V

a.c. power unit

This unit provides a source of low-voltage a.c. for the panel mounted modules and has two 12 V anti-phase outputs each rated at 1.25 A. Unlike the other panel set mounted modules, the a.c. power unit is housed in the rear of the rack system. The unit incorporates mains switching and fuses (mounted on the rear panel of the rack system) together with a single noise suppression capacitor.

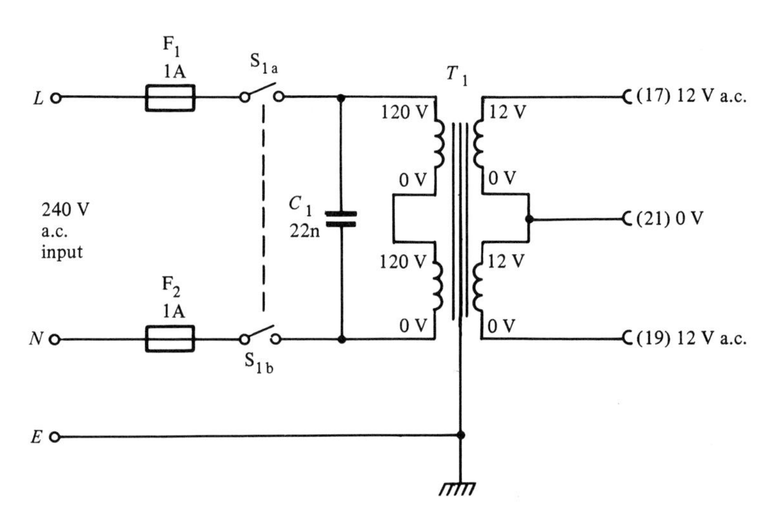

Figure 11.1 *a.c. power supply*

Components

S_1 d.p.d.t. miniature rocker mains switch.
F_1, F_2 1 A 20 mm quick blow fuses and holders.
C_1 22 n polypropylene capacitor rated at 1 kV d.c./350 V a.c.
T_1 30 VA open-lead toroidal transformer with 2 × 120 V primaries and 2 × 12 V secondaries (e.g. RS/Electromail 208–462)

Miniature subrack system (e.g. RS/Electromail 501–323) fitted with an appropriate number of card guides and 21-way DIN 41617 sockets.
IEC mains lead and connector.

d.c. power supply

The d.c. power supply is fitted in a 6TE panel set and not only feeds its outputs to the rack system bus (for use by other modules) but also provides the following fixed outputs at the front panel; + 12 V, + 5 V, − 5 V, and − 12 V. Each output is nominally rated at 250 mA but this figure can be exceeded depending upon the demands of other modules present in the rack. See Plate 16.

Components

Resistor

R_1 1 kΩ 0.25 W 5% carbon film.

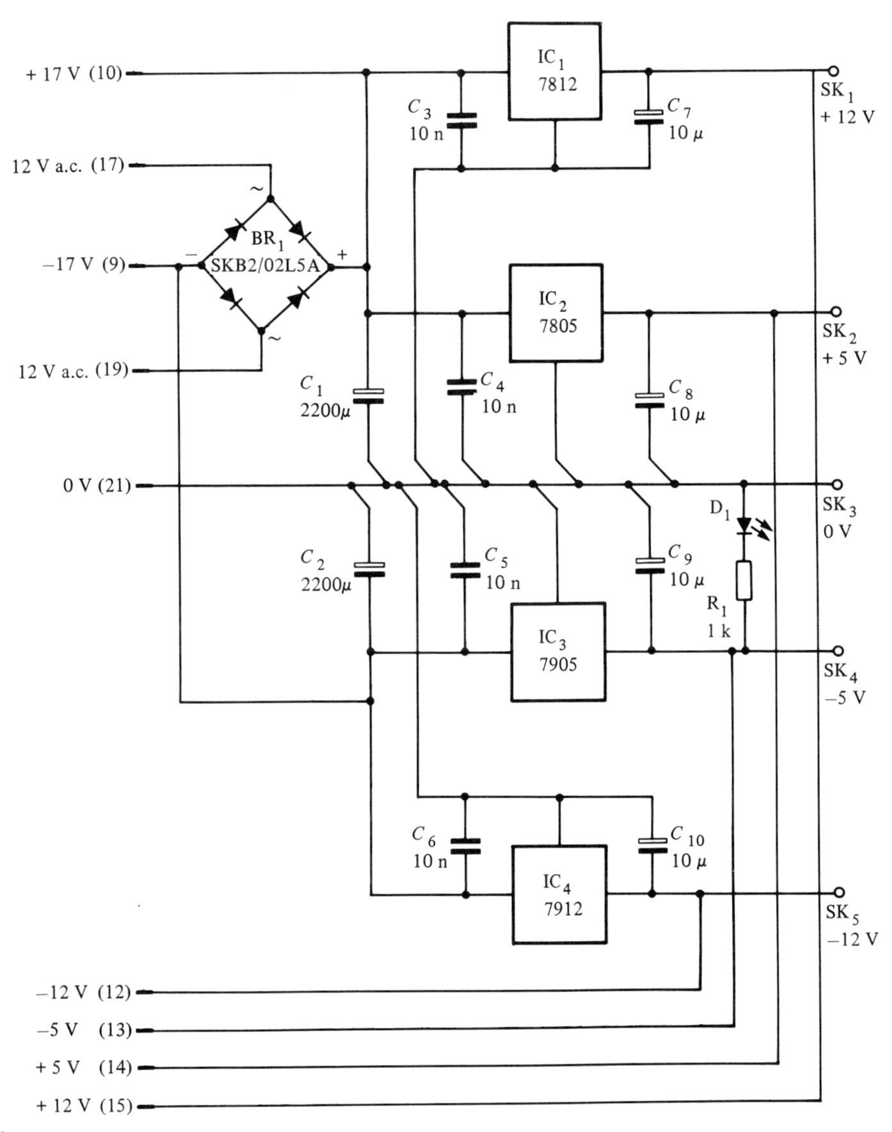

Figure 11.2 *d.c. power supply*

Capacitor

C_1, C_2	2200 μF 25 V axial lead electrolytic.
C_3 to C_6	10 nF sub-miniature polyester.
C_7 to C_{10}	10 μF 16 V radial lead electrolytic.

Semiconductors

BR_1 SKB2/02L5A.
IC_1 7812.
IC_2 7805.
IC_3 7905.
IC_4 7912.
D_1 Red LED

Miscellaneous

SK_1 to SK_5 2 mm sockets (appropriate colours).
6TE panel set.
Two heatsinks 7.1° C/W (modified).

Variable d.c. power supply

This unit is housed in a 12TE panel set and acts as a general purpose bench power supply having an output voltage which is variable over the range 2.7 V to 16.5 V. The power supply incorporates current limiting which is adjustable from 10 mA to 1.1 A. If desired, the output voltage may be displayed using the digital panel meter module described on page 239.

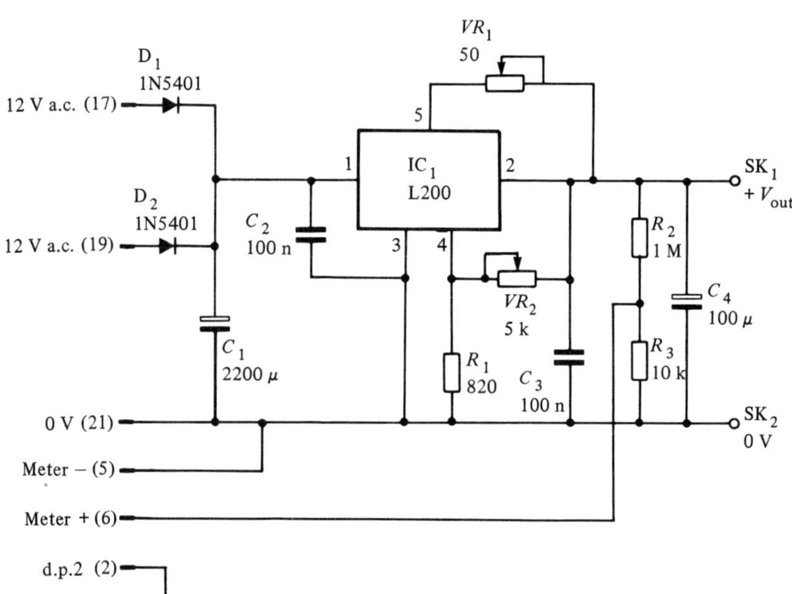

Figure 11.3 *Variable d.c. power supply*

Components

Resistors

R_1 820 Ω.
R_2 1 MΩ.
R_3 10 kΩ.

All fixed resistors are 0.25 W 5% carbon film.

VR_1 50 Ω wirewound.
VR_2 5 kΩ wirewound.

Capacitors

C_1 2200 μF 25 V axial lead electrolytic.
C_2, C_3 100 nF sub-miniature polyester.
C_4 100 μF 25 V axial lead electrolytic.

Semiconductors

D_1, D_2 1N5401.
IC_1 L200.

Miscellaneous

SK_1, SK_2 2 mm sockets (red and black).
12TE panel set.
TO220 heatsink rated at 9.9 °C/W.
Knobs (two required).

Digital panel meter

This unit provides a digital display for use in conjunction with the following modules:

Variable d.c. power supply (page 238)
Transistor tester (page 240)
Digital ohmmeter (page 242)

The digital panel meter is based on a readily available LCD digital meter module and it is housed in an 18TE panel set. The display functions with a basic sensitivity of 199.9 mV. Decimal point switching and input voltage conditioning are both carried out externally. The unit must be used in conjunction with the d.c. power supply module described on page 236.

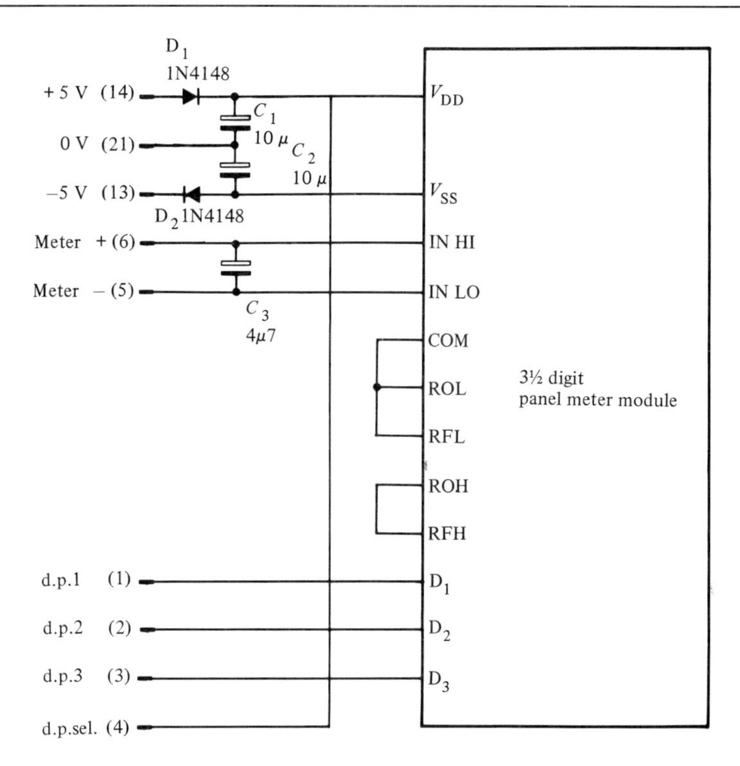

Figure 11.4 *Digital panel meter*

Components

Capacitors

C_1, C_2 10 μF 16 V radial lead electrolytic.
C_3 4.7 μF 35 V tantalum electrolytic.

Semiconductors

D_1, D_2 1N4148.

Miscellaneous

LCD panel meter module (e.g. RS/Electromail 258–041) and mounting bezel (e.g. RS/Electromail 258–057).
18TE panel set.

Transistor tester

This module functions on a simple static transistor tester which can be used to determine the current gain of n-p-n and p-n-p transistors. The following ranges are provided: 0 to 1999, 0 to 199.9, and 0 to 19.9 (at nominal base currents of 1 μA, 10 μA, and 100 μA respectively).

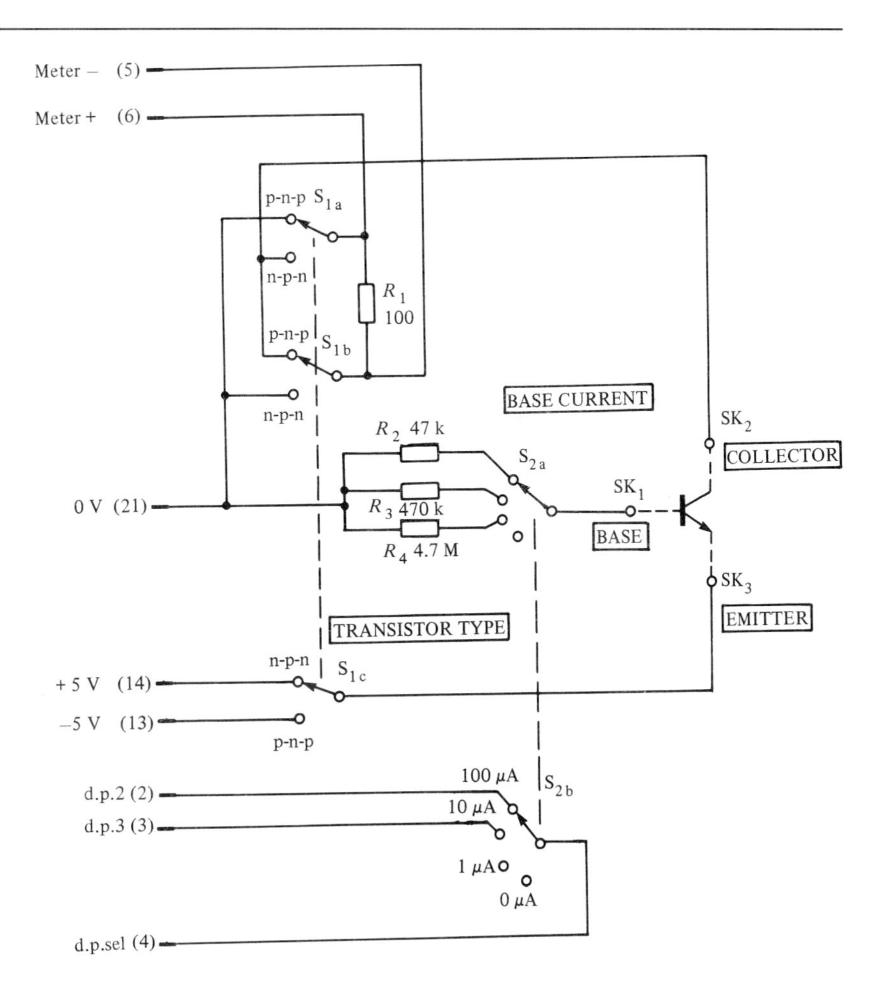

Figure 11.5 *Transistor tester*

The instrument also measures leakage current (base terminal left open circuit) in the range 0 to 1.999 mA and with a resolution of 1 μA. The transistor tester must be used in conjunction with the digital panel meter described on page 239 and the d.c. power supply described on page 236. The unit is housed in a 12TE panel set.

Components

Resistors

R_1 100 Ω.
R_2 47 kΩ.
R_3 470 kΩ.
R_4 4.7 MΩ.

All resistors are 0.25 W 5% carbon film.

Miscellaneous

S_1 3P 2 W (4P 3 W with rotation limit adjusted and one pole unused).
S_2 2P 4 W (3P 4 W with one pole unused).
Knobs (two required).
SK_1 to SK_3 1 mm sockets (appropriate colours).
12TE panel set.

Digital ohmmeter

This unit is housed in a 12TE panel set and, when used in conjunction with the digital panel meter, provides direct digital readings of resistance values. The unit is thus ideal for those requiring a more accurate means of measuring resistance than that provided by most conventional analogue multimeters. The instrument provides the following resistance ranges: 0 to 1.999 MΩ, 0 to 199.9 kΩ, 0 to 19.99 kΩ, 0 to 199.9 Ω. The digital ohmmeter must be used in conjunction with the d.c. power supply module.

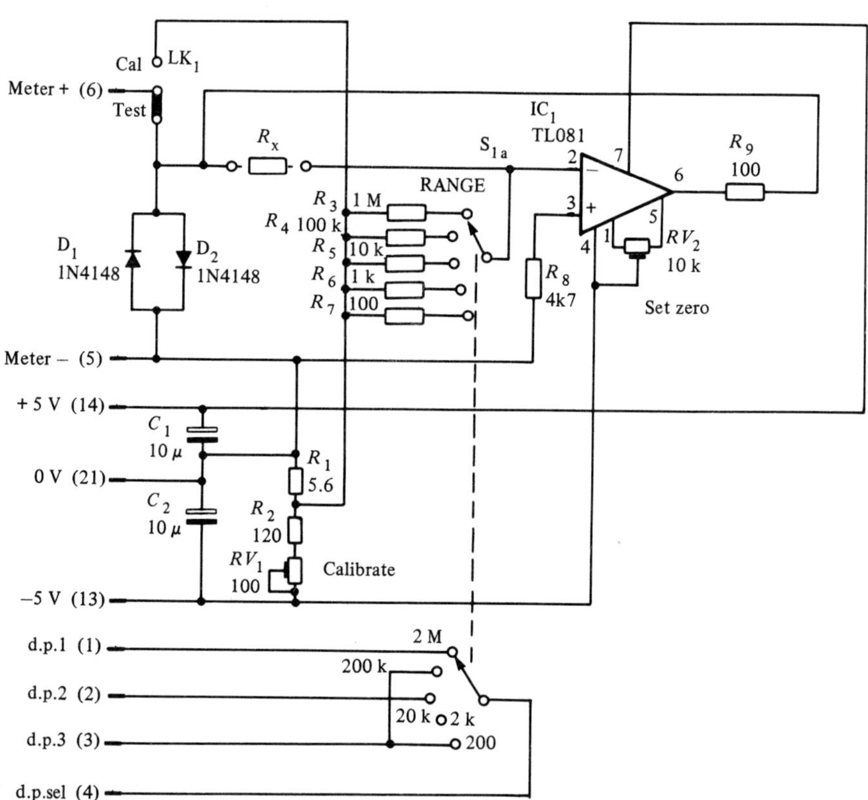

Figure 11.6 *Digital ohmmeter*

Components

Resistors

R_1 5.6 Ω.
R_2 120 Ω.
R_3 1 MΩ.
R_4 100 kΩ.
R_5 10 kΩ.
R_6 1 kΩ.
R_7 100 Ω.
R_8 4.7 kΩ.
R_9 100 Ω.

With the exception of R_3 to R_7 which are 1% metal film, all fixed resistors are 0.25 W 5% high-stability carbon film.

RV_1 100 Ω miniature horizontal skeleton preset.
RV_2 10 kΩ miniature horizontal skeleton preset.

Capacitors

C_1, C_2 10 μF 16 V radial lead electrolytic.

Semiconductors

D_1, D_2 1N4148.
IC_1 TL081.

Miscellaneous

S_1 2 P 5 W rotary switch (2 P 6 W type with rotation stop suitably adjusted).
12TE panel set.
Knobs (two required).
8-pin low-profile DIL IC socket.
SK_1, SK_2 2 mm sockets (black and red).
Jumper link and matching 3-way pin strip.

Function generator

This module acts as simple signal generator having sine, square and triangle outputs with frequency adjustable over the following frequency ranges: 1 Hz to 100 Hz, 100 Hz to 1 kHz, 1 kHz to 10 kHz, 1 kHz to 100 kHz. An output stage described on page 245. The function generator derives its impedance 100 Ω) but a wider range of outputs may be obtained using the output stage described on page 245. The function generator derives its power from the d.c. power supply and requires an 18TE panel set.

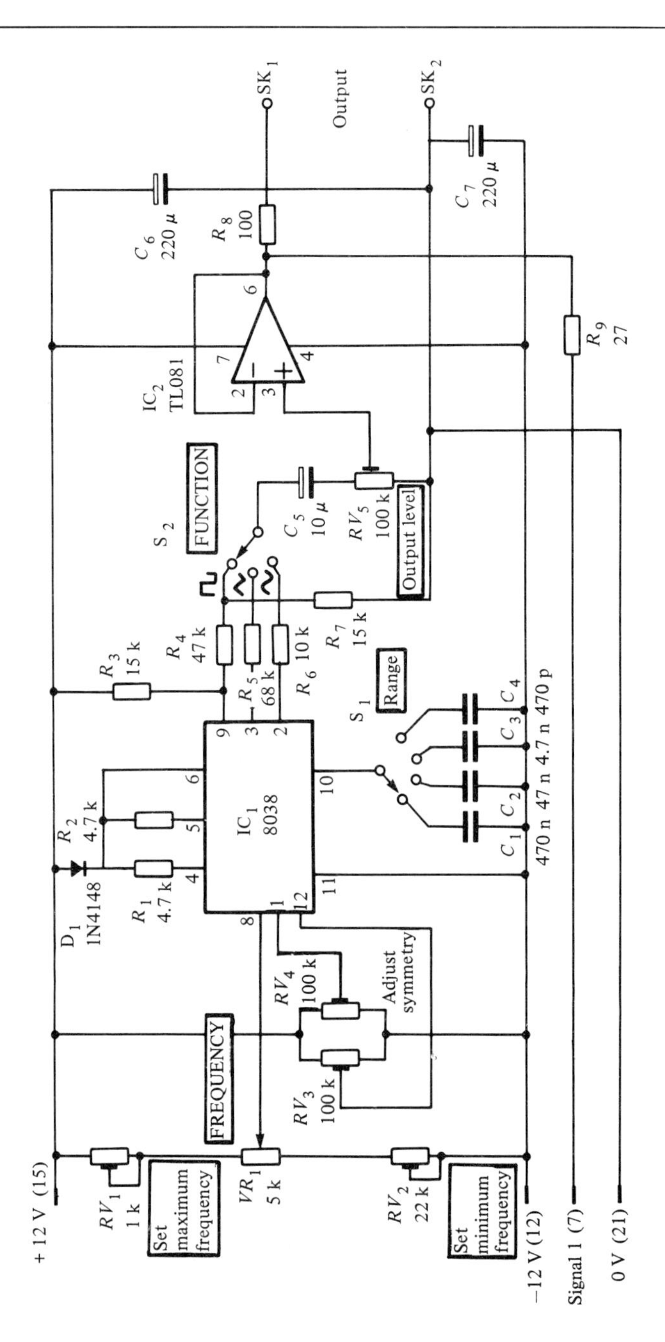

Figure 11.7 *Function generator*

Components

Resistors

R_1, R_2 4.7 kΩ.

R_3, R_7 15 kΩ
R_4 47 kΩ.
R_5 68 kΩ.
R_6 10 kΩ.
R_8 100 Ω.
R_9 27 Ω.

All fixed resistors are 0.25 W 5% carbon film.

Capacitors

C_1 470 nF polyester.
C_2 47 nF polyester.
C_3 4.7 nF polystyrene.
C_4 470 pF polystyrene.
C_5 10 μF 16 V radial lead electrolytic.
C_6, C_7 220 μF 16 V radial lead electrolytic.

Semiconductors

IC_1 8038
IC_2 TL081
D_1 1N4148

Miscellaneous

S_1 1P 4 W rotary switch (3 P 4 W type with two poles ignored).
S_2 4P 3 W rotary switch (4 P 3 W type with three poles ignored).
8-pin low profile DIL IC socket.
14-pin low profile DIL IC socket.
SK_1, SK_2 2 mm sockets (white and black).
Knobs (three required).
18TE panel set.

Output stage

This unit is designed to be used in conjunction with the function generator (page 243) and pulse generator (page 248) and provides outputs which are adjustable in the following ranges: 10 mV to 100 mV peak-peak, 100 mV to 1 V peak-peak, and 1 V to 10 V peak-peak. The unit has an output impedance of less than 1 Ω and requires a 12TE panel set. See Plate 17.

Components

Resistors

R_1 9kΩ (2 × 18 kΩ in parallel).
R_2 900Ω (2 × 1.8 k Ω in parallel).
R_3 100 Ω.

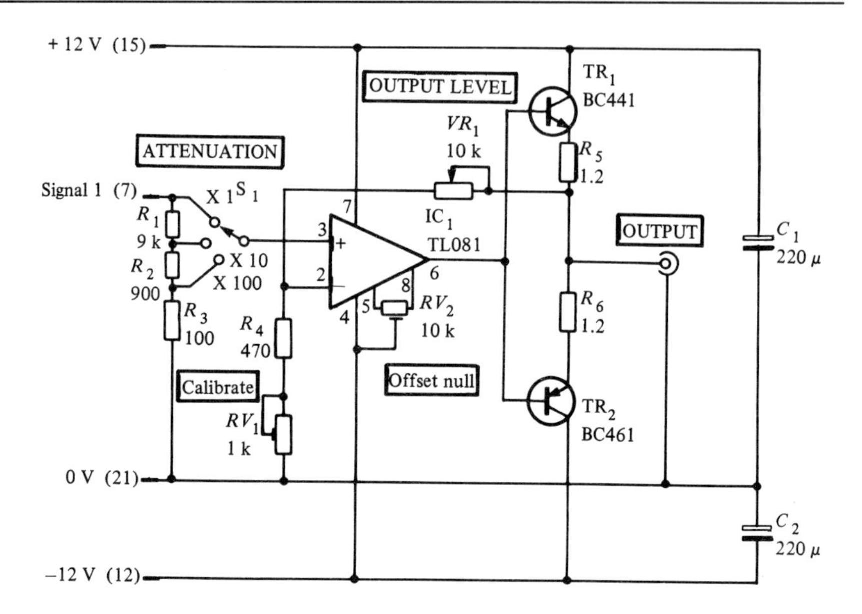

Figure 11.8 *Output stage*

All fixed resistors are 0.5 W 1% metal film.

RV_1 1 kΩ miniature horizontal mounting preset.
RV_2 10 kΩ miniature horizontal mounting preset.
VR_1 10 kΩ wirewound.

Capacitors

C_1, C_2 220 μF 16 V radial lead electrolytic.

Semiconductors

TR_1 BC441.
TR_2 BC461.
IC_1 TL081.

Miscellaneous

S_1 1 P 3 W rotary switch (4 P 3 W type with three poles ignored).
Push-fit TO5 heatsinks rated at 48 °C/W (two required).
12TE panel set.
8-pin low-profile DIL IC socket.
50 Ω BNC coaxial socket.
Knobs (two required).

Crystal calibrator

This module provides highly accurate and stable signals which can be used to calibrate equipment such as oscilloscopes, frequency meters,

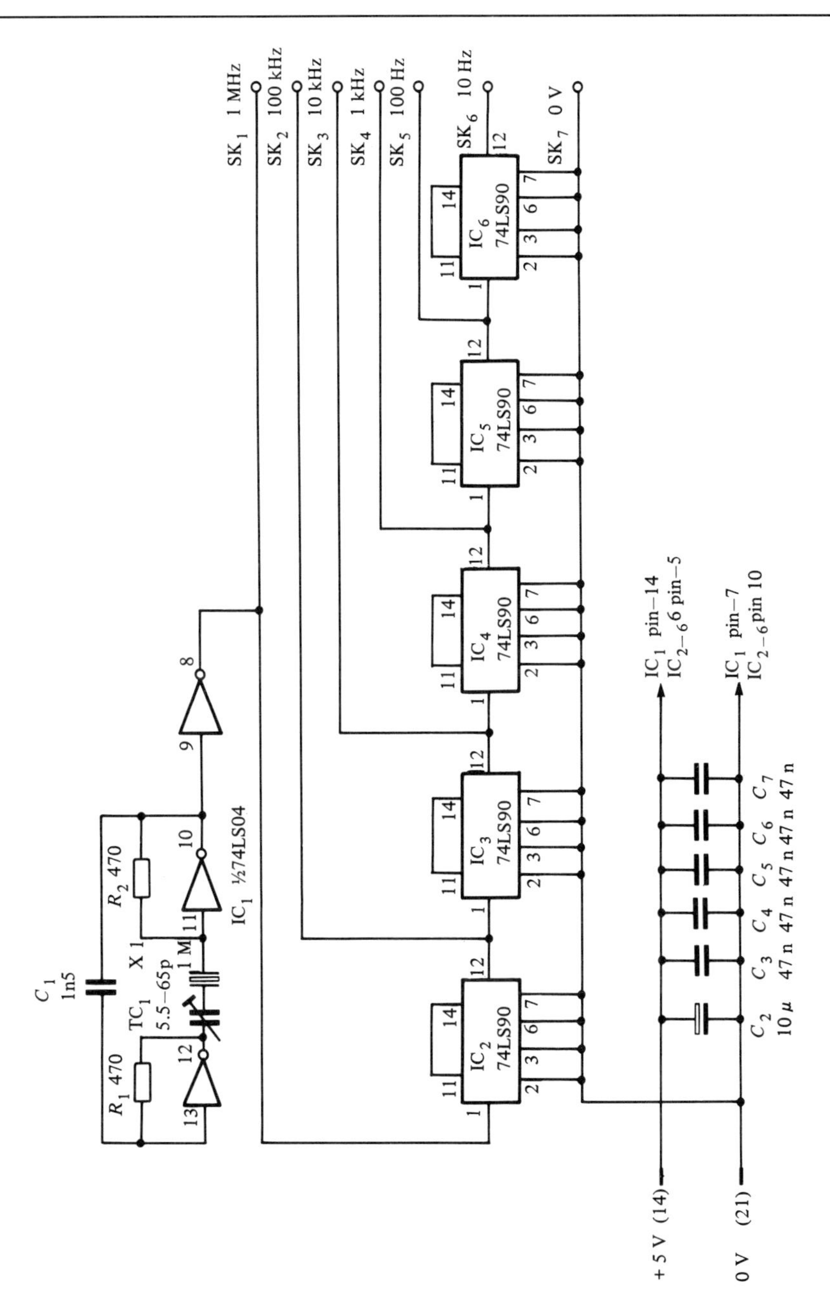

Figure 11.9 *Crystal calibrator*

counters, and radio receivers. The following fixed-frequency TTL compatible outputs are provided: 10 Hz, 100 Hz, 1 kHz, 10 kHz, 100 kHz, and 1 MHz. The unit can also be used as a general purpose clock generator for use with logic circuitry. The crystal calibrator obtains its power from the d.c. power supply module described on page 236 and requires a 6TE panel set.

Components

Resistors

R_1, R_2 470 Ω 0.25 W 5%.

Capacitors

C_1 1 nF 5 miniature plate ceramic.
C_2 10 μF 16 V radial lead electrolytic.
C_3 to C_7 47 nF disc ceramic.
$TC1$ 5.5 pF to 65 p trimmer.

Semiconductors

IC_1 74LS04.
IC_2 to IC_6 74LS90.

Miscellaneous

X1 1 MHz HC33/U or HC6/U crystal.
14-pin low-profile DIL IC socket.
SK_1 to SK_7 2 mm sockets (appropriate colours).
6TE panel set.

Pulse generator

This unit provides single-shot and repetitive pulses having widths selected in decade ranges from 5 s to 500 ns. When repetitive operation is selected, p.r.f.s ranging from 0.05 Hz to 50 kHz are available. An LED is incorporated to provide the user with a visible warning that the selected duty cycle is invalid. The unit provides complementary TTL compatible outputs at the front panel. A variable output may be obtained using the output stage described on page 245 (note, however, that performance will be impaired for pulse widths of less than 5 μs, or so). The pulse generator should be used in conjunction with the d.c. power supply, and is mounted in an 18TE panel set.

Components

Resistors

R_1, R_3 10 kΩ.
R_2, R_4 1 kΩ.
R_5 4.7 kΩ.
R_6, R_{11} 470 Ω.
R_7, R_{12} 100 Ω.

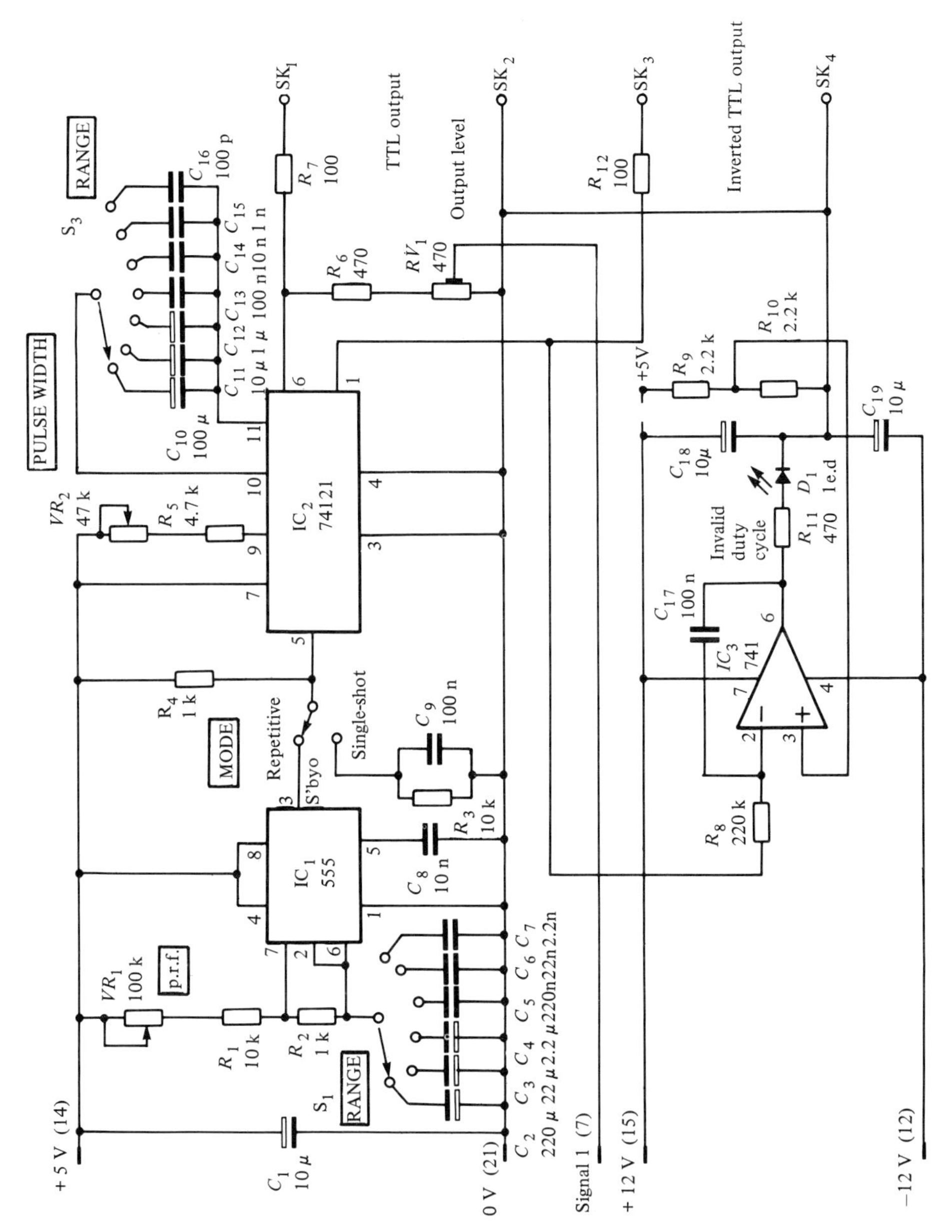

Figure 11.10 *Pulse generator*

R_8	220 kΩ.
R_9, R_{10}	2.2 kΩ.

All fixed resistors are 0.25 W 5% carbon film

RV_1	470 Ω miniature horizontal skeleton preset.
VR_1	100 kΩ linear carbon.
VR_2	47 kΩ linear carbon.

Semiconductors

IC_1 555.
IC_2 74121.
IC_3 741.
D_1 Red LED.

Miscellaneous

S_1 1 P 6 W (1 P 12 W type with rotation stop adjusted).
S_2 s.p.d.t. biased toggle with centre off position (e.g. RS/Electromail 317–184 d.p.d.t. type with one pole ignored).
S_3 1 P 7 W (1 P 12 W type with rotation stop adjusted).
SK_1 to SK_4 2 mm sockets (two red, two white).
18TE panel set.
14-pin low-profile DIL IC socket.
8-pin low-profile DIL IC sockets (two required).

Appendix 1 Selected symbols from BS 3939

General qualifying and supplementary symbols

Direct current or steady voltage

Alternating: general symbol

Indicates suitability for use on either direct or alternating supply

Variability: general symbol

Pre-set adjustment

Inherent non-linear variability

Mechanical coupling: general symbol

Example

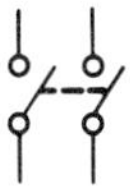

Unclassified symbols

Primary or secondary cell

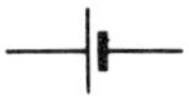

Notes:

1 The long line represents the positive pole and the short line the negative pole.

2 This symbol may also be used to indicate a battery. The nominal voltage should then be indicated on the diagram.

Battery of primary or secondary cells

Alternative symbol

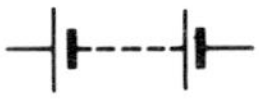

Example: 12 V battery

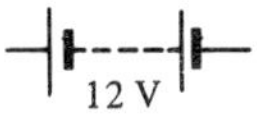

12 V

Earth

When it is necessary to distinguish between earth connections, the symbols may be annotated or numbered.

Example: signal earth

Signal

Frame or chassis not necessarily earthed

Positive polarity

Negative polarity

Impedence

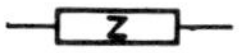

Z

Rectifier

Signal lamp: general symbol

Electric bell: general symbol

Electric buzzer: general symbol

Filament lamp

Cold cathode discharge lamp (e.g. neon lamp)

Conductors and connecting devices

Conductor or group of conductors: general symbol

Jumper (cross connection or temporary connection)

Two conductors: single-line representation

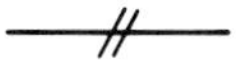

Two conductors: multi-line representation

Single line representation of n conductors

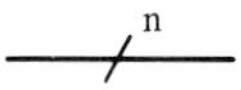

n

Note: The stroke may be omitted if there is no risk of confusion

General symbol denoting twisting of conductors

Example: two conductors twisted

General symbol denoting cable

Example: four conductors in cable

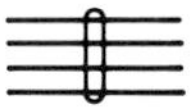

Crossing of conductor symbols on a diagram (no electrical connection)

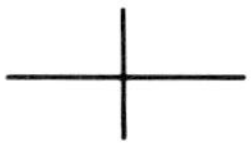

Junction of conductors

Double junction of conductors

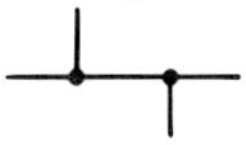

Terminal or tag: general symbol

Link normally closed;

with two readily separable contacts

with two bolted contacts

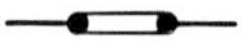

Example: hinged link, normally open

Plug (male)

Socket (female)

Coaxial plug

Coaxial socket

Multi-pole plug and socket devices including link inserts, coaxial or otherwise, may be made up from the standard symbols as illustrated below:

Example: multi-pole plug and socket, 5-pole

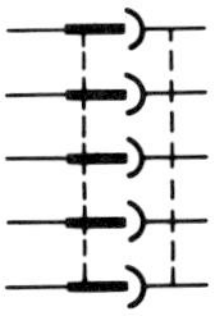

Or thus:

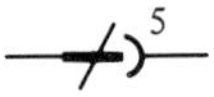

5

Jack sleeve (bush)

Jack spring

Examples: Three-pole concentric plug and jack

Three-pole concentric plug and break jack

Fuse: general symbol

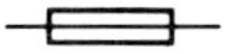

The supply side may be indicated by a thick line, thus:

Alternative general symbol

Resistors

Fixed resistor:
general symbol

Alternative general
symbol

Variable resistor:
general symbol

Resistor with moving
contact (rheostat)

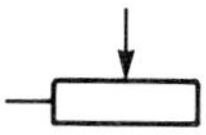

Voltage divider with
moving contact
(potentiometer)

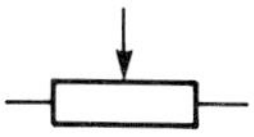

Resistor with pronounced
negative resistance-
temperature coefficient,
e.g. thermistor

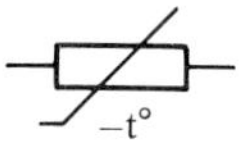

−t°

Capacitors

Capacitor:
general symbol

Polarized capacitor:
general symbol

+

Polarized electrolytic
capacitor

+

Variable capacitor:
general symbol

Capacitor with pre-set
adjustment

Inductors and transformers

Winding (i.e. of an
inductor, coil or
transformer): preferred
general symbol

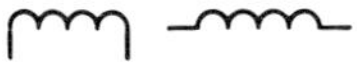

Winding with tappings

Example:
with two tappings

Note 1: If it is desired to
indicate that the winding
has a core it may be shown
thus:

Example: inductor with
core (ferromagnetic
unless otherwise
indicated)

Note 2: To the previous
symbol a note or the
chemical symbol of the
core material may be
added

Example: inductor
with a ferromagnetic
dust core

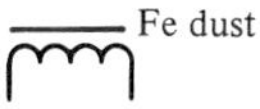

Fe dust

Transformer with ferro-
magnetic core
(Not in BS 3939)
If it is necessary to 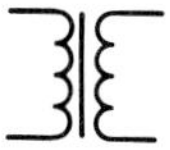
indicate that the core
has a gap it may be
shown thus:

Example; transformer
with gap in ferro-
magnetic core

Transformer general
symbols
complete form,
i.e. air cored

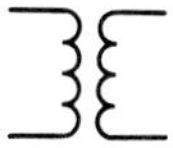

Rotating machines

Direct current motor: general symbol

Alternating current generator: general symbol

Alternating current motor: general symbol

Transducers

Microphone

Earphone (receiver)

Loudspeaker

Headgear receiver, double (headphones)

Measuring instruments and clocks

Indicating instrument, or measuring instrument: general symbol

Ammeter

Voltmeter

Wattmeter

Ohmmeter

Wavemeter

Oscilloscope

Galvanometer

Thermocouple (the negative pole is represented by the thicker line)

Contacts and relays

Make contact (normally open): general symbols

Break contact (normally closed): general symbols

Changeover contact, break before make

Changeover contact, make before break

Time-delayed contacts

Examples:
Make contact with delayed make

Push-button switches, non-locking

Make contact

Break contact

Control or protection relay: general symbol

Single element relay

Alternative symbol

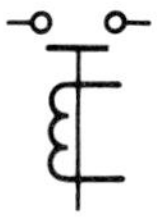

Relay coil: general symbol

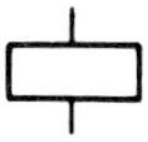

Relay contact-units;

Make contact-unit

Break contact-unit

Changeover (break before make) contact-unit

Changeover (make before break) contact-unit

Changeover (both sides stable) contact-unit

Semiconductor devices

p-n diode: general symbol

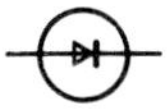

Alternative symbol

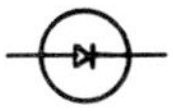

Note: The use of the envelope symbol is optional. It may be omitted if no confusion would arise. The envelope has to be shown if there is a connection to it.

Note: (not in BS 3939). Separate envelopes should not be shown for diodes and transistors forming part of a complex semiconductor device such as an integrated circuit.

p-n diode used as capacitive device (varactor) Envelope symbol has no physical counterpart (not in BS 3939).

Tunnel diode

Unidirectional breakdown diode, voltage-reference diode (voltage-regulator diode) e.g. zener diode

Thyristor: general symbol

Reverse-blocking triode thyristor, n-gate (anode-controlled)

Bidirectional triode thyristor

p-n-p transistor (also p-n-i-p transistor if omission of the intrinsic region will not result in ambiguity)

n-p-n transistor

n-p-n transistor with collector connected to envelope

Unijunction transistor with p-type base

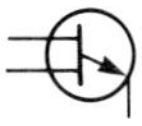

Unijunction transistor with n-type base

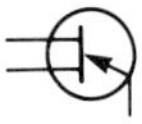

Junction-gate field-effect transistors (JUGFET)

With n-type channel

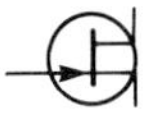

With p-type channel

Insulated-gate field-effect transistors (IGFET)

Depletion-type single-gate 3-terminal

n-channel

p-channel

Depletion-type, two-gate n-channel, 5-terminal with substrate connection brought out

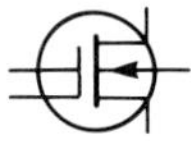

Enhancement-type, single-gate p-channel, 4-terminal with substrate connection brought out

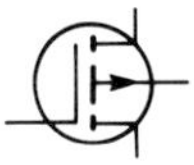

p-n diode, light-sensitive (photo-conductive cell with asymmetric conductivity)

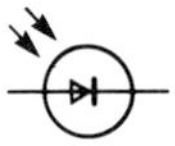

Photo-conductive cell with symmetrical conductivity

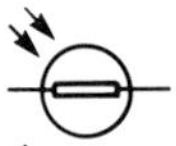

Photo-voltaic cell

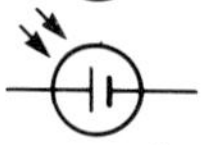

Light-generating semiconductor diode

Light-emitting semi-conductor diode, LED

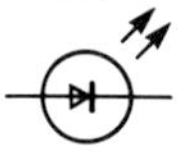

Binary logic elements

Elementary combinative elements

AND &

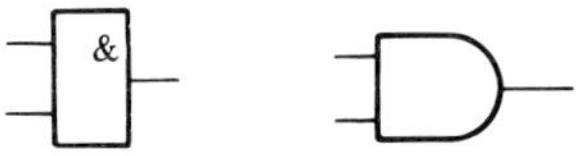

OR ≥ 1

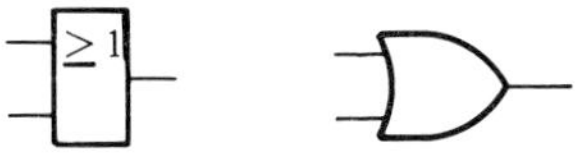

NOT 1

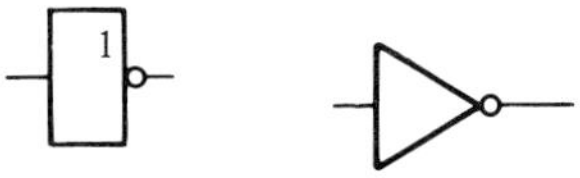

Complex combinative elements

Exclusive OR = 1

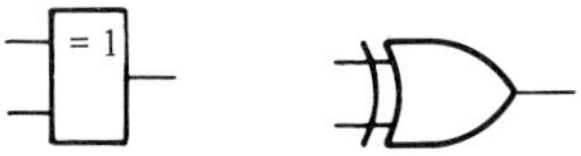

Logic identity (exclusive-NOR) =

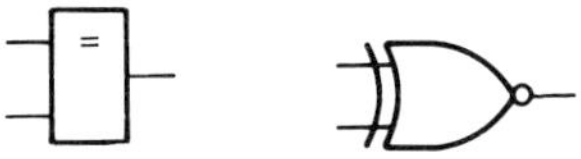

Examples of the use of combinative elements

AND with negated output (NAND) &

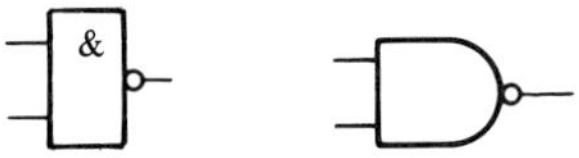

OR with negated output (NOR) ≥ 1

Bistable elements: general symbol Flip-flops (not in BS3939) Q $\bar{Q}$

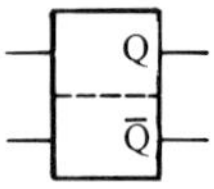

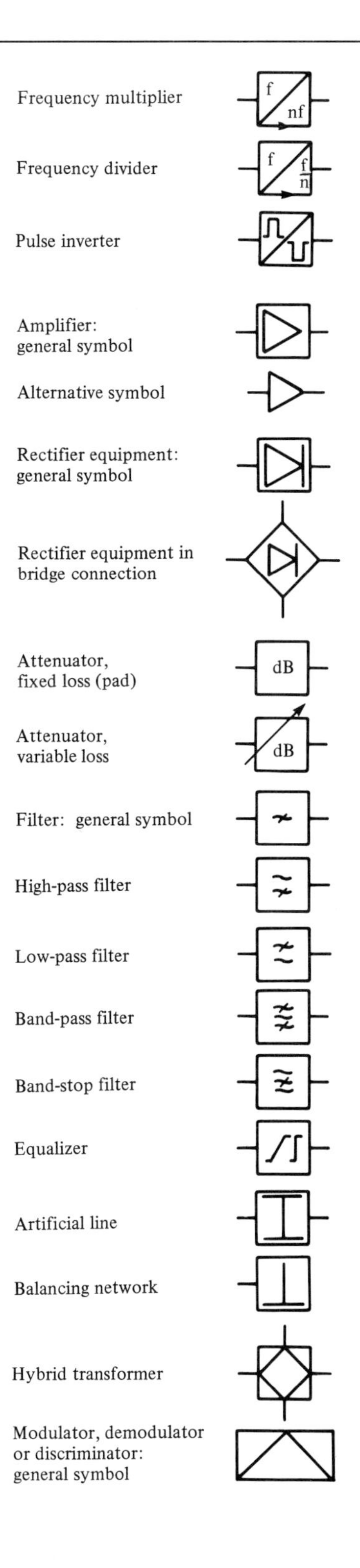
Frequency multiplier
f
nf
Frequency divider
f
f
n
Pulse inverter
Amplifier:
general symbol
Alternative symbol
Rectifier equipment:
general symbol
Rectifier equipment in
bridge connection
Attenuator,
fixed loss (pad)
dB
Attenuator,
variable loss
dB
Filter: general symbol
High-pass filter
Low-pass filter
Band-pass filter
Band-stop filter
Equalizer
Artificial line
Balancing network
Hybrid transformer
Modulator, demodulator
or discriminator:
general symbol

Appendix 2 Pin connecting data

Regulators

78L series

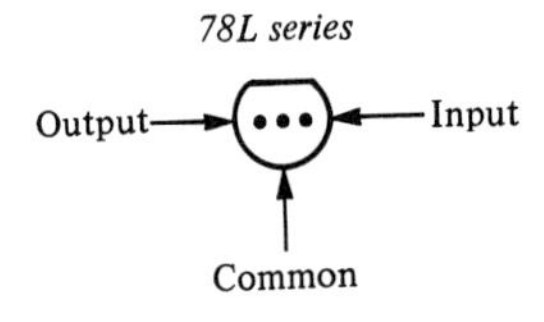

79L series

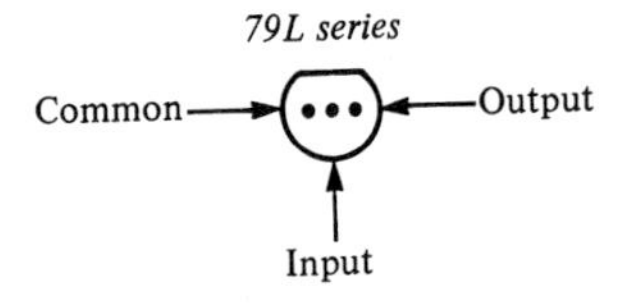

78 series

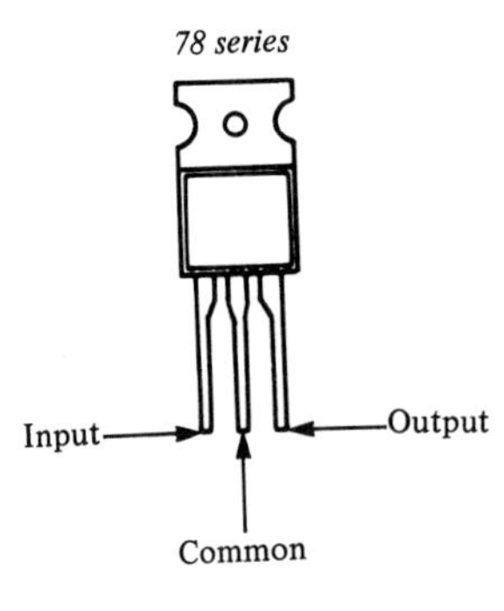

79 series

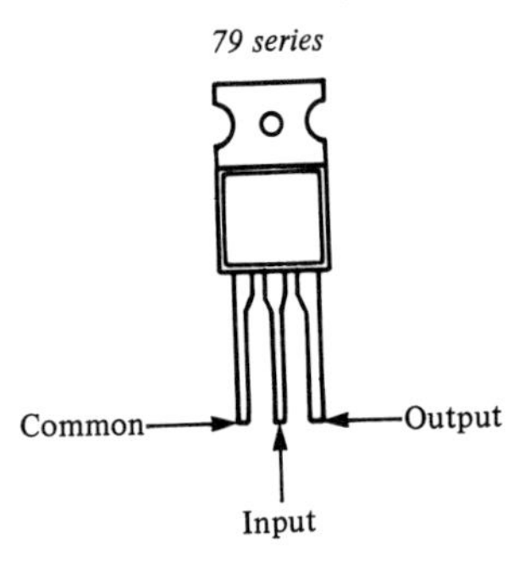

317L

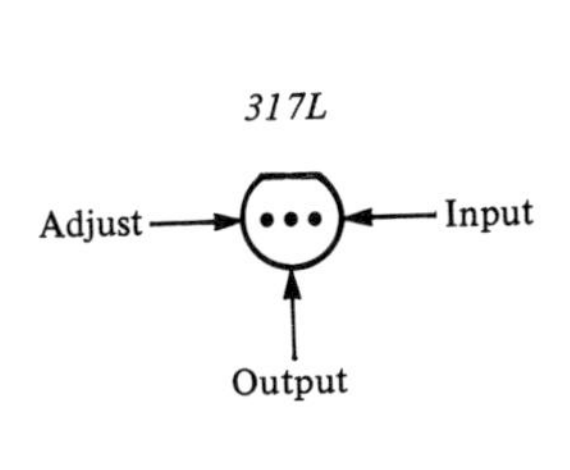

317M

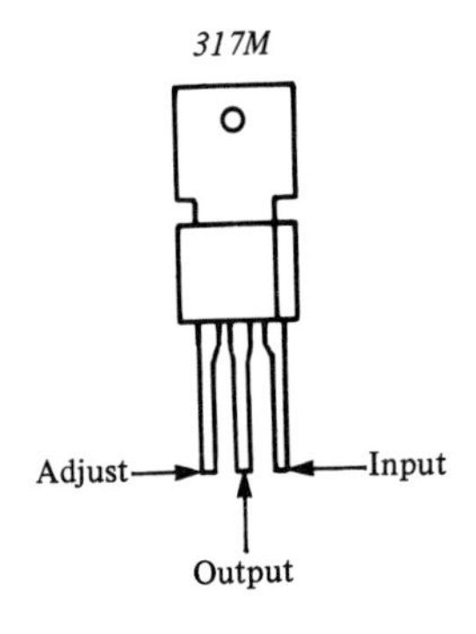

317K/338K

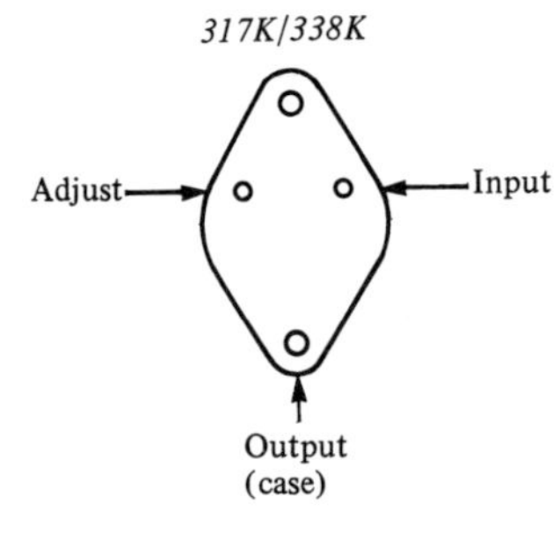

L200

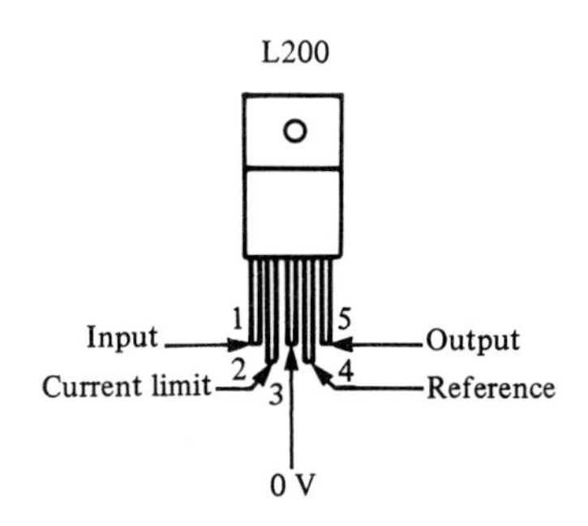

4195

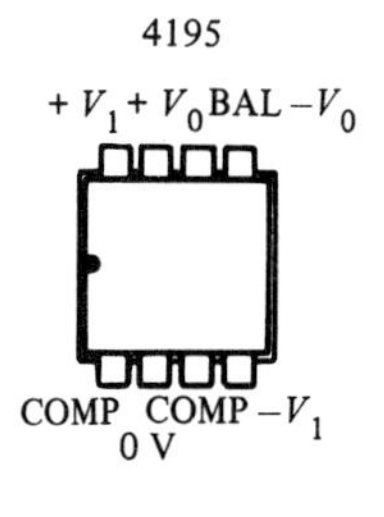

Transistors

BC184L/BC212L (TO92)

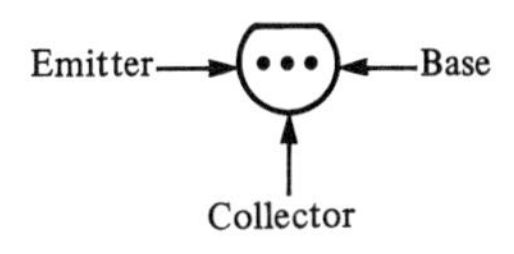

2N3819/BF244 (TO92)

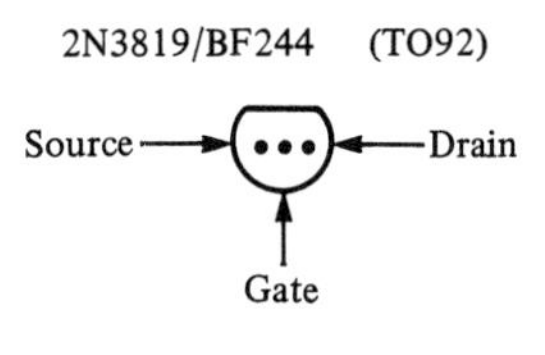

BC441/BC461/2N3053 (TO39)

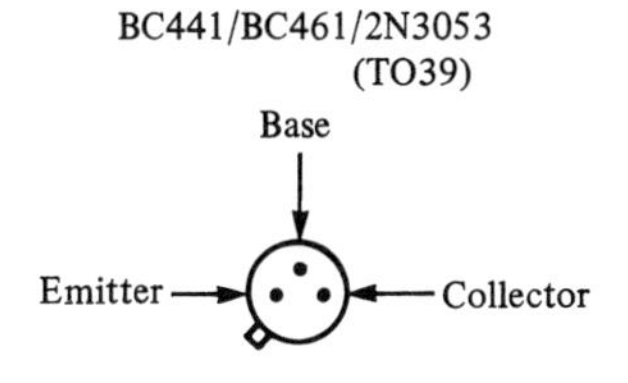

(TO220) TIP 31A

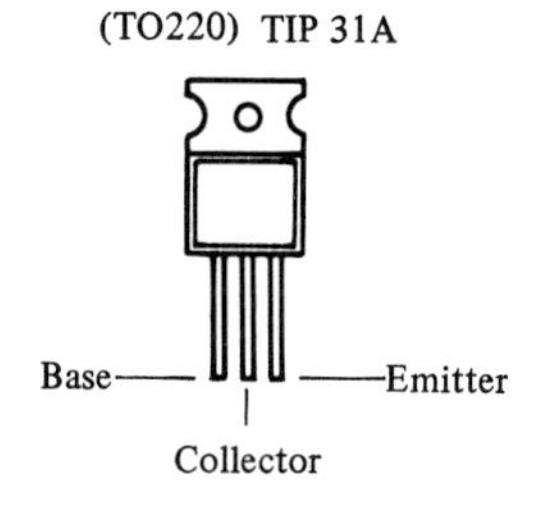

BC108/BC109/BC179 (TO18)

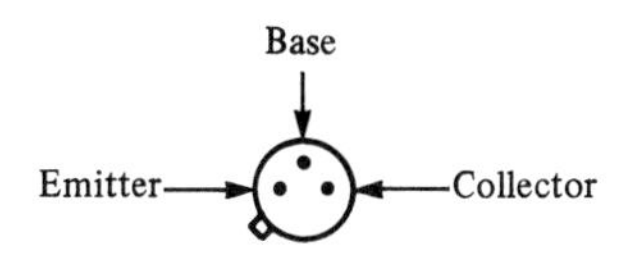

BC548 (TO92)

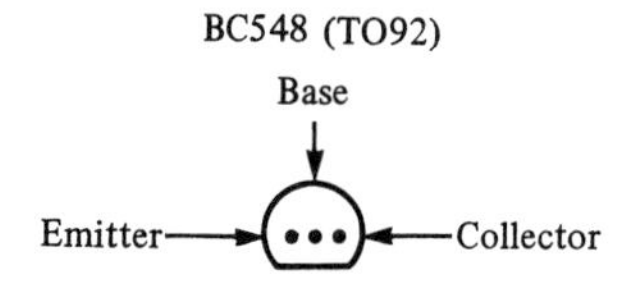

3N201 (TO72)

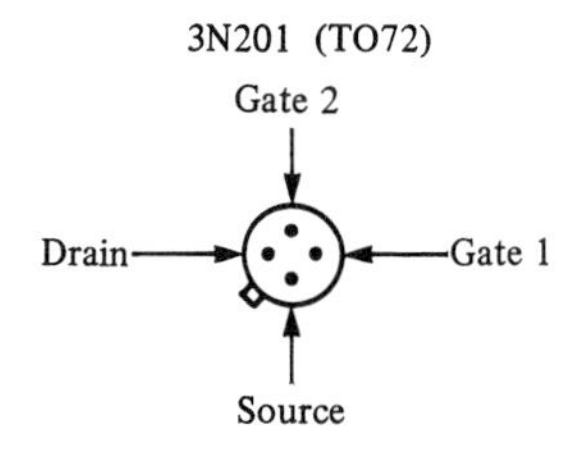

TIP3055/TIP2955/TIP141/TIP146 (TAB)

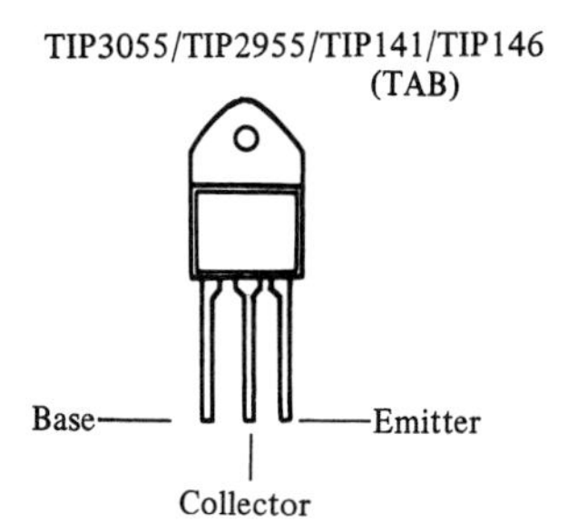

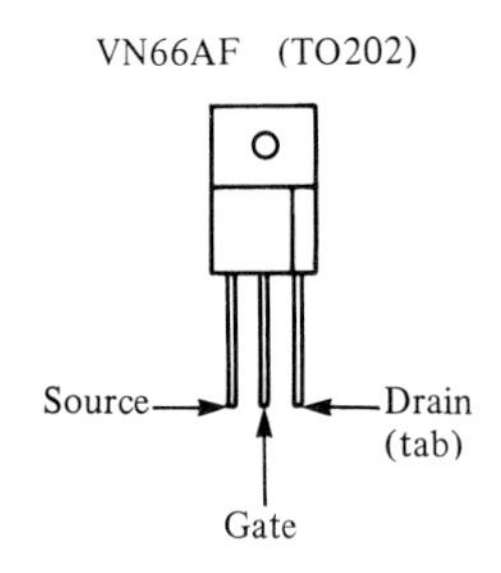

Diodes

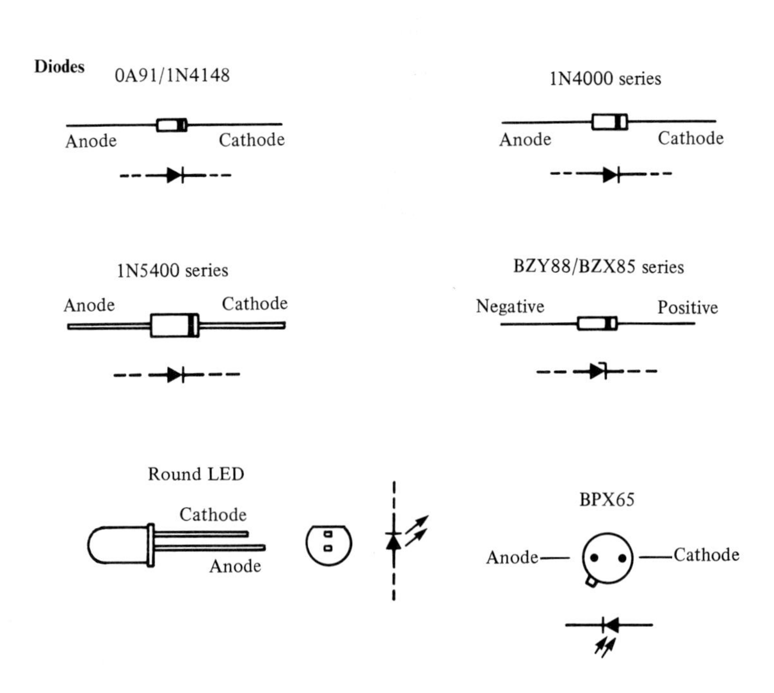

Semiconductor temperature sensor

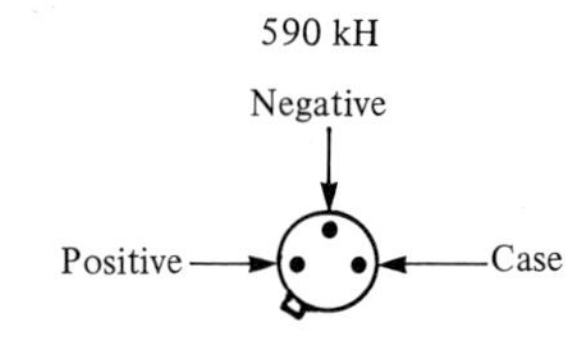

Plate 1 *Digital multimeter offering 0.05% d.c. accuracy and true RMS a.c. measurements up to 100 kHz*
(Courtesy Philips Test and Measuring Instruments)

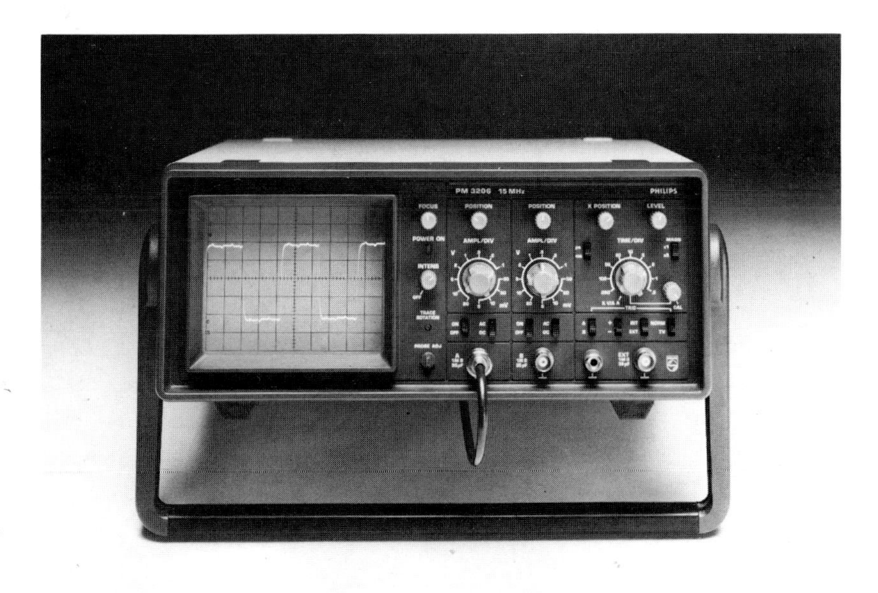

Plate 2 *Low-cost 15 MHz dual beam oscilloscope*
(Courtesy Philips Test and Measuring Instruments)

Plate 3 *Dual channel digital storage oscilloscope suitable for analysis of repetitive signals to well over 60 MHz*
(Courtesy Philips Test and Measuring Instruments)

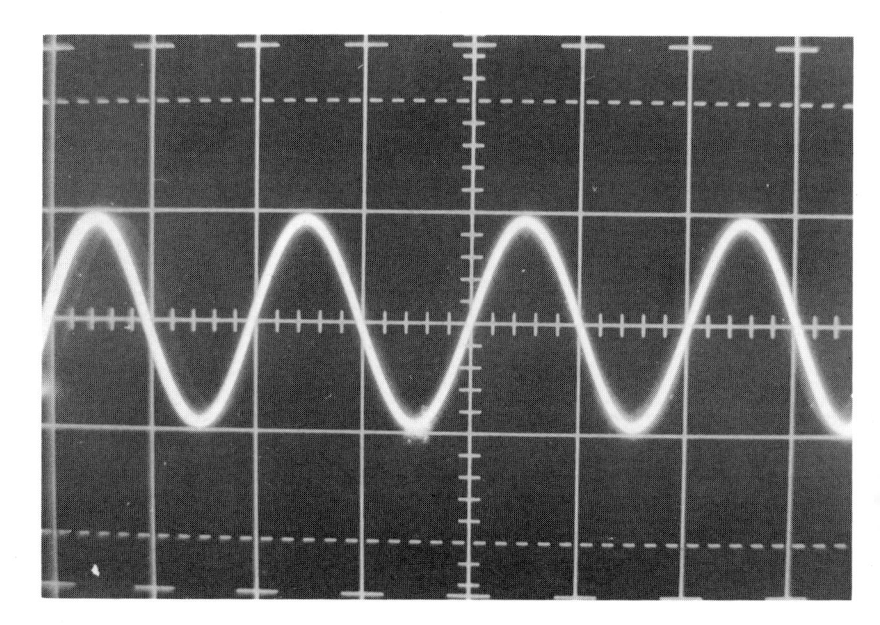

Plate 4 *1.9 v peak-peak 1 kHz sine wave displayed on an oscilloscope having timebase setting of 500 us/cm and vertical attenuator setting of 1 V/cm*

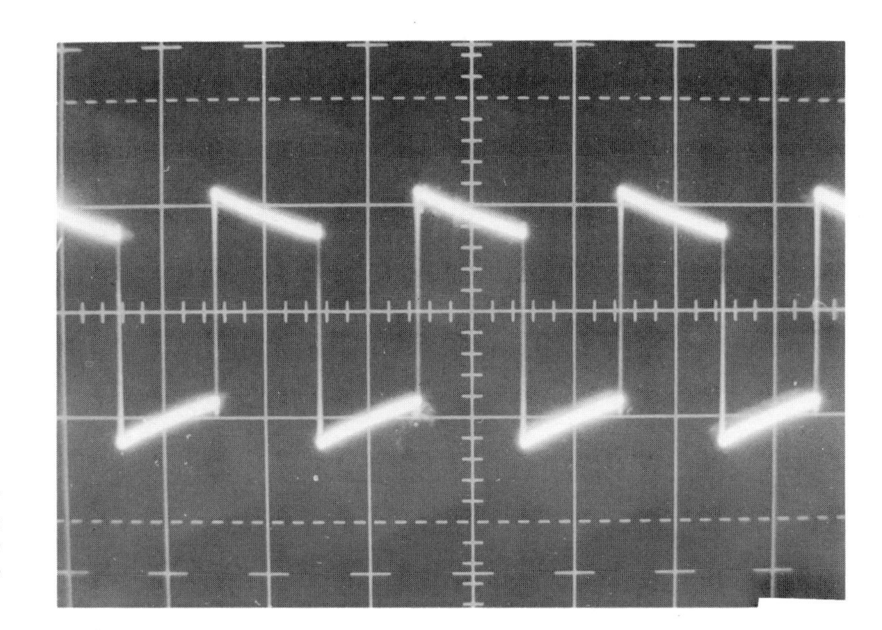

Plate 5 *Square wave with fast rise time (indicating good high frequency response) and slight sag (indicating restricted low frequency response)*

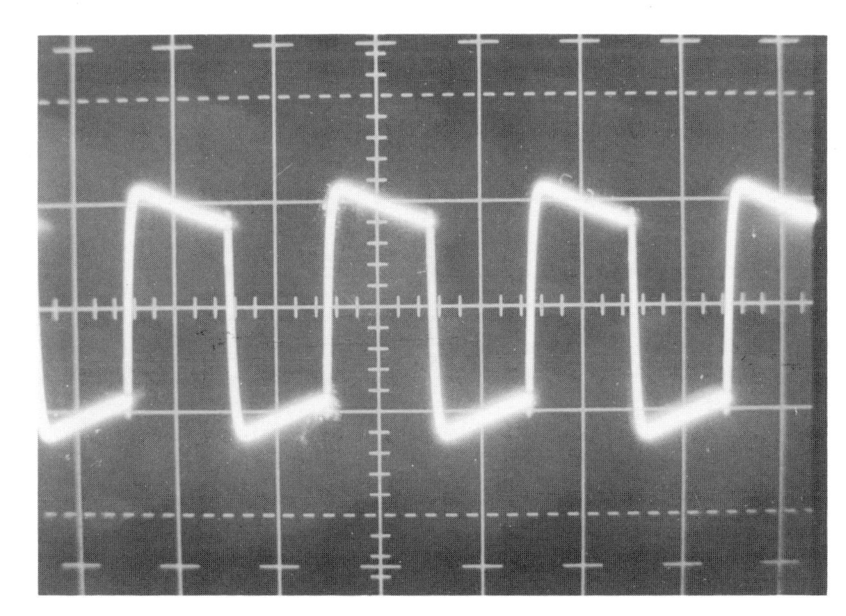

Plate 6 *Square wave with poor rise time (indicating restricted high frequency response and slight sag (indicating restricted low frequency response)*

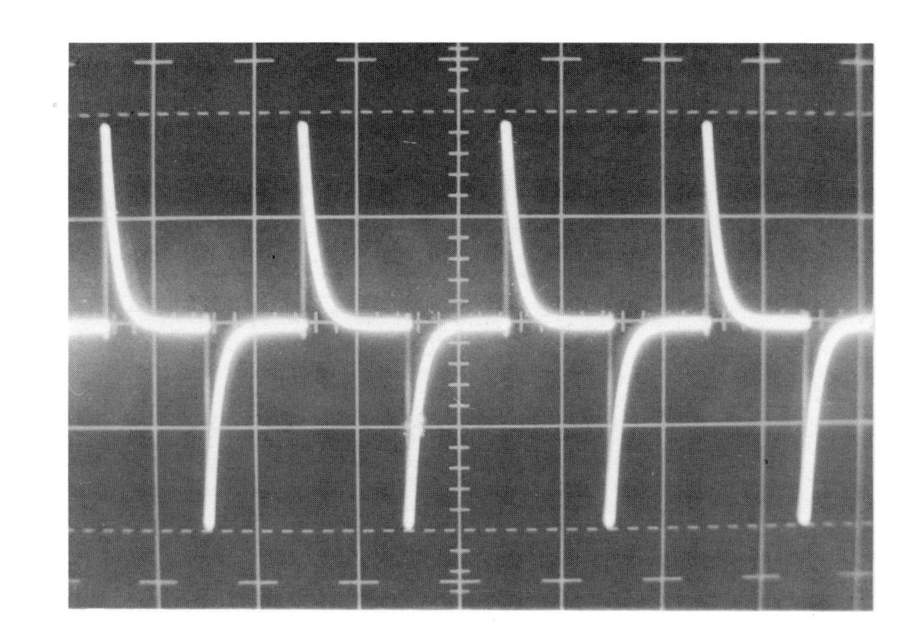

Plate 7 *Severely differentiated square wave (indicating good high frequency response coupled with very poor low frequency response)*

Plate 8 *Carefully locate the component pads on the underside of the PCB*

Plate 9 *Apply soldering iron to pads, heat and remove solder using a desoldering tool*

Plate 10 *Gently withdraw the component from the upper side of the PCB*

Plate 11 *Insert the replacement component, carefully checking correct lead orientation*

Plate 12 *Solder component leads to the pads*

Plate 13 *Trim the leads using a pair of side cutters*

Plate 14 *Example of an incorrectly made joint; large solder blobs, excessive flux due to dirty component leads, and solder bridges between adjacent tracks*

Plate 15 *Various heatsinks (including TO3, TO5 push-fit, and TO220/TAB types). The largest heatsink shown is rated at 6.8° C/W while the smallest shown is rated at 48° C/W*

Plate 16 *Completed d.c. poser supply module*

Plate 17 *Completed output stage module*

Thyristors and triacs

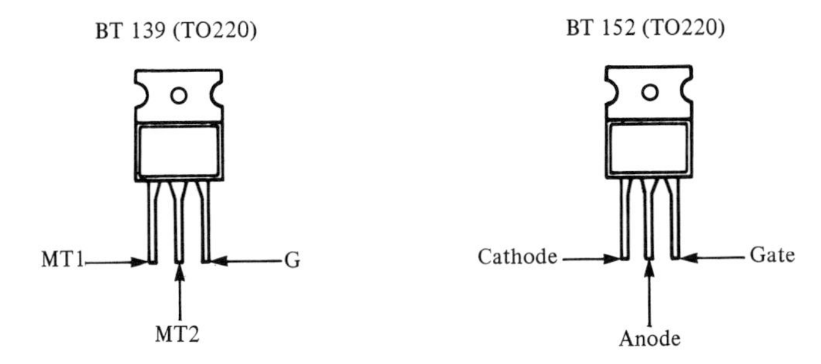

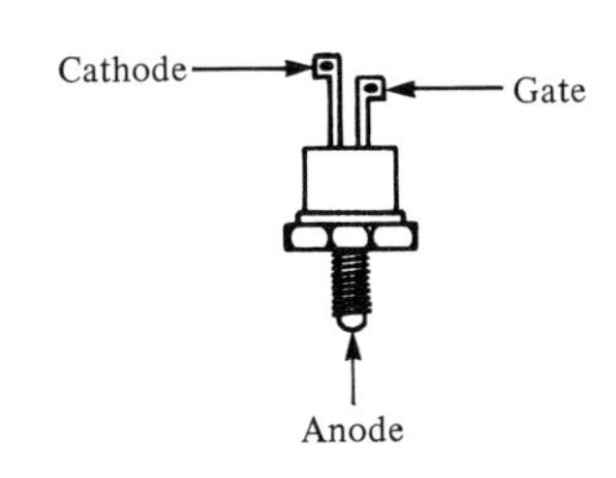

Solid state ray

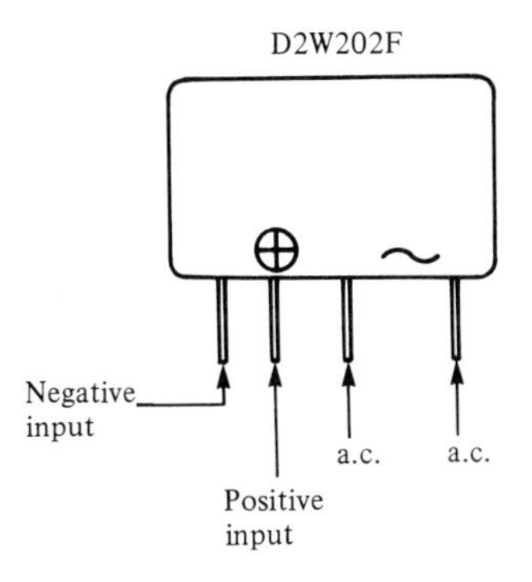

Dual comparator

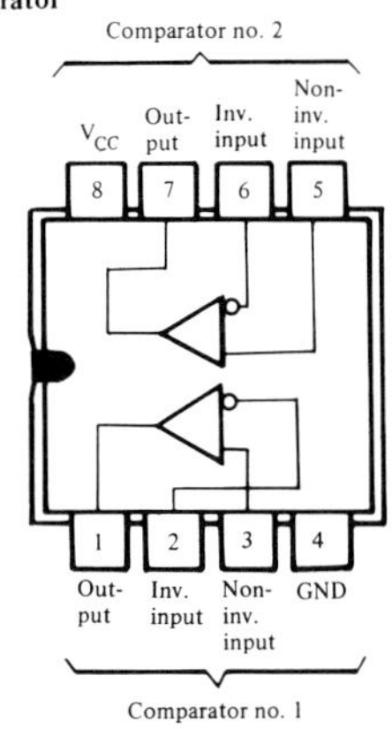

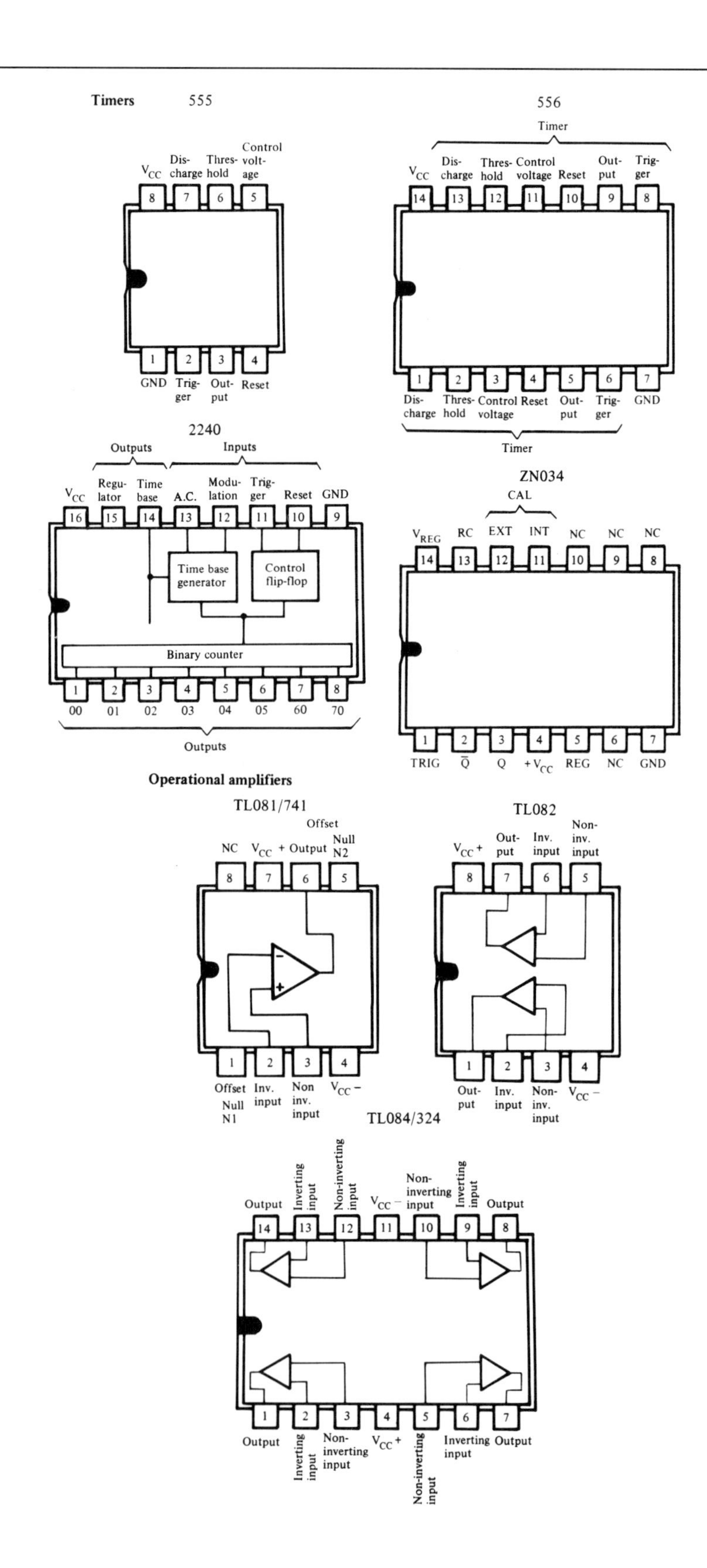
Timers
555
Control volt-age
Dis-charge
Thres-hold
V_{CC}
8 7 6 5
1 2 3 4
GND
Trig-ger
Out-put
Reset
556
Timer
V_{CC}
Dis-charge
Thres-hold
Control voltage
Reset
Out-put
Trig-ger
Dis-charge
Thres-hold
Control voltage
Reset
Out-put
Trig-ger
GND
Timer
2240
Outputs
Inputs
V_{CC}
Regu-lator
Time base
A.C.
Modu-lation
Trig-ger
Reset
GND
Time base generator
Control flip-flop
Binary counter
00 01 02 03 04 05 60 70
Outputs
ZN034
CAL
V_{REG}
RC
EXT
INT
NC
NC
NC
TRIG
$\bar{Q}$
Q
$+V_{CC}$
REG
NC
GND
Operational amplifiers
TL081/741
NC
V_{CC} +
Output
Offset Null N2
Offset Null N1
Inv. input
Non inv. input
V_{CC} −
TL082
V_{CC} +
Out-put
Inv. input
Non-inv. input
Out-put
Inv. input
Non-inv. input
V_{CC} −
TL084/324
Output
Inverting input
Non-inverting input
V_{CC} −
Non-inverting input
Inverting input
Output
Output
Inverting input
Non-inverting input
V_{CC} +
Non-inverting input
Inverting input
Output

Power operational amplifier

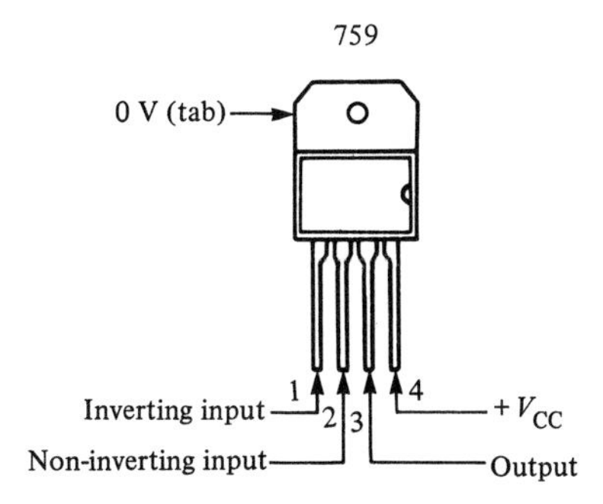

Linear integrated circuits

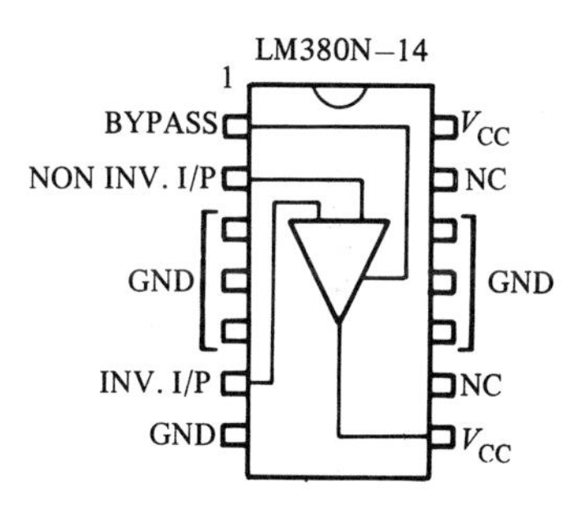

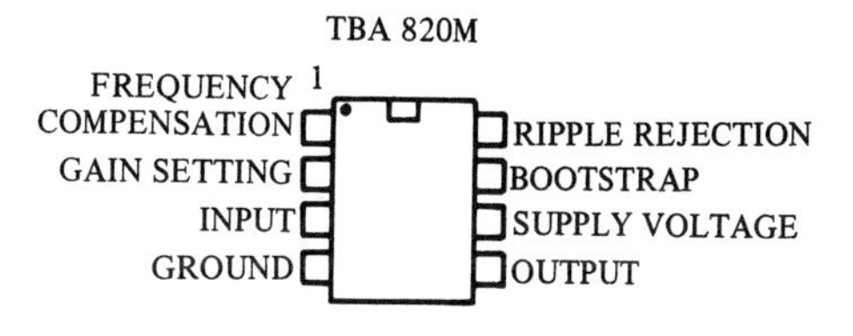

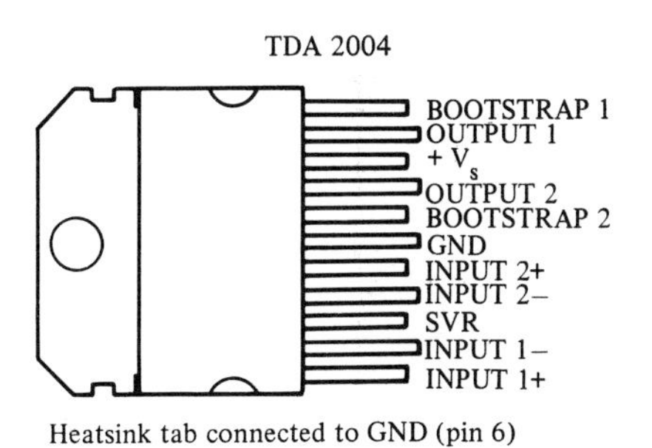

Heatsink tab connected to GND (pin 6)

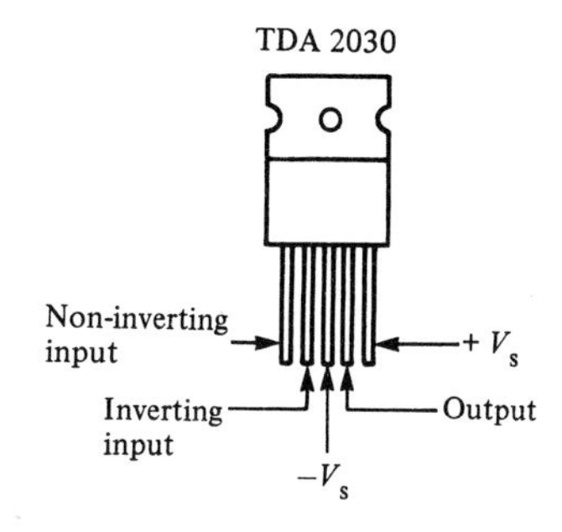

Common TTL integrated circuits

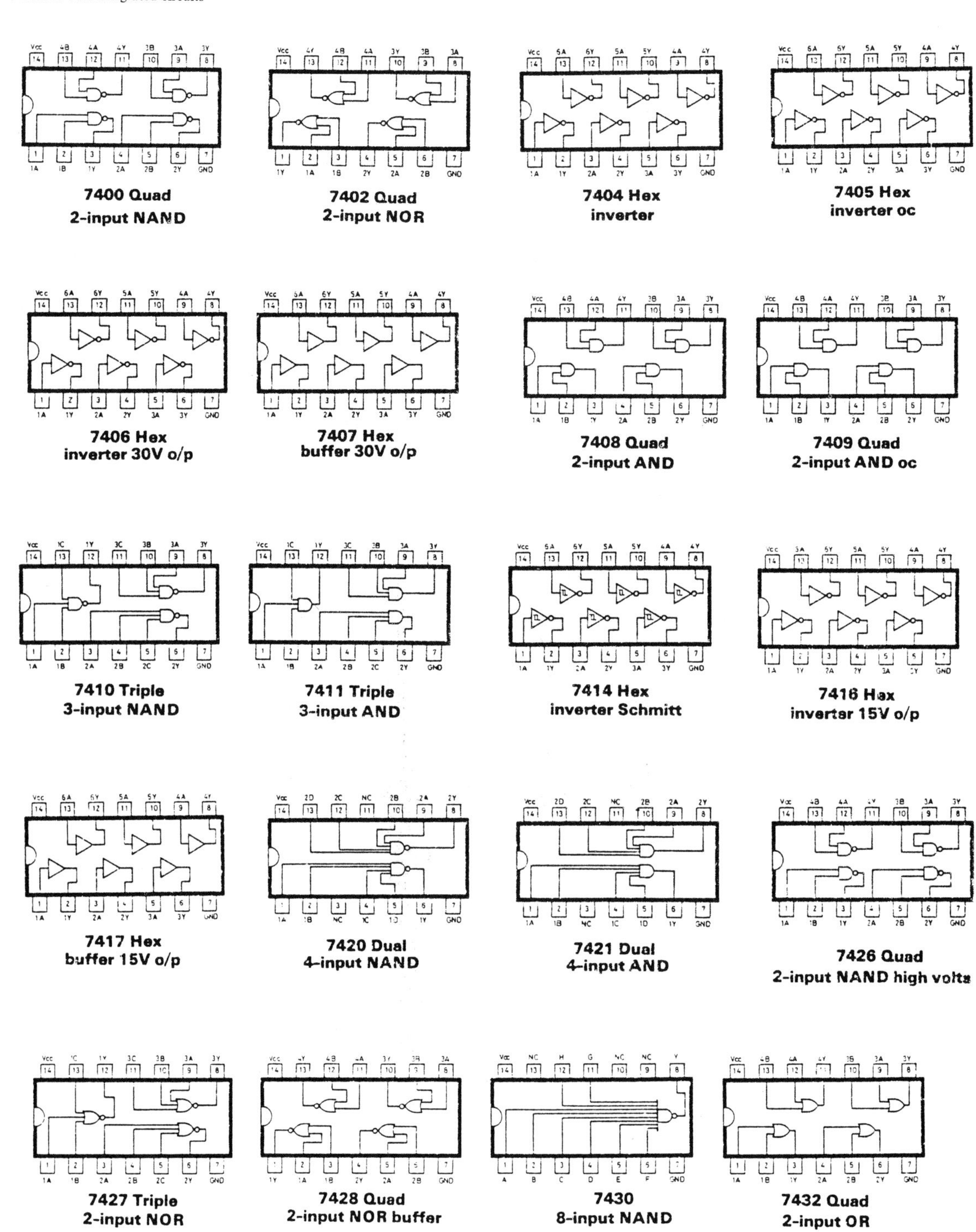

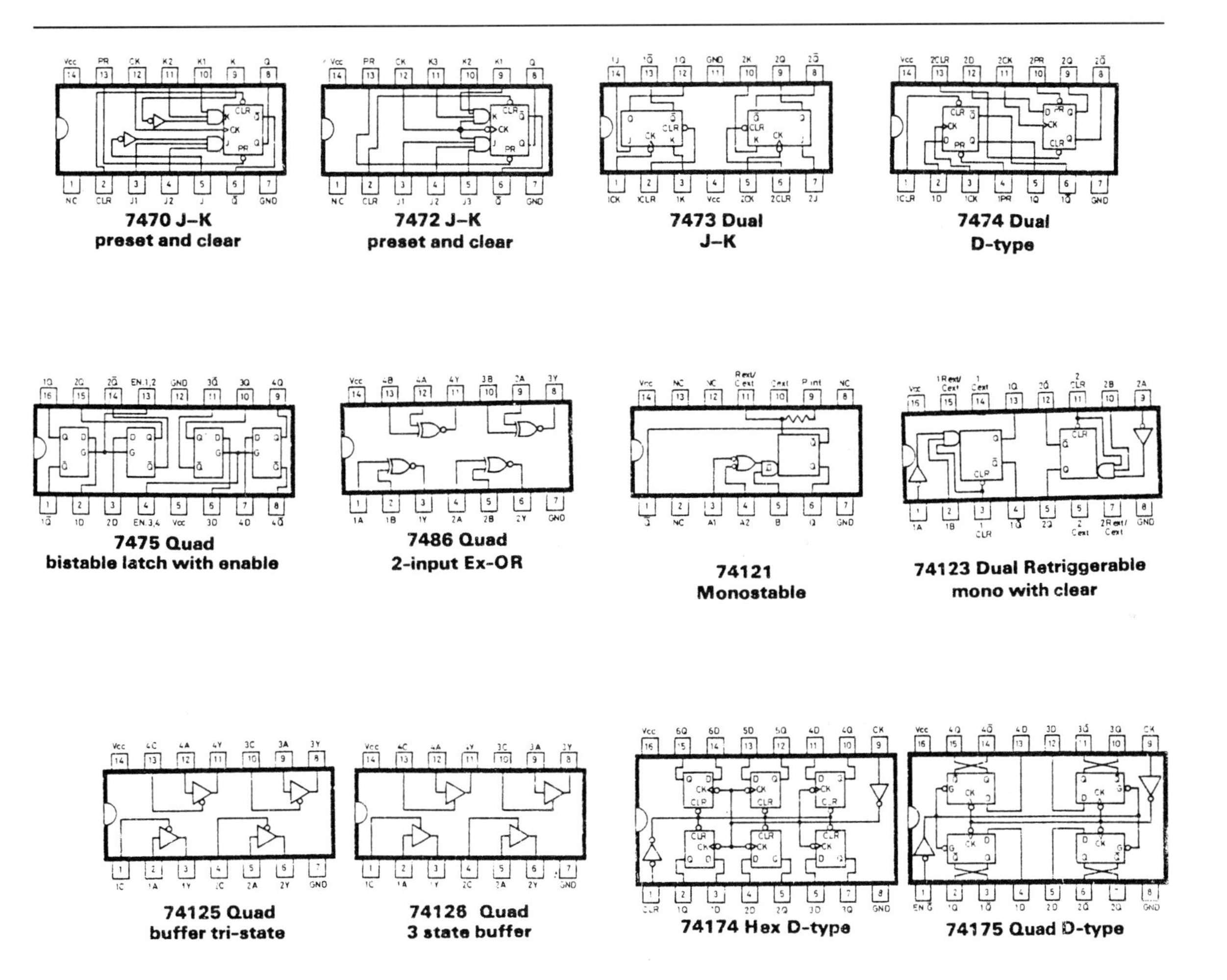

7470 J–K preset and clear

7472 J–K preset and clear

7473 Dual J–K

7474 Dual D-type

7475 Quad bistable latch with enable

7486 Quad 2-input Ex-OR

74121 Monostable

74123 Dual Retriggerable mono with clear

74125 Quad buffer tri-state

74126 Quad 3 state buffer

74174 Hex D-type

74175 Quad D-type

Appendix 3 Transistor mounting details

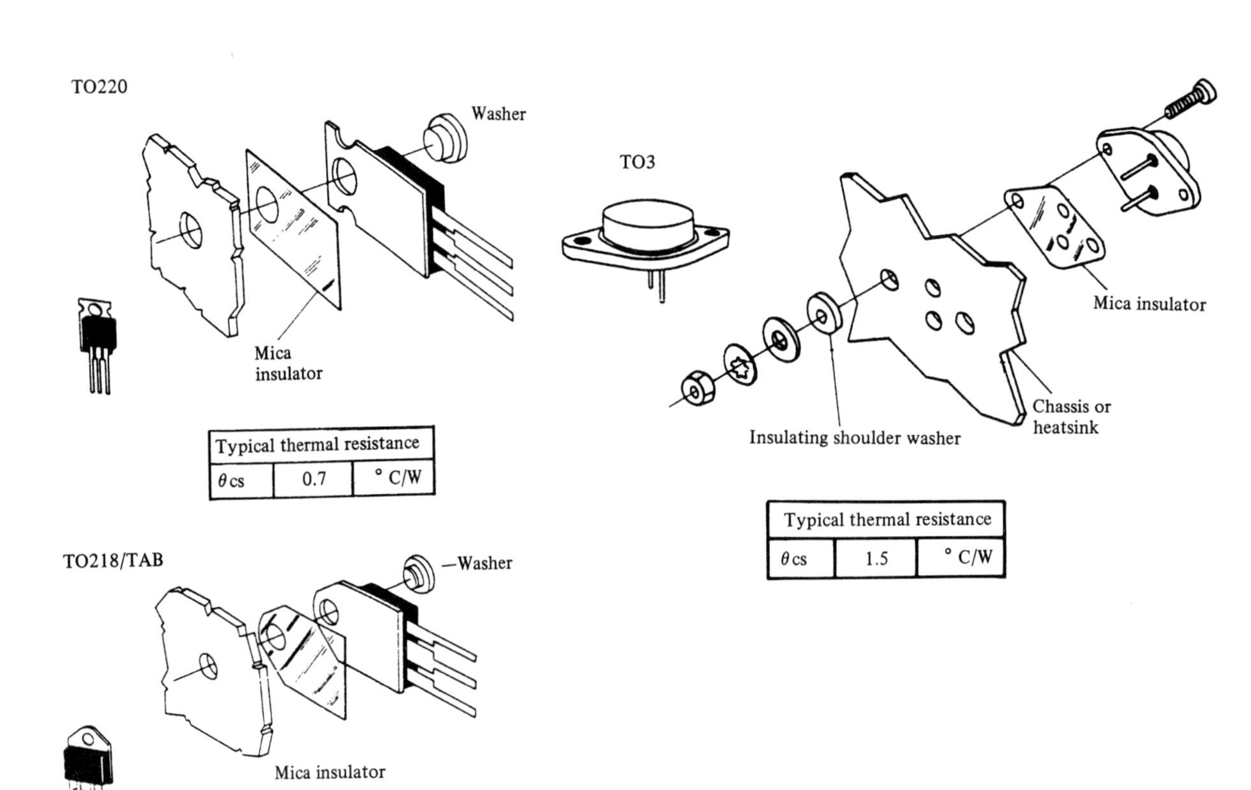

TO220

Typical thermal resistance		
θcs	0.7	° C/W

TO3

Typical thermal resistance		
θcs	1.5	° C/W

TO218/TAB

Typical thermal resistance		
θcs	0.6	° C/W

Appendix 4 Useful addresses

Component suppliers

Electromail,
PO Box 33,
Corby,
Northants,
NN17 9EL
Tel: 0536–204555

Supplies a vast range of 'industry standard' components and devices. Extensive catalogue available at moderate cost.

Maplin Electronic Supplies,
PO Box 3,
Rayleigh,
Essex,
SS6 8LR
Tel: 0702–554161

Reasonably priced catalogue (available direct or from most branches of WH Smith) catering mainly for the hobbyist and home constructor. Retail shops in London, Birmingham, Manchester, Southampton and Southend.

Magazines

Everyday Electronics and Electronics Monthly,
Editorial Offices and Subscriptions,
6 Church Street,
Wimborne,
Dorset,
BH21 1JH
Tel: 0202–881749

Monthly magazine especially suited to the beginner. Educational features are included on a regular basis as are projects suitable for the newcomer.

Practical Electronics,
Editorial Offices,
16 Garway Road,
London,
W2 4NH

Subscriptions,
Lumpett House,
Farndon Road,
Market Harborough,
Leicestershire,
LE16 9NR

Monthly magazine devoted to all facets of electronics.

Index